NORTH-HOLLAND RESEARCH MONOGRAPHS

FRONTIERS OF BIOLOGY

VOLUME 23

Under the General Editorship of
A. NEUBERGER
London
and
E. L. TATUM
New York

NORTH-HOLLAND PUBLISHING COMPANY
AMSTERDAM · LONDON

EXOBIOLOGY

Edited by

CYRIL PONNAMPERUMA

Laboratory of Chemical Evolution, Department of Chemistry,
University of Maryland, College Park, Maryland

1972

NORTH-HOLLAND PUBLISHING COMPANY

AMSTERDAM · LONDON

Library of Congress Catalog Card Number: 71–146196
ISBN North-Holland: 0 7204 7123 0
ISBN American Elsevier: 0 444 10110 1

PUBLISHERS:
NORTH-HOLLAND PUBLISHING COMPANY – AMSTERDAM
NORTH-HOLLAND PUBLISHING COMPANY, LTD. – LONDON

SOLE DISTRIBUTORS FOR THE U.S.A. AND CANADA:
AMERICAN ELSEVIER PUBLISHING COMPANY, INC.
52 VANDERBILT AVENUE
NEW YORK, N.Y. 10017

PRINTED IN THE NETHERLANDS

General preface

The aim of the publication of this series of monographs, known under the collective title of '*Frontiers of Biology*', is to present coherent and up-to-date views of the fundamental concepts which dominate modern biology.

Biology in its widest sense has made very great advances during the past decade, and the rate of progress has been steadily accelerating. Undoubtedly important factors in this acceleration have been the effective use by biologists of new techniques, including electron microscopy, isotopic labels, and a great variety of physical and chemical techniques, especially those with varying degrees of automation. In addition, scientists with partly physical or chemical backgrounds have become interested in the great variety of problems presented by living organisms. Most significant, however, increasing interest in and understanding of the biology of the cell, especially in regard to the molecular events involved in genetic phenomena and in metabolism and its control, have led to the recognition of patterns common to all forms of life from bacteria to man. These factors and unifying concepts have led to a situation in which the sharp boundaries between the various classical biological disciplines are rapidly disappearing.

Thus, while scientists are becoming increasingly specialized in their techniques, to an increasing extent they need an intellectual and conceptual approach on a wide and non-specialized basis. It is with these considerations and needs in mind that this series of monographs, '*Frontiers of Biology*' has been conceived.

The advances in various areas of biology, including microbiology, biochemistry, genetics, cytology, and cell structure and function in general will be presented by authors who have themselves contributed significantly to these developments. They will have, in this series, the opportunity of bringing together, from diverse sources, theories and experimental data, and of integrating these into a more general conceptual framework. It is unavoidable, and probably even desirable, that the special bias of the individual authors will become evident in their contributions. Scope will also be given for presentation of new and challenging ideas and hypotheses for which complete evidence is at present lacking. However, the main emphasis will be on fairly complete and objective presentation of the more important and more rapidly advancing aspects of biology. The level will be advanced, directed primarily to the needs of the graduate student and research worker.

Most monographs in this series will be in the range of 200–300 pages, but on occasion a collective work of major importance may be included exceeding this figure. The intent of the publishers is to bring out these books promptly and in fairly quick succession.

It is on the basis of all these various considerations that we welcome the opportunity of supporting the publication of the series '*Frontiers of Biology*' by North-Holland Publishing Company.

E. L. TATUM
A. NEUBERGER, *General editors*

Preface

Are we alone in the universe? This question looms larger than ever today. If we can sample the alien dust of another world, what hidden secrets may we not discover? If we scan the radio waves from a distant galaxy, what may we not hear if we listen closely? If we probe into the chemistry of life's origin, what may we not discern about its possibility elsewhere in the universe. The search for life beyond the earth is the driving force of the new science of exobiology.

It was over a decade ago, that Joshua Lederberg of Stanford University addressed an international meeting of the committee of space research and used the term 'exobiology' for the study of extraterrestrial life. During the years that have elapsed, since the COSPAR meeting of Nice in 1957, Exobiology has come of age. We have not yet discovered life elsewhere in the universe, but each day appears to bring us nearer to our goal.

Three approaches present themselves to us in our search for extraterrestrial life: listening to signals from intelligent beings, the landing of instruments in remote parts of the universe, and the recapitulating of the manner in which life began on the earth. The latter aspect is one of paramount importance for laboratory investigations. Our primary objective, thus, becomes the understanding of the origin of life in the universe. This is the scientifically broader question before us. If we can understand how life began on the earth, we can argue that the sequence of events which lead to the appearance of terrestrial life may be repeated in the staggering number of planetary systems of our universe.

In this volume I have endeavored to gather together within the pages of a single volume, a number of contributions from the acknowledged authorities in the field. The discussions in depth are intended for the researcher and advanced student in this new and fascinating branch of science.

I am greatly indebted to all the contributors for their willingness to co-operate in this effort, and for their patience during the interminable round of communications between publisher, editor, and author. To Dr. Linda Caren I owe a special debt of gratitude for her invaluable editorial assistance in the preparation of this book.

CYRIL PONNAMPERUMA

List of contributors

WILLIAM A. BONNER, *Department of Chemistry, Stanford University, Stanford, California 94305.*

SHERWOOD CHANG, *Exobiology Division, National Aeronautics and Space Administration, Ames Research Center, Moffet Field, California 94305.*

MARGARET DAYHOFF, *National Biomedical Research Foundation, Silver Spring, Maryland 20910.*

BERTRAM DONN, *Laboratory for Space Sciences, Goddard Space Flight Center, Greenbelt, Maryland 20771.*

NORMAN GABEL, *Research Department, Illinois State Psychiatric Institute, Chicago, Illinois 60612.*

HAROLD P. KLEIN, *Exobiology Division, National Aeronautics and Space Administration, Ames Research Center, Moffet Field, California 94305.*

KEITH A. KVENVOLDEN, *Exobiology Division, National Aeronautics and Space Administration, Ames Research Center, Moffet Field, California 94305.*

LYNN MARGULIS, *Department of Biology, Boston University, Boston, Massachusetts 02215.*

A. I. OPARIN, *A. N. Bach Institute of Biochemistry of the Academy of Sciences of the USSR, Moscow, USSR.*

CYRIL PONNAMPERUMA, *Laboratory of Chemical Evolution, Department of Chemistry, University of Maryland, College Park, Maryland 20742.*

BERNARD PULLMAN, *Institut de Biologie Physico-Chimique, Université de Paris, Paris, France.*

S. ICHTIAQUE RASOOL, *Goddard Institute of Space Studies, New York, New York 10025.*

CARL SAGAN, *Center for Radiophysics and Space Research, Cornell University, Ithaca, New York 14850.*

WILLIAM SCHOPF, *Department of Geology, University of California – Los Angeles, Los Angeles, California 90024.*

DINESH O. SHAH, *Laboratory of Surface Chemistry, Lamont-Doherty Geological Observatory, Columbia University, Palisades, New York 10964.*

PETER SYLVESTER-BRADLEY, *Department of Geology, University of Leicester, Leicester, England.*

CARL WOESE, *Department of Microbiology, University of Illinois, Urbana, Illinois 61801.*

Contents

Chapter 3.   *The geology of juvenile carbon, by P. C. Sylvester-Bradley*   62

Chapter 4.   *Primordial organic chemistry, by Norman W. Gabel and*
Cyril Ponnamperuma   95

Contents

Contents

PART I

Origin of life on Earth

C. Ponnamperuma (ed.), Exobiology. © North-Holland Publishing Company

CHAPTER 1

The appearance of life in the universe

A. I. OPARIN

A fundamental task of space biology is the discovery and investigation of extraterrestrial life, of life which has appeared, developed and now exists independently of our planet. The discovery of such life anywhere in the universe certainly is of paramount importance. Man everywhere, although in varying manner, has strived to define his place on earth and also in the vast universe. Just now we find ourselves at a stage when a deep revision is taking place of our basic concepts of the foundation of science and human knowledge.

According to legend the ancients, with a layout that was colossal in regard to both resources and energies, attempted to erect the grandiose tower of Babel. With the aid of this, one was to be able to reach the firmament, considered to be supported from within by a rigid dome extending to the earth. At present, in the era when man is already travelling in the cosmos, we look at the sky with a sensation altogether different from that of our grandfathers and even our fathers. We are passionately aware of the fact that we are not alone in the universe. We are destined not to spare effort and material to obtain direct proof of this. As man lives not 'by bread alone', the solution of this problem is of absolute necessity to him.

One-hundred to two-hundred years ago the world presented itself to a man of science as a grandiose mechanism, created to the exact law of conservation of matter and energy. In that scheme, all is unchangeable and all phenomena occur in a closed cycle, unchanged by restoring themselves to the original position. It was, therefore, said that anything in the world could be easily calculated and, even predicted in the same manner as we predict the course of day and night or winters and summers – phenomena at the basis of which lies the revolution of the earth or its orbiting around the sun.

However, in modern times that belief in the unchangeability of all that exists yields its place, in the minds of people, to another principle that has already been proclaimed by the great dialectic of ancient Greece, Heraclitus ($\pi\alpha\nu\tau\alpha$ $\rho\epsilon\iota$) – 'everything is in flux'.

We live in a world that perpetually changes, goes through evolution. That process of development has a progressive character and causes the appearance of each new and more complex form of real actuality. Striving to grasp only a single glimpse of the general view of that universal progress from evolution of elements to the appearance of conscious beings and of the human community, we must clearly understand that this evolution does not appear to be in a straight line. At different times and on different cosmic objects this may have been accomplished in varied manners. Therefore, we are able to demonstrate evolution in a schematic manner as a system of parallel or divergent lines some branches of which can bring out the complex and perfect form of organization and motion of matter. About many of these forms we still know nothing; in fact we cannot even guess about their existence. However, independently of the degree of perfectness which they may have attained in the process of their development, in the majority of cases, they may have no relation with life.

Life is the result of only one out of all the many-numbered branches of the evolution of matter which we have already mentioned. Characteristic to it are the specific routes of appearance and perfection. In our present day, we only know life by one singular example, only by our terrestrial life and from it we must formulate our opinion about other possible forms of biological organization. Hence, the earth and events which took place on it must to a considerable degree serve as a model for our paramount ideas concerning life in the universe.

Earlier, the emergence of life on earth was considered to be the result of a lucky accident. This could not give anything for a theory concerning life outside of our planet except the calculation of the probability of such an event which gave hardly any satisfaction to our mind.

However, we have now a basis for looking at the emergence of our life not as a 'lucky happenstance', but rather a phenomenon entirely regulated by laws and an unalienable and integral part of the general evolutionary development of our planet. If we accept the laws of nature to be universal we should expect that an analogous evolution of matter had to lead to the origin of life, on other planets of our universe.

The search for life outside of the earth is only a part of the more fundamental question of the origin of life in the universe. The study of the origin of life on earth presents itself as the investigation of just one example of events that must have occurred in the cosmos innumerable number of times. Therefore, the explanation of the question of how life sprang up on earth must of necessity give strong proof in favor of the theory of its existence elsewhere in the universe.

According to contemporary theory, the onset of the rise of life on earth was proceeded by a process by which carbon compounds become successively more complex, and the formation from these of a polymolecular system. It is possible to break down this process into several stages:

(1) Appearance of hydrocarbons and cyanides and their immediate derivatives in cosmic space and during the formation of the earth and the subsequent development of its crust, atmosphere, and hydrosphere.

(2) Conversion on the earth's surface of the initial carbon-containing compounds into more and more complex organic substances – monomers and polymers – appearance of the so-called 'primordial soup'.

(3) The self-formation, in this soup, of polymolecular open systems capable of mutually interacting with the environment and capable of growth and multiplication on the basis of this interaction – appearance of probionts.

(4) The further evolution of 'probionts', the development of more perfect metabolism, more perfect molecular and super molecular structures accomplished through the basis of prebiological selection – the appearance of the primordial organisms.

It is easy to see that the first stage indicated to us bore a very pronounced chemical character. The path of processes of this phase was completely and entirely subordinated to the laws of physics and chemistry general to all nature. Therefore, we could, a priori, expect the discovery of such processes in this or another form not on earth alone but also on many other various heavenly bodies. However, on our planet at one special stage in the development of matter emerged the transition from chemical into biological evolution. After this change, new and much less universal laws became predominant.

As an exclusively specific feature of anything living we find a superbly complete adaptation or, as it is frequently called, purposiveness of the entire organization of the living bodies. This purposiveness is directed towards their permanent self-preservation and self-perpetuation, within a given set of conditions of the external medium. There is also an adaptation of the construction of the separate parts of living bodies (molecules, organoids, cells, tissues and organs) towards the functions they carry out in the life processes.

For a long period of time, the essence of the purposiveness appeared to be mystical and supernatural. In it one saw the purpose-oriented execution of an outline for the living existence from some sort of spiritual authority governing life. Darwin gave a rational account for the formation of this purposiveness on the basis of natural selection, a specific biological law.

However, the ideas of Darwin covered the completely formed living organisms only, already possessing highly-organized metabolism and intracellular structure. In this respect, Darwinism stands before us like the glittering top of an iceberg of which nearly nine-tenths are hidden from our view beneath the water. Prebiological and early bioevolution, in the course of which fundamental features and qualities were formed characteristic of each living species, lasted throughout the flow of far longer time and was enriched by events no less dramatic than the widely studied Darwinian development of life. However, with the transition from chemical to biological evolution, the principle of natural selection came to act with the result that one and the same complex of organic substances, on different celestial bodies, may have undergone very different pathways in their development.

The appearance of compounds for the primary step of biological evolution: hydrocarbons, cyanides, and their immediate derivatives occurred in inconceivably distant time, many billions of years ago, long before the formation of the solar system. This was a result of both the extremely large predominance of hydrogen in the cosmos and the appearance of carbon even before the formation of heavy elements necessary for planetary systems, during the steady state of stellar evolution, independent of the flare-up of supernovae. This is evident from the study of spectra of all classes of stars, including older types, which are reckoned to be 15–20 billion years.

Therefore, the compounds of carbon and hydrogen, are extremely widespread throughout the cosmos; they are present both on the surface of stars at temperatures of several thousand degrees and under very high gravitation and also in interstellar gas and dust at extremely low gravity and a temperature close to absolute zero. Evidence for this is available from our present-day cloud of interstellar material as well as by examining the spectra of comets, cosmic bodies formed under conditions almost like those of the interstellar medium. These investigations indicate that comets abound in light hydrocarbons and the cyanides.

Of basic interest from the point of view of the problem discussed here are the data obtained through the study of meteorites. In the first place, up to the collection of samples from the lunar surface, they were the only extraterrestrial bodies we had access to. Secondly, meteorites are in their composition very similar to that of the concentration of cosmic dust, the planetesimals from which the earth and the earthlike planets have been formed.

In the composition of certain meteorites, the so-called 'carbonaceous chondrites', have been detected not only the initial simple compounds of carbon and hydrogen but also their far more complex derivatives, different organic substances that can be formed by abiogenic pathways independent of life.

Therefore, in the opinion of J. D. Bernal (1965), the second phase of the evolution of carbon-containing compounds – their conversion to all more complex organic materials – began long before the formation of the earth. According to Bernal, those conversions became complete abiogenically on the surface of particles of cosmic dust or in planetesimal environments under the action of short-wave ultraviolet rays or cosmic radiation. In this fashion, the earth obtained certain quantities of organic substances, already in prepared form as early as in the process of its build up as a planet, and later on was fed with them through the fall of meteorites and cometary material on its surface.

However, the principal mass of organic substance required for the development of life appears to have been formed endogenously on earth upon formation of the terrestrial crust, hydrosphere and secondary atmosphere. From this point of view exceedingly interesting are the numerous contemporary attempts to detect organic materials of abiogenic origin in some geologic samples. However, the treatment hitherto of geochemical data is very complicated because of the possibility of contamination by material of recent biological origin (Schopf, this volume, page 16).

According to the now most widely spread concept, our planet was formed by the accumulation and concentration of cold solid bodies or planetesimals. Such gases as molecular hydrogen, helium, etc. might still be preserved in absorbed form in the solid samples and thence have contributed to the primary atmosphere. However, that atmosphere was not long-lasting, as it consisted of gases that were not retained by the earth's gravity.

Having become a compact mass, the earth kept on developing. This evolution was mainly related to the thermal history of our planet. Under the influence of heat formed at the expense of gravitational energy and the decay of radioactive elements a partial melting of the primary rocks took place resulting in the formation of the earth's crust, hydrosphere, and atmosphere.

The principal mass of water initially was present in bound form in hydrated minerals. Therefore, at the outset, there was less water on the surface of the earth than now, and the formation of the oceans took place very gradually, in conjunction with the buildup of the earth's crust. With that development also coincided the formation of the secondary atmosphere. The latter was basically different from the present-day one. It had reducing character – being void of free oxygen and contained besides water vapor such compounds as gaseous hydrocarbons, ammonia, and hydrogen sulfide. Thanks to the absence of molecular oxygen, it was totally transparent to shortwave ultra-violet rays as no ozone screen which blacks the access to these rays to the earth's surface today, could be formed.

Reproducing the parameters of the earth at that time under laboratory conditions, numerous investigators in different countries of the world, S. Miller, C. Ponnamperuma, J. Oró, T. Pavlovskaya, A. Pasynsky and others, have convincingly demonstrated the inevitability of the synthesis of manifold organic compounds in the secondary atmosphere and the hydro-sphere of the earth (Gabel and Ponnamperuma 1968; Lemmon 1970). In some cases their chemical nature is very close to those substances which we nowadays isolate from contemporary living things, in particular amino acids, sugars, purine and pyrimidine bases, nucleotides, organic acids, as well as their various polymers, including a number of protein-like and nucleic acid-like substances among them. S. Fox (1970), on the basis of his model experi-ments, even suggests that the protein-like polymers formed abiogenetically under described conditions would possess a definite intramolecular organiza-tion, a periodic sequence of amino acid residues in a polypeptide chain that provided them a certain specific catalytic activity.

The substances formed in this manner on the terrestrial surface could certainly undergo a secondary degradation under the action of such factors as, for example, short wave ultraviolet radiation. But an important portion could escape such an influence as a result of the migration of the polymer molecules or their complexes from the surface to deeper water layers, or by absorption on soil particles etc. In this fashion in the water of the terrestrial hydrosphere, various organic substances were accumulated forming a com-plex water solution of these substances which we are now accustomed to call the 'primordial soup'.

We can today in a sufficiently established manner demonstrate the charac-ter of those chemical conversions that took place within that 'soup' and even the succession in the origin of the varied organic substances. In general this corresponds to the order of changes which takes place in a simple aqueous solution of these materials and it is certainly fundamentally different from all that which we today find in living nature. Both the complex order of chemical reactions characteristic of biological metabolism and any kind of biological structure were absent. In particular even though they possess certain intramolecular organization and catalytic action, the protein-like polymers of the 'primordial soup' are principally different from protein of present-day organisms. The structure and the enzymatic action of the latter are found everywhere completely adapted to perform the function which they carry out in living entities.

Such adaptativeness or 'purposiveness', of organization even at the molecular level was certainly absent from the simple aqueous solution of chemical compounds. This 'purposiveness' could only arise during the process of the origin of life, at the transition from chemical to biological evolution, on the basis of the formation and natural selection of integral prebiological systems.

We know that life is not simply dispersed through space similar to substances in the 'primitive soup'. It manifested itself through organisms – discrete systems spatially separated from the exterior environment but also interacting with the latter in the manner of open systems, that is, systems whose stability and longevity are determined not by their state of rest, but by their continuous transformation of substances. The thermodynamics of this class of systems is notedly different from the classical. This results in a property characteristic of all organisms, that they yield not an increasing but a decreasing entropy.

In the process of the evolution of matter before the origin of organisms it was necessary that a self-formation preceded within the 'primordial soup', of organisms far more primitive, than living, multimolecular open systems, and only as a result of their gradual evolution could the predecessors of all living things on earth appear.

One is able not only to imagine such systems initial for biological evolution but even to reproduce in modelling experiments a great variety of them (e.g. Goldacre's (1958) vesicles, Fox's (1963) microspheres, the coacervates of Bungenberg de Jong (1936) and many others.

For the further evolution of these systems it was important that they interacted with their environment in the fashion of open systems and that their stability bore not a statical but a dynamic stationary character. From that point of view, coacervate droplets appear especially suitable, but not the only possible models, for the reproduction of the phenomena that took place in the distant past.

Under conditions similar to the 'primordial soup', the spontaneous formation of coacervate drops takes place readily through combination of various yet not specific polymers, for example, polypeptides or polynucleotides, to form systems visible under the microscope. These are separated from the surrounding solution by a clearly defined surface, but are capable of selective uptake of various substances, amino acids, sugars, mononucleotides, different salts, from the surrounding 'soup' and of transferring to the external solution the products of chemical reactions occurring within the complexes.

The rate of these reactions can be strongly enhanced by the inclusion of various catalysts, inorganic, organic, or enzymatic, in the coacervate drops. Here further processes of degradation as well as synthesis of polymers can take place forming these drops at the expense of the energy-rich substances entering the complexes from the exterior solution. If the speed of the latter processes exceeds the rate of breakdown, the complexes gain not only dynamic stability but also the capability of growth. They increase before our eyes in volume and mass. In this manner, a stream of matter and energy occurs in the complexes that in the simplest way models the metabolism of living things. We have reproduced in the coacervate drops, relatively complicated schemes of the metabolic pathways in which several mutually complementary reactions participate. In all those events, the growth rate of the drops was found to be directly related to their individual organization, and regulated primarily by the rates of various reactions taking place within them. From this point of view, some drops can be considered more, and others less accomplished, as to their organization.

Placing different coacervates in one and the same external medium, a solution of identical substances, we find a rapid growth of the more developed units. On the other hand the less accomplished systems are retarded or even degraded.

Model experiments of this nature demonstrate to us the elements of a new law which had to develop at the interface of chemical and biological evolution, the beginning of prebiological natural selection. This new law also lays at the basis of all further evolution of systems, which give rise to organisms, we conditionally can name 'probionts'. On their successive development we can speculate while still only hypothetically, and basing it principally on data of a comparative biochemical study of metabolism and structure in primitive organisms existing today.

The evolutionary development of 'probionts', heretofore is completely related to their catalytic apparatus as the paramount factor of organization of their metabolism which is based on the various reaction rates of their components. Certainly, at the stage of evolution investigated by us, inorganic salts and organic materials present in the primordial soup were the only catalysts available to the 'probionts'. The activity of their catalysts was probably very low. However, upon their definite mutual combination, that activity could be increased a hundred or a thousand times.

We can ourselves envision a very colossal number of different atomic groupings and combinations which in one way or another are capable of catalyzing the reactions necessary for the existence of probionts. However, as a result of the fact that natural selection all along rejected less perfected complexes and destroyed the 'probionts' which possessed these, only the very few most effective ones remained to us. These are widely known in biochemistry as coenzymes. Their number is relatively insignificant, but they are extremely universal catalysts for all living beings without exception. This fact points to their very early formation in the process of the development of life. The necessary constancy in the concentration of coenzymes in the growth and fission of 'probionts' under the influence of extraneous mechanical affects like ocean breakers for example, could be sustained because of the uptake of these compounds, or the formation of their components, from the exterior environment. We have something analogous in contemporary organisms which are compelled to obtain from the ambient environment vitamins, which play the role of coenzymes in metabolism. But, in the probionts the capability of synthesizing coenzymes for themselves was gradually worked out, which emancipated the systems under progress from their large dependence upon the environment.

However, the probiont metabolism gradually became more complex and called for a more perfect combination of a great number of metabolic reactions to form chains and cycles or a coordinated network of biochemical reactions. For that kind of coordination the catalytic action and specificity of coenzymes was insufficient. In the process, therefore, of further evolution, the coenzymes were supplemented by the whole arsenal of far more powerful catalysts, the enzymes, which were proteins whose secondary and tertiary structure was extremely well adapted to their functions. In this way the era of the coenzymes was displaced by a new epoch when protein substances with specific intramolecularly organized structures came to play a decisive role.

The initially formed protein-like polymers with their random arrangements of amino acid residues could serve as material for the formation of coacervate drops and probionts, but they were very poor catalysts or were entirely deprived of that function.

In the probionts, polymerization of amino acids could certainly take place giving rise to those combinations of residues that were able to play the role of enzymes. In ordinary unorganized polymerization, however, that preeminence would be quickly lost, since there would be no difference from the wide growth of the probiont. Therefore, the development of those organizations that fixed the steadfastness of the primary structure of the newly-synthesized polymers was very important. In this organization, an extremely important role was discharged on the part of the polynucleotides.

In contemporary organisms the synthesis of enzymatic proteins is realized by an extremely complex and perfect mechanism with the aid of which amino acids sequentially join in a polypeptide chain in the order that is dictated by a specific, strictly controlled combination of mononucleotide residues in the molecules DNA and RNA. A mechanism of this kind could only appear in a process of a continuous evolution of probionts and subsequently formed living systems. But already at very much earlier stages of development, the polynucleotides occluded in the probionts could exhibit a certain effect on the polymerization of amino acids in these systems. The intramolecular structure of these primary polynucleotides was still not very perfected. It varied greatly during the process of probiont growth. Each variant which appeared in this manner could to a certain degree consolidate itself within a given growing system as a result of complementariness of polynucleotides, and besides it influenced the order of arrangement of amino acid residues in the polypeptide synthesized in this system.

If the combination of amino acid residues, developed in this way, was advantageous from the point of view of enforcing the catalytic activity of polypeptides, then the system which gave rise to this combination would appear superior in its far more rapid proliferation and multiplication. In the opposite event it was annulled by natural selection. Thus, gradually the intramolecular structure of the proteinoid polypeptides and the polynucleotides participating in their synthesis in parallel, became ever more ordered and more adapted to the functions which those polymers carried out in the growing and multiplying systems.

Yet, one must clearly imagine that there were no polynucleotides capable of replication or polypeptides having some specificity in their sequence under the influence of these polynucleotides which entered the selective process but rather a whole system, although primitive but more or less perfect in metabolism either corresponding or not to a given environment. The role of the nucleic acids, in this instance, was that they spatially fixed the constancy of the synthesis of catalytically advantageous amino acid combinations in the growing and multiplying systems and served as a stabilizing factor in their evolution. Thus, only at a sufficiently late stage of evolution arose a new epoch when the systems reached a previously unprecedented height of that exact self-perpetuation which characterizes our whole present-day world of living things.

We can follow the further development of living systems and the improvement of their metabolism and supermolecular structures to a certain extent on the basis of comprehensive comparative biochemical investigations. Data obtained in this fashion clearly state that some forms of organization of metabolism, in other words, some combinations of biochemical reactions developed already at the initial steps of the development of life and therefore, are characteristic for all contemporary organisms (without exclusion). On the other hand, others appeared significantly later as supplements to already existing metabolic mechanisms.

The sole source of nourishment for the primitive organisms might have been the organic substances of the 'primordial soup'. In agreement with this was the ability to utilize organic food which is at the very basis of life, and is characteristic without exception of all living organisms of our planet.

The absence of free oxygen in the secondary atmosphere of the earth and in the hydrosphere, caused the anaerobic character of the energy metabolism of the first organisms. And indeed, the data of comparative biochemistry clearly show that anaerobic metabolism lies at the basis of energetics of all contemporary organisms without exception, higher plants and animals capable of respiration, among them.

In the process of the development of life the supply of abiogenically generated organic substances on the earth surface gradually exhausted itself, because the development of life must have proceeded faster than the formation of these materials and they came to be deficient. This change in the environment brought about the appearance of the organisms which, thanks to the ability acquired by them to utilize light energy, were able to synthesize anew organic materials out of inorganic carbon compounds, from the carbon dioxide in the atmosphere. Indeed, similarly, such processes went on in the 'primitive soup' in a simple manner; coacervates, probionts, or prime organisms were able to a certain degree to synthesize organic substances under the conditions of the reducing atmosphere at the expense of the energy of short-wave ultraviolet light. But in the earth's atmosphere, the gradual formation of free oxygen in an abiogenic route proceeded, although very slowly, whereupon an ozone screen developed that barred the access of shortwave ultraviolet rays to the earth's surface.

Parallel with this transition there must have occurred a change which enabled the primitive organisms to utilize the longwave light which is issued in such abundance from the sun to our planet. The main difficulty, however, was the fact that each separate quantum of visible light carries with it only a relatively minute amount of energy. Therefore, for the successful realization of biologically important photochemical reactions, the participation of photosensitizers in the latter is needed.

On the foundation of the data of comparative biochemistry and model experiments, it is possible to follow through the development of such photosensitizers as porphyrins and their magnesium derivatives, and their inclusion in the systems capable of photosynthesis. The pathways of successive perfection of this system emerged on one side through the selection of more and more effective pigments such as porphyrins and on the other, through the complication and functional adaptation of the supermolecular structure of the photosynthetic apparatus. Thus, arose photosynthesis – novel and very perfect way of synthesis of organic substances which emerged in place of the earlier, very slow, and imperfect abiogenic synthesis. Therefore, in the further development of life, photosynthesis assumed a foremost, monopolistic significance in the formation of organic substances on the surface of the earth. Its origin changed the existing condition of life here. Part of the organisms came to build the organic compounds necessary for themselves, another part retained the previous heterotrophic form of nourishment, utilizing organic materials that were developed on some biogenic route with the help of photosynthesis. Thus, two branches of the living world appeared: plants and animals.

However, the origin of photosynthesis did not only produce a plentiful supply of organic matter, but it also led up to the rapid formation of free oxygen in the terrestrial atmosphere. This changed the whole character of chemical processes being accomplished here, and allowed the majority of living things to improve its energetic metabolism significantly, building upon the previous anaerobic mechanism, new supplementary systems of oxygen respiration and in this way entirely utilizing energy hidden in organic substances. Parallel to the perfection of the metabolism took place the evolution of spatial organization of living bodies. Its origin and development was intimately tied to the evolutional development of functions related to the structure.

If anaerobic fermentation may be achieved also in a homogeneous solution, photosynthesis and respiration already require for their realization very complex structures. The very most primitive structural formations seem to be proteinlipid membranes that we may obtain already at the stage of coacervate formation and which are characteristic of all organisms without exception. However, the building of such structures as chloroplasts, mitochondria, or cell nuclei, took place only in the process of the gradual evolution of living systems. In this manner we see that the very long time-span of the evolution of primitive organisms in principle was linked to the improvement of the metabolism and intramolecular organization of biologically important substances.

But at a certain stage of the development of life, subcellular and cellular structures appeared, the course of evolution of which can already be followed on the basis of paleontological data, attributed to the earliest times in the history of our planet.

What general conclusions may we draw on the basis of all the materials presented here with respect to possible pathways of the origin and development of extraterrestrial life? The same initial carbonaceous material that was needed for the origin of our terrestrial life and life analogous to that formed on earth, may be present on other heavenly bodies. Therefore, the absence of life cannot serve as an impediment to its formation. The whole question centers around how this material evolved further on one or another cosmic object. It is clear that a successive evolution, analogous to ours, of carbon compounds could only exist within a relatively narrow framework of exterior conditions: temperature, gravitation, magnetic field, illumination, hydration, and so on. It appears that only planetary systems are able to satisfy these conditions and according to the calculations of Harlow Shapley (1960), even in this kind of system only one or perhaps two planets may be located in a zone where the parameters of temperature and radiation appear suitable for the origin of life. The extent of hydration of the planet is also of great importance. The abiogenic synthesis of organic matter and its polymers can take place, as demonstrated by the works of S. Fox, in entirely dry conditions under the influence of high temperatures. The finding of such compounds in meteorites, demonstrate this possibility but in meteorites the evolution of these compounds did not reach as far as it did on earth.

It is important to have in view that the above-mentioned external sources of energy not only cause the synthesis of organic substances but also their decomposition. Thermodynamic equilibrium here does not in general contribute to the advantage of the synthesis. Thus, the organic materials that arose abiogenically on the surface of this or that cosmic body did not of necessity have to evolve further, as on earth. They could also have disappeared, degrading themselves under the impact of the very influences which formed them. In support of this idea is the circumstance that the preliminary analyses of materials obtained from the lunar surface unprotected from cosmic interactions, did not reveal here perceptible quantities of organic matter (Chang and Kvenvolden, this volume, page 400).

The accumulation of these substances necessary for further evolution could only proceed if that subsequent to their formation, they could in some measure escape the effects of the destructive radiation or high temperature. On the terrestrial surface, the hydrosphere, the great masses of free water, played an important role in this respect. Related to the aqueous environment were also all successive advances of the systems emerging for the development of life. Therefore, the presence of water had a decisive significance for all pathways of evolution even during its chemical stage.

Upon the transition to the biological era determined by natural selection based on mutual interaction of the systems and the surrounding media, that significance grew still more. Hence, the pathways of evolution could strongly differ on different cosmic objects even under relatively similar external circumstances.

The epoch of interplanetary cosmic travel into which mankind just now enters opens for the science of life new vistas and distant perspectives. It may offer to us an insight into life and the pathways of its origin and development in forms that may be distinct from the terrestrial. Yet, even if, to our great disappointment, we do not discover life on our neighboring planets, we will still learn very much that is new about the pathways of evolution of organic matter which lay at the foundation of the origin of our terrestrial life.

References

BERNAL, J. D., 1965, Molecular matrices for living systems; In: The origins of prebiological systems and of their molecular matrices, S. W. Fox, ed. (Academic Press, New York) 65.

BUNGENBERG DE JONG, H. G., 1936, La coacervation, les coacervats et leur importance en biologie. VI. Generalités et coacervats complexes and VII. Coacervats auto-complexes. (Hermann, Paris).

FOX, S. W., 1965, Simulated natural experiments in spontaneous organization of morphological units from proteinoid; In: The origins of prebiological systems and of their molecular matrices, S. W. Fox, ed. (Academic Press, New York) 361.

FOX, S. W., 1970, Experiments related to the simultaneous origin of protein and nucleic acid; In: Molecular evolution, Vol. 1, proceedings of the third international conference on the origin of life, Pont-à-Mousson, April 1970, R. Buvet and C. Ponnamperuma, eds. (In press).

GOLDACRE, R. J., 1958, Surface films, their collapse on compression, the shapes and sizes of cells, and the origin of life; In: Surface phenomena in chemistry and biology, J. F. Danielli et al., eds. (Pergamon Press, New York) 278.

LEMMON, R. M., 1970, Chemical evolution; Chem. Rev. 70, 95.

PONNAMPERUMA, C. and N. W. GABEL, 1968, Current status of chemical studies on the origin of life; Space Life Sci 1, 64.

Shapley, H., 1960, On the evidences of inorganic evolution; In: Evolution after Darwin, Vol. I., S. Tax, ed. (University Press, Chicago) 23.

C. Ponnamperuma (ed.), Exobiology. © North-Holland Publishing Company

CHAPTER 2

Precambrian paleobiology

J. WILLIAM SCHOPF

1. *Precambrian paleobiology*

The nature of the Precambrian biota – its antiquity, composition and evolution – and the well-known faunal discontinuity near the beginning of the Paleozoic, have long been recognized as particularly puzzling problems in paleontology. The evolutionary continuum well-documented in Phanerozoic sediments and the diversity and complexity of the early Paleozoic biota augur well for a substantial period of Precambrian evolutionary development. Until recently, however, evidence of this development remained largely undeciphered; the nature of Precambrian life was a fertile subject for speculation, essentially unfettered by the poorly known fossil record. The past few years have witnessed a renewed interest in these classic problems and a marked proliferation of available data; this increased activity has resulted in the emergence of a new subdiscipline of paleontological science, that of Precambrian paleobiology.

Although widely regarded as a new area of emphasis, Precambrian paleobiology is firmly rooted in the pioneering studies of the early 1900's by C. D. Walcott and J. W. Gruner, and to a major extent it represents a variation, rather than an innovation, on their original theme. Walcott was an acknowledged leader in the early search for Precambrian fossils and was one of the first to stress the probable algal origin of Precambrian laminated stromatolites (Walcott 1883, 1899, 1912, 1914). This interpretation, however, and reports by Gruner (1922, 1923, 1924, 1925), who claimed to have discovered filamentous microfossils in Precambrian cherts, were viewed with varying degrees of skepticism by contemporary paleontologists (e.g. Hawley 1926). Subsequent investigations have shown that, in part, this skepticism was well-founded and that certain of these early interpretations were erroneous (e.g. Tyler and Barghoorn 1963). Nevertheless, the fundamental association of cherts, stromatolites and Precambrian microfossils suggested by the studies of Walcott and Gruner has been fully confirmed, and with minor modification has formed the basis of the productive investigations of recent years.

Surprisingly, perhaps, these early studies excited little sustained interest; although several other occurrences of possible microfossils were reported subsequently from Precambrian cherts (e.g. Moore 1918; Ashley 1937; Cahen et al. 1946), it was not until 1954, with the publication of a short note by S. A. Tyler and E. S. Barghoorn describing microorganisms from stromatolitic cherts of the Gunflint Iron-Formation, that the potentialities of the Precambrian began to be widely appreciated. In the 15 years that have followed this report, Precambrian paleobiology has 'come of age' and the field has developed a distinctive interdisciplinary flavor – a merging of techniques and data from diverse branches of chemistry, geology and biology – to a degree previously unknown in paleontological science. As the result of this increased activity, numerous sediments containing diverse types of structurally preserved microorganisms are now known from the Precambrian, and organic geochemical studies have yielded putative evidence of the physiology and biochemical complexity of early life. Based on these new data it has become possible to outline, in the broadest of terms, major events in Precambrian biological history.

2. *Limitations of the Precambrian fossil record*

At the outset of this discussion it should be stressed that in spite of the recent progress noted above, the Precambrian biota remains very incompletely known; inferences here drawn from the fragmentary data available should be regarded as being of a most tentative sort.

The deficiences in the early fossil record are of varied sources. To some degree they reflect a traditional dogma of paleontology that the 'Precambrian is unfossiliferous'; until recently, this view effectively limited inquiry into the paleobiology of very ancient sediments. Certain of these gaps, however, may be inherent to the field: the Precambrian, encompassing the earliest seven-eighths of geologic time, presents problems of a rather different sort from those normally encountered in studies of the Phanerozoic 'overburden'.

At present, fewer than three dozen occurrences of cellularly preserved microorganisms are known from the Precambrian, spanning a segment of biologic history more than four times as long as that encompassed by the entire Phanerozoic. The most important of these fossiliferous deposits, and the evidence of early biologic activity they contain, are listed in text-fig. 1. Unfortunately, relatively few of these sediments contain communities of microorganisms preserved in situ on which inferences of evolutionary status and paleoecology might be based most reliably. Furthermore, of the few such assemblages known, all are preserved in inorganically precipitated primary cherts; although these siliceous sediments have provided the geologic setting responsible for cellular preservation of delicate microorganisms, such cherts reflect unusual ecologic conditions and almost certainly contain a restricted, rather atypical sample of the total biota.

A variety of age effects, essentially inherent to the Precambrian, also complicate the interpretation of the early record. As might be expected, there generally appears to be an inverse correlation between the age of a sedimentary unit studied and the fidelity of organic preservation observed; thus, as the record of biologic activity is traced back into the earliest Precambrian, the morphological fossil record becomes increasingly difficult to decipher. This trend is paralleled by the geochemical degradation of chemical fossils; many biogenic organic compounds detected in Tertiary sediments (e.g. proteins, optically active amino acids, unsaturated fatty acids, predominance of *n*-alkanes with an odd number of carbon atoms etc.) are of decreased abundance in Paleozoic deposits and may be completely absent from sediments of greater geologic age. This loss of information is further compounded by the effects of biologic evolution, so that not only are the earliest organisms relatively poorly preserved, but they are also of limited diversity and of such simple morphology that their biological affinities are difficult to determine. As a result of these age effects, the oldest fossil-like microstructures now known (Engel et al. 1968; B. and L. Nagy 1969) are of uncertain biogenicity.

Younger Precambrian microorganisms, although generally better preserved and morphologically more complex than those of the Early Precambrian, rather commonly lack modern and fossil morphological counterparts; the phylogenetic position of such forms is highly conjectural. As M. F. Glaessner (1968) has suggested, such organisms might represent phyletic side branches, only remotely related to members of well-known systematic categories. In addition, certain evolutionary transitions (e.g. the development of the eucaryotic cell and the origin of chemosynthetic bacteria) were apparently the result of intracellular changes in ultrastructure and biochemistry, not readily preservable in the fossil record and not originally evinced by changes in organismal morphology. Other transitions (e.g. from unicellular algae to primitive protozoans) may have been reflected initially in behavioral patterns, rather than in obvious differences of form. It is doubtful that the earliest appearance of such evolutionary innovations would be recognized in the geologic record unless chemical fossils were detected indicating the evolution of new, phylogenetically restricted, biosynthetic pathways. A systematic search for chemical fossils of this type (e.g. sterol derivatives perhaps indicating the presence of eucaryotic organisms; derivatives of polyunsaturated fatty acids suggesting the occurrence of oxygen-producing photosynthesizers) has yet to be made, however, and at present the newly derived stock might not be recognized until relatively advanced forms, morphologically comparable to their modern descendents, appeared in the record.

Finally, there are several restrictions imposed by the geologic record itself: in general, Precambrian sediments are of rather limited areal distribution and of moderate- to high-grade metamorphism. Moreover, the oldest known rocks (ca. 3.5×10^9 years) are comparable in age to the earliest known fossils (more than 3.1×10^9 years). A variety of evidence suggests that these early organisms were physiologically rather advanced; if so, their biochemical complexity would seem to imply the existence of a substantial period of prior evolutionary development. Direct evidence of the beginnings and earliest evolution of living systems may not be detectable unless very ancient sediments, perhaps 3.5 to 4.25 billion years in age, are discovered.

3. *Chemical fossils*

As an outgrowth of early interest in the chemistry of coals and crude oils and the biologic and geologic processes producing these materials, and gaining general acceptance among paleontologists with the studies of P. H. Abelson (1954, 1959) on amino acids in fossil shells and bones, organic geochemistry has come to play an increasingly significant role in paleobiological investigations. Nowhere is this more apparent than in studies of Precambrian sediments from which a diverse suite of chemical fossils have been reported during the past decade (e.g. *n*-alkanes, isoprenoids, steranes, fatty acids, amino acids, porphyrins, sugars and ratios of the stable isotopes of carbon and of sulfur).

The techniques used to isolate and characterize chemical fossils have been reviewed by T. C. Hoering (1967a), who also discussed Precambrian studies carried out prior to 1967; reviews by M. Calvin (1969) and W. Van Hoeven (1969) include certain of the more recent data. Text-fig. 1 summarizes the geologic distribution of the major categories of chemical fossils reported from the Precambrian. Before considering the implications of these distributions, however, the definition of such fossils and the problems inherent to their interpretation must first be examined.

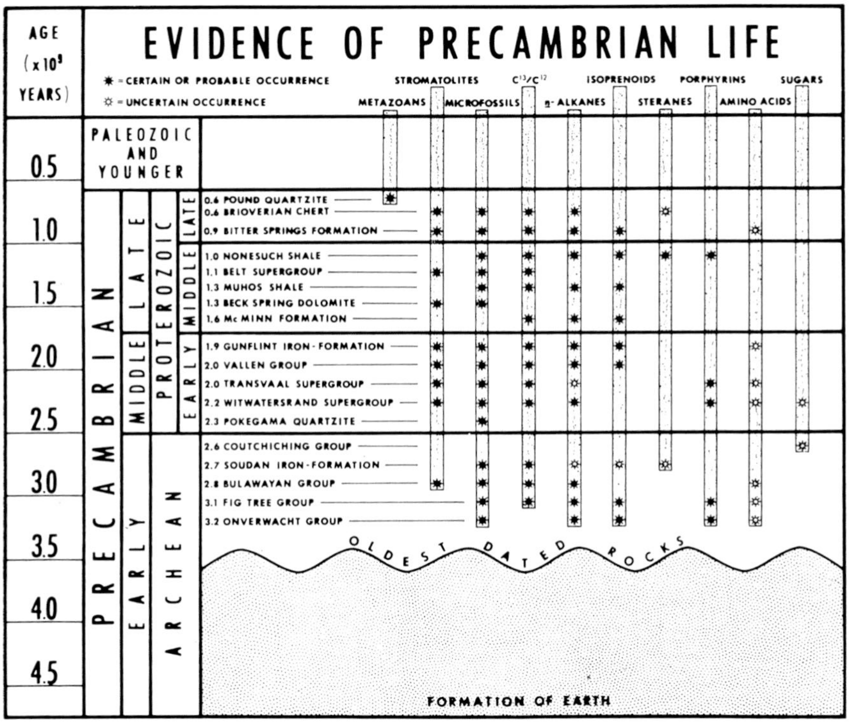

Text-fig. 1. Histogram showing distribution of organic geochemical and morphological evidence of biologic activity reported from Precambrian sediments.

3.1. Definitions

In its broadest sense, the term 'fossil' may embrace *any* evidence of early life (cf. Barghoorn et al. 1965). The distinction here posed, viz. between 'morphological' and 'chemical' fossils, is used to differentiate the morphological evidence of previously existent organisms (the paleontologist's 'fossil') from the organic compounds and/or isotopic fractionation effects derived from such organisms. In a sense, this distinction may be somewhat superficial since some morphological fossils are predominantly or entirely composed of organic matter, and since chemical fossils exhibit specific, albeit molecular, morphology. Although other terms might therefore be proposed, this particular nomenclature seems to be in vogue and is gaining wide acceptance (particularly among organic geochemists). In any case, such a distinction is generally useful, for the techniques of detection and the problems encountered in interpretation are inherently different for these two categories of biological remnants.

3.2. General limitations

Unlike analyses of Phanerozoic molluscs (Abelson 1963; Hare and Mitterer 1967) and vascular plants (Swain et al. 1968), chemical studies of Precambrian life have involved the analysis of extracts of whole-rocks, rather than of individual fossil organisms. Since sediments containing essentially monospecific assemblages (e.g. some diatomites and tasmanite coals) are unknown in the Precambrian, such analyses cannot be correlated with particular taxa and are therefore of relatively limited phylogenetic usefulness. Moreover, although it is generally possible to show that these extracted materials are indigenous to a sediment, rather than being the result of contamination introduced in the laboratory, it is substantially more difficult to establish that they are syngenetic with Precambrian sedimentation. This problem arises because the concentrations of organic matter extracted from early sediments are almost always on the order of a few parts per million or less (for exceptions, see Hoering 1967b) and the deposits are generally permeable to some degree; a minute amount of secondarily emplaced material (e.g. carried by connate water) can greatly influence analytical results. In addition to demonstrating the indigenousness of such chemical fossils, therefore, other criteria (i.e. physical or chemical 'tests') are clearly needed to establish a Precambrian age. If it could be shown, for example, that the $C^{13}:C^{12}$ ratios of syngenetically deposited organic matter of differing geologic age vary in a systematic fashion, as some data seem to suggest (Smith et al. 1970), or if the chemistry of organically preserved microorganisms could be correlated with their micromorphology (Schopf 1968b, 1970), this problem might be resolved.

3.3. Uncertain occurrences

In the absence of such tests, however, some indication of the probable age of extractable compounds can be obtained from a consideration of their geochemical stabilities. Recent studies seem to indicate that certain amino acids (Abelson 1959) and sugars (Vallentyne 1963; Van Hoeven 1969) are relatively unstable in the geologic environment. To explain the unexpected occurrence of these compounds in extracts of ancient sediments, it has been suggested that they may have been protected from diagenetic alteration by being chemically bound to their encompassing matrix (e.g. Degens et al. 1964; Degens 1965; Oberlies and Prashnowsky 1968; Schopf et al. 1968). Although the possible stabilizing effect of mineral substrates has not been investigated fully, preliminary studies suggest that such effects may be minimal (Abelson and Hare 1968a,b; Smith et al. 1970).

Certainly, the available data provide insufficient grounds for concluding that some amino acids and sugars cannot survive from the Precambrian. Nevertheless, the syngenetic nature of amino acids (Harrington and Toens 1963; Prashnowsky and Schidlowski 1967; Schopf et al. 1968; Oberlies and Prashnowsky, 1968; Pflug et al. 1969; Kvenvolden et al. 1969) and of sugars (Swain et al. 1966; Swain et al. 1967; Prashnowsky and Schidlowski 1967; Oberlies and Prashnowsky 1968; Van Hoeven 1969) detected in extracts of Precambrian sediments has not been demonstrated; based on stability considerations, it seems likely that in part, these compounds are of secondary, relatively recent origin.

In addition to the geochemically less stable compounds discussed above, a few hydrocarbon occurrences, denoted by the open symbols in text-fig. 1, are regarded as being of questionable nature. Insufficient data are available to fully evaluate the report of sterol-like compounds isolated from Late Precambrian Brioverian cherts by Roblot et al. (1966). Carbon isotopic data (Hoering 1967a) suggest that soluble hydrocarbons extracted from sediments of the Middle Precambrian Transvaal Supergroup may not be syngenetic with deposition. Similar reasoning (Hoering 1967a,b), and mineralogic evidence of an elevated thermal history (J. F. Machamer, cited by Cloud et al. 1965), suggest that the *n*-alkanes, isoprenoids, steranes and fatty acids reported from the Early Precambrian Soudan Iron-Formation (Meinschein 1965; Burlingame et al. 1965; Oró et al. 1966; Johns et al. 1966; Van Hoeven et al. 1969; Calvin 1969) are partially, or entirely, of recent origin.

3.4. Probable occurrences

Ignoring these uncertain occurrences, it is evident from text-fig. 1 that several types of geochemically stable organic compounds are detectable in Precambrian sediments. Although it should be realized from the arguments presented above that the Precambrian age of these components has not been firmly established, much of the available evidence is consistent with this interpretation. For example, (i) the distributions of *n*-alkanes in these sediments, lacking or exhibiting only slight (Barghoorn et al. 1965; Van Hoeven et al. 1969) odd-carbon number preference, is suggestive of considerable geologic age; (ii) the carbon isotopic data (Hoering 1967a) do not seem indicative of secondary emplacement; (iii) paleontologic studies demonstrate that syngenetically deposited organic matter is present in these sediments; and (iv) stability considerations (Abelson 1959) indicate that these compounds (denoted by the solid symbols in text-fig. 1) could reasonably be expected to survive in a geologic setting for several billion years.

On the other hand, it is evident that some portion of these geochemically stable materials might be of post-depositional origin. The difficulties in assessing the degree to which secondary emplacement has occurred are clearly illustrated by variations in analytical results reported from different laboratories in studies of sediments from the same locality. For example, analyses of fossiliferous black cherts from the Gunflint Iron-Formation (from the Schreiber Beach locality of Barghoorn and Tyler 1965) reported by Smith et al. (1970) yielded only 0.029 to 0.039 ppm of total alkanes (including surface contaminants), as compared with yields of about 5 ppm reported by Oró et al. (1965) and 10 to 100 ppm reported by Van Hoeven et al. (1969); the distributions of *n*-alkanes observed were generally similar, but Oró et al. reported no evidence of odd-carbon number preference, Van Hoeven et al. detected a predominance of *n*-alkanes with an odd number of carbon atoms in the C_{23}-C_{29} range, and Smith et al. showed odd-carbon preference in the C_{27}-C_{33} range. Finally, Oró et al. (1965) reported that 'less than 1 per cent of the hydrocarbons recovered from the chert' could reasonably be attributed to surface contamination, as evidenced by extraction of untreated rocks with a benzene-methanol solvent solution; Smith et al. (1970), however, report values of 82 and 89 per cent for two separate analyses of similarly extractable material. These wide variations in analytical results presumably reflect local differences in the quality of organic preservation and/or levels of in situ contamination, and they present major problems in interpreting the origin(s) of these components.

It is possible that the problems of interpretation illustrated by these analyses are practically, if not theoretically, insuperable and that the extractable components of Precambrian sediments – including those compounds of established geochemical stability – can provide no convincing evidence of early biochemical processes. Whether this conclusion will prove applicable generally remains to be demonstrated; at present, the data suggest that some portion (and perhaps the majority) of these stable compounds, particularly that extracted from the interior of rock specimens, is actually of Precambrian age. In the absence of more definitive criteria to indicate whether these components are syngenetic with Precambrian sedimentation, the only apparent recourse is to use the morphological fossil record as a basis for evaluating the chemical evidence. This solution, of course, is notably imperfect, for it relegates extractable chemical fossils to a subsidiary position, to be accepted only if they are consistent with previous interpretations based on morphological evidence. Nevertheless, such an approach provides a rationale for considering the totality of data available, rather than completely disregarding extractable organic materials as being indigenous, but possibly not syngenetic.

3.5. Biogenicity

If it is assumed, therefore, that the occurrences regarded as 'certain or probable' in text-fig. 1 have been correctly interpreted, four categories of chemical fossils (n-alkanes, isoprenoids, porphyrins and $C^{13}:C^{12}$ values) extend into the Early Precambrian. With the exception of the carbon isotopic values, representatives of each of these categories have been synthesized abiotically in experiments designed to simulate conditions that may have produced organic compounds on the primitive earth. [In principle, comparable carbon isotopic ratios might also be abiotically produced, although this has yet to be demonstrated experimentally.] The environments inducing these reactions (starting materials, energy sources etc.) vary widely – some being more plausible in geologic terms than others – and, considering the arsenal of materials available to the synthetic chemist, it should not be surprising that compounds of these types can be produced by solely abiotic processes.

Nevertheless, the distribution of molecular species in the Precambrian sediments listed in text-fig. 1 seems indicative of biogenicity; while not inconceivable, the possibility of an abiotic origin for these compounds seems relatively remote. It may be noted, for example, that normal alkanes predominate over all other possible isomers in each of these deposits, a feature typical of biological materials, and occur in marked excess over the thermodynamically more stable isoalkanes and cycloalkanes characteristic of many abiotic syntheses. A similar case can be made for the presumed biogenicity of phytane ($C_{20}H_{42}$), an isoprenoid hydrocarbon isolated from the eight sediments indicated. According to L. and M. Fieser (1961, page 105), there are 366,319 possible structural isomers of this compound (i.e. structural arrangements of 20 carbon atoms and 42 hydrogen atoms, not including stereoisomers); only two or three of these isomers commonly occur as products of living systems (Calvin 1969) and these appear to be the same isomers detectable in ancient sediments. If these chemical fossils had been synthesized by a random abiotic process, thousands of isomers, many of which are energetically similar, might be expected in these deposits. This specificity, coupled with the detailed similarities of distribution between Precambrian chemical fossils and compounds extracted from extant organisms and Phanerozoic sediments, has persuaded most investigators that an abiotic origin for these compounds is most unlikely (Hoering 1967a; Calvin 1969).

3.6. Chemical evidence of photosynthesis

Assuming that the relatively stable chemical fossils reported from Precambrian sediments are of biological origin and are the same age as the sediments from which they have been extracted, what inferences may be drawn regarding the biochemistry of the primitive biota? Based on analogy with extant organisms, and on the apparent universality of biochemical processes in living systems, the most obvious inference is that these compounds, or their geochemical precursors, were produced by a series of enzymatically catalyzed intracellular reactions. The presence of n-alkanes and isoprenoids (Oró and Nooner 1967; Meinschein 1967; Hoering 1967a; MacLeod 1968; Calvin 1969) and of porphyrins (Kvenvolden et al. 1968; Kvenvolden and Hodgson 1969) in sediments of the Early Precambrian Swaziland Supergroup (Fig Tree and Onverwacht Groups) would appear to imply, therefore, that organisms exhibiting nucleic acid-directed enzyme synthesis and complex biosynthetic pathways were extant more than 3000 million years ago.

The probable relationship of these chemical fossils to the physiology of early organisms is more conjectural; but if it is assumed that they played functional roles comparable to those of their modern analogues, biological sources can be suggested based on the distribution of these analogues in the extant biota. Thus, since compounds containing isoprene derivatives (e.g. chlorophylls, carotenoids, phycobilins, steroids, vitamin A) are limited in distribution to photosynthetic organisms (including photosynthetic bacteria) and their evolutionary descendents, the presence of isoprenoid hydrocarbons in the Early Precambrian Swaziland Supergroup would seem to indicate that photosynthetic autotrophy had evolved by this point in earth history. The occurrence of porphyrins in these sediments is consistent with this interpretation since porphyrin-containing respiratory pigments (e.g. chlorophylls, cytochromes*) are present in all photosynthetic organisms; moreover, bacterial and algal chlorophylls (as well as those of higher plants) contain a phytol side chain, thought to be the geochemical precursor of pristane and phytane, the predominant isoprenoids extracted from ancient sediments (Oró et al. 1965; Calvin 1969).

The carbon isotopic composition of organic matter in the Swaziland sediments seems similarly suggestive of photosynthetic activity. Studies by R. Park and S. Epstein (1960, 1961) have shown that the two stable carbon isotopes C^{12} and C^{13} undergo kinetic fractionation during the process of green-plant photosynthesis; a series of steps, the most important of which involves the enzyme-mediated carboxylation of ribulose/1.5 diphosphate to yield 3-phosphoglyceric acid, produces organic matter enriched in the lighter isotope C^{12}, relative to its concentration in the atmospheric carbon dioxide reservoir. Carbon isotopic ratios have been measured in a wide variety of naturally occurring materials produced by both biologic and inorganic processes, and of both modern and geologic age (for a summary of these values see Degens 1965, page 282). The $C^{13}:C^{12}$ ratios of apparently syngenetic, non-extractable organic matter in sediments of the Fig Tree Group (Hoering 1967a) and the Onverwacht Group (D. Oehler 1969, personal communication) fall well within the range of values characteristic of organic materials of known photosynthetic origin or derivation (e.g. petroleum, coal) and differ markedly from values typical of inorganically produced carbon compounds. These ratios presumably reflect the occurrence of carbon isotopic fractionation; as such, they constitute suggestive evidence of algal, or possibly bacterial, photosynthesis.

In summary, the reported occurrence of isoprenoids and porphyrins in sediments of the Swaziland Supergroup, and the carbon isotopic ratios of organic matter deposited apparently syngenetically with these sediments, are consistent with, and seem suggestive of, the presence of autotrophic photosynthesizers of bacterial and/or algal affinities. For reasons detailed above,

---

* It should be noted that although *Desulfovibrio*, an anaerobic sulfate-reducer, contains cytochromes, it seems an unlikely source for Early Precambrian porphyrins since $S^{34}:S^{32}$ measurements (Smith et al. 1970) suggest that sulfate-reducing bacteria may not have evolved until much later in geologic time. The cobalt-containing tetrapyrroles (vitamin B_{12}) of anaerobic methane bacteria represent a more plausible source for ancient porphyrins, but such bacteria apparently lack isoprenoids.

this interpretation should be regarded as tentative, to be accepted only if it is consistent with evidence from the morphological fossil record.

4.   The Early Precambrian record

It is perhaps a truism that the most interesting and challenging scientific problems involve questions of broadly encompassing import for which solutions have yet to be found, but which have the appearance of being soluble, at least in principle. The origin of biological systems is one such problem. A variety of approaches have been applied to this question, the most promising of which has involved laboratory syntheses of organic 'biologic' compounds in nonbiologic systems (for a recent review of these experiments see Ponnamperuma and Gabel 1968). To provide a putative solution to this problem, however, such experiments must do more than 'merely' spontaneously assemble a group of abiotically produced organic molecules into a complex, self-replicating living system – a feat that may prove attainable within the next few decades – for it is crucial that this endpoint be reached through a series of steps that are plausible in terms of the Early Precambrian environment (and even then, only one of the potential pathways to biological complexity will have been traced, with no firm guarantee that living systems originated in precisely this manner).

The plausibility of such experiments (concentrations and types of starting materials, energy sources etc.) can be determined only by referral to the geologic record, and the limitations it sets on the nature of the primitive environment. In this regard, it should be recognized that Early Precambrian sediments provide no evidence for the existence of the methane-ammonia atmosphere commonly postulated for the primitive earth (Abelson 1966). On the contrary, certain of the oldest sedimentary sequences now known (e.g. the Swaziland Supergroup with an apparent age of about 3.2 billion years) contain inorganically precipitated carbonate units (Ramsay 1963) indicating that carbon dioxide, and not methane, may have been the dominant form of atmospheric carbon at this point in geologic time. Nevertheless, it is possible, and perhaps likely, that a highly reduced atmosphere was present very early in earth history, prior to the deposition of the oldest known rocks some 3.5×10^9 years ago (Donn et al. 1965). Thus, the geologic evidence suggests that if living systems originated in the presence of a methane-ammonia atmosphere, an assumption basic to most experimental approaches to the problem (Ponnamperuma and Gabel 1968), this signal event must have occurred during the earliest third of earth history.

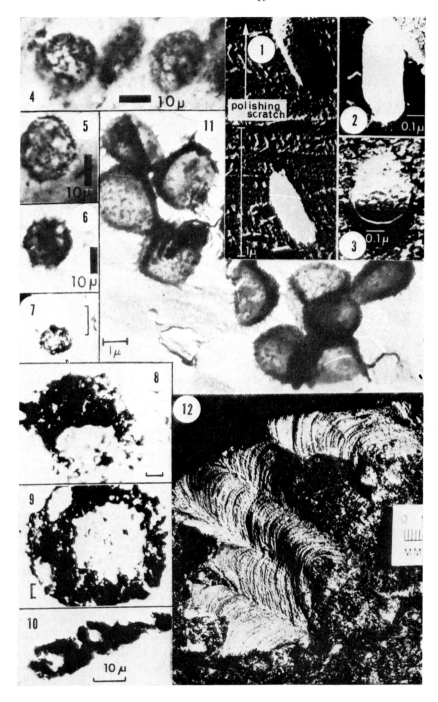

4.1. Oldest known microfossils

Based on comparative studies of the morphology and biochemistry of extant organisms, it seems probable that unicellular, microscopic, anaerobic pro-caryotes were among the earliest living systems; presumably, these earliest forms were heterotrophic, using abiotically produced organic matter as a carbon source. Morphological evidence supporting the major portion of these suppositions recently has been secured from carbonaceous sediments of the Swaziland Supergroup occurring in the eastern Transvaal near Bar-berton, South Africa. This evidence indicates that unicellular microorgan-isms, almost certainly of procaryotic affinities, were extant more than 3000 million years ago; it seems apparent, however, that these forms were relative-ly advanced, suggesting that direct evidence of life's beginnings will be dis-covered only in still older deposits.

In figs. 1–3 are shown electron micrographs of bacteriumlike microfossils (*Eobacterium isolatum* Barghoorn and Schopf) organically preserved in primary black cherts of the Fig Tree Group, the middle unit of the Swaziland Supergroup, exceeding 3.1×10^9 years in age (Ulrych et al. 1967; Allsopp et al. 1968). Although the probable physiological characteristics of these

---

Figs. 1–12. Microfossils, microstructures and stromatolites from the Early Precambrian of Africa. Except as otherwise noted, line for scale in each figure represents 10 microns. Figs. 1–3. Transmission electron micrographs showing *Eobacterium isolatum* Barghoorn and Schopf in surface replicas of black chert from the Fig Tree Group (ca. 3.1×10^9 years old), eastern Transvaal (Barghoorn and Schopf 1966): (1) organic cell (white, below) and its imprint in the rock surface (above); (2) rod-shaped cell; (3) transverse section showing organic cell wall and cell lumen infilled with silica.

Figs. 4–6. Optical photomicrographs showing *Archaeosphaeroides barbertonensis* Schopf and Barghoorn in petrographic thin sections of black chert from the Fig Tree Group (Schopf and Barghoorn 1967): (4) spheroids exhibiting irregularly reticulate surface tex-ture; (5) medial optical section; (6) unicell containing coalified organic matter interpreted as degraded remnants of original cytoplasm.

Figs. 7–10. Optical photomicrographs showing possibly biogenic microstructures in petro-graphic thin sections (figs. 7, 10) and in ground preparations of sediments from the Onver-wacht Group (ca. 3.2×10^9 years old), eastern Transvaal (B. and L. Nagy 1969; Engel et al. 1968; Nagy and Urey 1969).

Fig. 11. Transmission electron micrograph showing spinous spheroids ('Mikrosporen') in a surface replica of stromatolitic dolomite from the Bulawayan Group (ca. 2.8×10^9 years old), Rhodesia (Oberlies and Prashnowsky 1968).

Fig. 12. Optical photograph of finely laminated algal stromatolites from the Bulawayan Group, collected by K. A. Kvenvolden at Huntsman Quarry near Turk Mine, 33 mi NNE of Bulawayo, Rhodesia.

minute, rod-shaped unicells are quite uncertain, their size, shape, cell wall ultrastructure and general organization seem indicative of procaryotic (eubacterialean) affinities (Barghoorn and Schopf 1966).

Optical photomicrographs of a second species from the Swaziland sediments, *Archaeosphaeroides barbertonensis* Schopf and Barghoorn, are shown in figs. 4–6. These organically preserved unicells, exhibiting a granular surface texture (fig. 4) and commonly containing clumps of coalified organic matter (fig. 6), have been detected in thin sections and in hydrofluoric acid-resistant residues of the Fig Tree cherts and shales (Pflug 1966a, 1967; Schopf and Barghoorn 1967; Engel et al. 1968). In general morphology, *A. barbertonensis* is similar to modern coccoid blue-green algae (Chroococcales) and is particularly comparable to algal microfossils of younger Precambrian age. [Compare figs. 4–6 with figs. 16, 38, 43, 44 and with figs. 12 and 13 in Cloud and Licari (1968) showing Early Precambrian unicells from the Soudan Iron-Formation.] It appears probable, therefore, that *A. barbertonensis* was a photosynthetic autotroph; if so, this organism would seem a likely source for the chemical fossils (isoprenoids, porphyrins and carbon isotopic ratios) suggestive of photosynthetic activity detected in the Fig Tree sediments. Furthermore, its general morphology seems more suggestive of algal than of bacterial affinities, perhaps indicating that oxygen-producing green-plant photosynthesis had evolved by this stage in biologic history. The evidence for this interpretation is not entirely compelling, and the chemical data seem equally consistent with the presence of photosynthetic bacteria (not producing oxygen as a by-product); it may be noted, however, that the reactants used by algal photosynthesizers, carbon dioxide and water, were readily available in the Swaziland environment, and some atmospheric oxygen was apparently present, possibly, but not certainly of photosynthetic origin (Schopf and Barghoorn 1967).

Figs. 7–10 show microstructures from the Onverwacht Group (about 3.2 billion years in age), the lowest unit of the Swaziland Supergroup, recently reported by A. E. J. Engel, B. Nagy and their coworker (Engel et al. 1968; Nagy and Urey 1969; B. and L. Nagy 1969). Although roughly similar in morphology to *A. barbertonensis*, these microstructures are commonly cup-shaped, rather than spheroidal, and they exhibit a much broader range of morphologic variability. Because of this variability, the biogenic nature of these forms is regarded as uncertain (B. and L. Nagy 1969). However, this relatively inconstant morphology might be interpreted as the result of incomplete preservation and the presence of several morphologically similar species. Additional, more detailed studies are needed, particularly of apparently less distorted specimens, to confirm or refute the probable biogenicity of these objects.

In summary, the bacterium-like rods and alga-like spheroids of the Fig Tree Group, with a minimum age of about 3.1 billion years, constitute the oldest morphological evidence of living systems now known from the geologic record; the Onverwacht microstructures, of probable, but as yet uncertain biogenicity, are perhaps 100 million years older. The physiological characteristics of this assemblage are conjectural, but morphologic comparisons with extant organisms suggest that the spheroidal microfossils were photosynthetic autotrophs, an interpretation in agreement with the available geochemical evidence. Whether exhibiting bacterial or algal photosynthetic pathways, the biochemical complexity of these forms would seem to imply the occurrence of a substantial period of prior evolutionary development – perhaps several hundred million years in duration – apparently indicating that biological systems originated extremely early in earth history. Moreover, if these organisms were physiologically comparable to extant algae, their presence would have resulted in the production of free atmospheric oxygen, significantly altering the presumed anoxic primitive environment and initiating the transition from an oxygen-deficient to the highly oxygenic atmosphere of later geologic time.

4.2. Oldest known stromatolites

Additional evidence of Early Precambrian biologic activity was described in 1940 by A. M. Macgregor who reported the occurrence of the oldest stromatolitic structures now known from sediments of the Bulawayan Group of Rhodesia. These finely laminated calcareous stromatolites (fig. 12), comparable in gross morphology to the layered bioherms of modern blue-green algal communities, are apparently on the order of 2.8 billion years in age (Nicolaysen 1962) and it is possible that they are only slightly younger than the Swaziland microfossils (Holmes 1965, page 371). The biogenicity of the Bulawayan stromatolites is indicated by their typically biohermal morphology, by their carbon isotopic composition (Hoering 1967a), and by their association with the spinose, organic spheroids shown in fig. 11 (Oberlies and Prashnowsky 1968). In addition, based on analogy with modern stromatolitic communities, their regularly laminated organization would appear to reflect a phototrophic response comparable to that exhibited by extant algal biocoenoses; since photosynthetic bacteria do not exhibit such laminated organization, these stromatolites constitute firm, if indirect evidence of

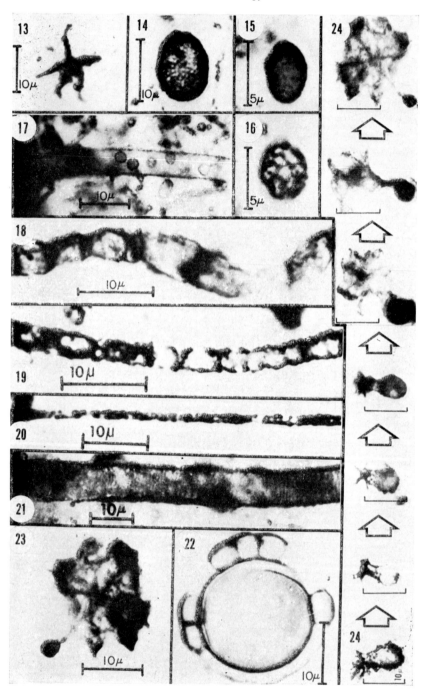

oxygen-producing photosynthesis. Moreover, this organization presumably reflects the presence of algal filaments, since the laminae of younger bioherms are produced primarily by the subparallel aggregation of filamentous, rather than of coccoid cyanophytes. This latter inference seems similarly suggested by the occurrence of threadlike microfossils (Oscillatoriaceae?) in black cherts of the Early Precambrian Soudan Iron-Formation of Minnesota (Gruner 1925, page 152), comparable in age to the Bulawayan stromatolites.

Based on the morphological and chemical fossils from the Bulawayan and Swaziland sediments, and on the occurrence of probable algae in the Soudan Iron-Formation (Gruner 1925; Cloud and Licari 1968), there seems little doubt that autotrophic, photosynthetic microorganisms were extant during the Early Precambrian. Although the Swaziland organisms may have exhibited bacterial pathways, rather than the more advanced process of algal photosynthesis, microscopic algae had apparently become established more than 2.8 billion years ago; it seems evident, therefore, that biologically generated oxygen has been added to the environment since very early in earth history.

5. The Middle Precambrian record

In marked contrast to the paucity of paleobiologic data available from the Early Precambrian, and the primitive, simple nature of the oldest known microfossils, Middle Precambrian sediments (i.e. between about 2.5 and 1.7 billion years in age) contain a wealth of evidence of biologic activity and a strikingly advanced biota. During recent years, Middle Precambrian microfossils have been reported from the Witwatersrand Supergroup of South

---

Figs. 13–24. Microfossils from the Middle Precambrian Gunflint chert. Except as otherwise noted, line for scale in each figure represents 10 microns.

Figs. 13–24. Optical photomicrographs showing microorganisms in petrographic thin sections of stromatolitic black chert from the Gunflint Iron-Formation (1.6–2.0×10^9 years old) of Ontario, Canada (Barghoorn and Tyler 1965): (13) *Eoastrion simplex* Barghoorn; (14) *Huroniospora microreticulata* Barghoorn; (15) *H. psilata* Barghoorn; (16) *H. macroreticulata* Barghoorn; (17) *Entosphaeroides amplus* Barghoorn; (18, 19) *Gunflintia grandis* Barghoorn; (20) *G. minuta* Barghoorn; (21) *Animikiea septata* Barghoorn; (22) *Eosphaera tyleri* Barghoorn; (23) *Kakabekia umbellata* Barghoorn; (24) specimens of *K. umbellata* arranged in an ontogenetic sequence proposed by Barghoorn and Tyler (1965).

Africa (Schidlowski 1966; Prashnowsky and Schidlowski 1967; Pflug et al. 1969), the Transvaal Supergroup of South Africa (Cloud and Licari 1968), the Vallen Group of Greenland (Bondensen et al. 1967), the middle (LaBerge 1967) and lower Belcher Group of Hudson Bay (Hofmann and Jackson 1969), the Gunflint Iron-Formation of southern Canada (Barghoorn and Tyler 1965; Cloud 1965), and several other localities (see Glaessner 1962, 1966). Of these occurrences, the most significant in terms of biological diversity and fidelity of preservation are the microbiotas of the Gunflint Iron-Formation (figs. 13–26) and the lower Belcher Group (figs. 27–36) of the Canadian Shield. Both of these assemblages are preserved in carbonaceous, microcrystalline cherts associated with, or in part comprising, laminated algal stromatolites; and in both deposits, the organisms are organically preserved in a three-dimensionally little altered condition. Consideration of these assemblages, and in particular the microbiota of the Gunflint cherts for which more detailed information is available, provides much insight into the evolutionary status of Middle Precambrian life.

As is noted above, the discovery of the Gunflint assemblage in 1953 by S. A. Tyler, and the subsequent announcement of its occurrence (Tyler and Barghoorn 1954) are in large measure responsible for the recent increase of interest in Precambrian paleobiology. Studies of the cherty Gunflint stromatolites, culminating in the description of several new genera of microscopic thallophytes (Barghoorn and Tyler 1965), clearly demonstrated, for the first time, that some and presumably all unmetamorphosed Precambrian stromatolites are of biologic origin; that the laminations of such stromatolites reflect the presence of filamentous blue-green algae aggregated in undulating sheet-like mats; and that primary carbonaceous cherts can provide a mineralogic matrix suitable for the organic preservation of delicate Precambrian microorganisms. Equipped with the knowledge of this association (an association that in retrospect seems evident from the early studies of Walcott and Gruner), a strategy emerged that has yielded high returns in recent years: the microscopic investigation of Precambrian, unmetamorphosed, primary black cherts, particularly those found in association with laminated stromatolitic structures. It is this strategy that has resulted in the discovery of virtually all of the important Precambrian microbiotas now known.

5.1. *Microbiota of the Gunflint chert*

The Gunflint Iron-Formation, thought to be between 1.6 and 2.0 billion years in age (Hurley et al. 1962; Davidson, according to Cloud, 1968, pages

10–11; Faure and Kovach 1969) is a moderately thick sedimentary sequence occurring in a latitudinally trending trough along the northern shore of Lake Superior in Ontario, Canada. The most fossiliferous chert horizon, the lowest member of the formation, varies from a few inches to about two feet in thickness, and is commonly stromatolitic throughout its lateral extent of some 150 miles. The Gunflint microbiota is particularly well-preserved in unmetamorphosed siliceous stromatolites near Schreiber, Ontario, at the easternmost exposure of the formation. Microorganisms are also prevalent, however, in more altered facies to the west, and spheroidal, apparently planktonic members of the assemblage are not restricted in distribution to stromatolites.

The Gunflint biota is composed predominantly of microscopic algae that formed laminar mats near the sediment-water interface of the shallow, agitated basin; at the Schreiber Beach locality, these mats enveloped boulders of an underlying conglomeratic unit producing mound-shaped biohermal masses. In a few cases, the cells of dead algal filaments became infilled with pyrite during diagenesis, producing a chain of crystals (fig. 25) that records the position and shape of individual algal cells, but not their detailed cellular structure. Most commonly, however, the living organisms were entrapped and embedded in rapidly deposited colloidal silica which infilled the cell lumina of the microorganisms while leaving their organic cell walls intact. Following lithification, this amorphous silica crystallized to form crypto-crystalline quartz, structurally preserving the organic microfossils.

Of the 12 species of microscopic plants recognized in the assemblage (Barghoorn and Tyler 1965), the most commonly occurring are unicellular spheroids (figs. 14–16) and septate filaments (figs. 18–20) primarily of cyano-phycean affinities. The ellipsoidal to spheroidal unicells (*Huroniospora* spp.), a few to several microns in diameter, exhibit a variety of surface textures (e.g. figs. 14–16). The variable size and surface texture of these microfossils suggests that several species are represented; their general morphology, however, and their occurrence in stromatolitic bioherms indicates that most of these forms can be referred to the Cyanophyceae (Chroococcales). The filamentous members of the biota (figs. 17–21) represent a similarly hetero-geneous mixture of apparently procaryotic microorganisms. Certain of the smaller septate forms (e.g. fig. 20) may be related to filamentous bacteria, such as those shown in fig. 37 from the 2.1 billion year-old Witwatersrand Supergroup of South Africa (Oberlies and Prashnowsky 1968); however, the occasional preservation of intercalary cells comparable to the reproductive heterocysts of Late Precambrian (figs. 46–48; Schopf and Barghoorn 1969)

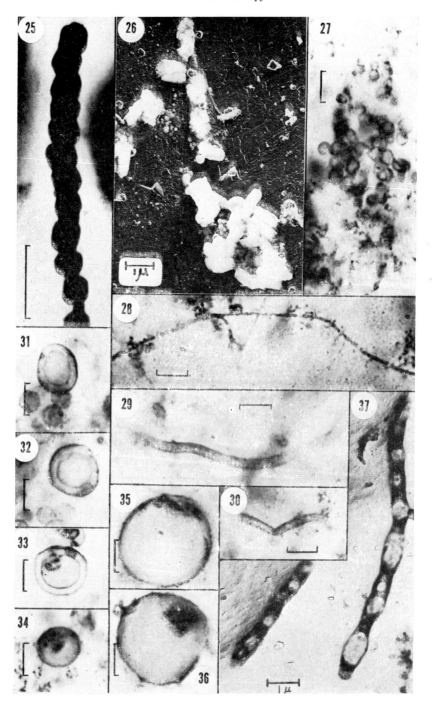

and extant blue-green algae (fig. 45) noted by G. Licari and P. E. Cloud, Jr. (1968) indicates that the majority of these microfossils (*Gunflintia* spp.) are of cyanophycean (nostocalean) affinities. The larger Gunflint filaments (figs. 17–19, 21, 25), of less common occurrence in the assemblage, are also comparable to blue-green algae (Oscillatoriaceae) although the endospore-containing organism shown in fig. 17 (*Entosphaeroides amplus* Barghoorn) is also comparable to certain modern iron-bacteria (e.g. *Crenothrix polyspora* Cohn).

Other organisms, of less certain biological relationships, are shown in figs. 13, 22–24. *Eoastrion* (fig. 13) is a minute actinomorphic organism composed of short filamentous appendages radiating from a central irregular disc. As Cloud (1965) has suggested, the size and general organization of this microfossil seems strikingly similar to that exhibited by *Metallogenium personatum* Perfil'yev, an iron- and manganese-oxidizing bacterium described from several Karelian lakes (Kuznetsov et al. 1963; fig. 57, page 172). *Kakabekia umbellata* Barghoorn (fig. 23) characteristically exhibits a tripartite organization consisting of an umbrella-like cap, a basal bulb, and a slender connecting stalk. In fig. 24 a series of individuals is shown arranged in the possible ontogenetic sequence suggested by Barghoorn and Tyler (1965); according to their scheme, the basal bulb decreases in size, producing a corresponding increase in size of the umbrella-like cap, during the development of the adult form. The biological relationships of *K. umbellata* are quite uncertain, and it is probably only superficially similar to an extant anucleate organism with which recently it has been compared (Siegel and

---

Figs. 25–37. Microfossils from the Middle Precambrian of Canada and South Africa. Except as otherwise noted, line for scale in each figure represents 10 microns.

Fig. 25. Algal filament replaced by pyrite, optically photographed in a petrographic thin section of black chert from the Gunflint Iron-Formation (1.6–2.0×10^9 years old) of Ontario, Canada (Barghoorn and Tyler 1965).

Fig. 26. Transmission electron micrograph showing rod-shaped bacteria in a surface replica of the Gunflint chert (Schopf et al. 1965).

Figs. 27–36. Optical photomicrographs showing microorganisms in a petrographic thin section of laminated dark gray chert of the lower Belcher Group (ca. 1.7×10^9 years old) from the Belcher Islands of Hudson Bay (Hofmann and Jackson 1969): (27) cf. *Huroniospora* spp.; (28) cf. *Biocatenoides* sp.; (29, 30) possibly aff. *Eomycetopsis*; (31–33) 'Type 4' spheroid of Hofmann and Jackson, possibly aff. *Eosphaera*; (34–36) 'Type 2' spheroid of Hofmann and Jackson exhibiting peripheral dark spot.

Fig. 37. Transmission electron micrograph showing filamentous bacteria from the 'Basel Reef' of the Witwatersrand Supergroup (ca. 2.1×10^9 years old) of South Africa (Oberlies and Prashnowsky 1968).

Siegel 1968). Modern and fossil morphological counterparts of *Eosphaera tyleri* Barghoorn, the double-walled spheroid shown in fig. 22, are apparently unknown; its general morphology suggests that it may have been a planktonic alga, possibly reproducing by means of the peripheral sporelike bodies.

Electron microscopic studies of the Gunflint assemblage have corroborated the presence of morphological features originally noted in optical studies (Cloud and Hagen 1965). In addition, electron microscopy of replicas of polished and etched surfaces of the fossiliferous chert has demonstrated the occurrence of the rod-shaped bacteria shown in fig. 26 (Schopf et al. 1965). Based on morphological comparisons with extant microorganisms, it seems likely that these microfossils are related to modern iron-bacteria (e.g. *Sphaerotilus*); their occurrence may indicate that the iron-bearing facies of the formation were deposited at least partially as the result of biologic activity.

To reiterate briefly, the Gunflint assemblage is composed predominantly of procaryotic microorganisms including representatives of several extant cyanophycean families (Chroococcaceae, Oscillatoriaceae Nostocaceae) and a variety of chemosynthetic bacteria. Other taxa, of less certain biological relationships are also represented; it remains to be established whether these forms were anucleate, eucaryotic or perhaps intermediate in cellular organization. It seems notable that assured eucaryotic (nucleated) organisms have not been identified in the biota. Moreover, structural features characteristic of higher thallophytes (e.g. tissue and sporangial formation) are apparently not represented, and organisms exhibiting true branching, a growth habit typical of eucaryotic plants (but also exhibited by one group of cyanophytes), are very rare. Furthermore, as described by Barghoorn and Tyler (1965), the one branching species known from the assemblage (*Archaeorestis schrei-berensis* Barghoorn) is generally preserved in a rather unusual manner (its tubular form being outlined by pyrite particles rather than organic matter) and it is possible, and perhaps likely, that the original morphology of this organism is not as faithfully preserved as that of other members of the biota. In any case, all of the Gunflint fossils that can be related with confidence to modern microorganisms appear to be of procaryotic (i.e. bacterial or cyanophycean) affinities. If this proves to be a general characteristic of the Middle Precambrian biota – a possibility that is apparently consistent with the limited data available – it would establish that the advanced, eucaryotic cell did not evolve until relatively late in geologic time.

5.2. Microbiota of the lower Belcher Group

In addition to the well-known Gunflint assemblage, other occurrences of microorganisms recently have been reported from the Middle Precambrian. The best preserved of these appears to be the microbiota of the lower Belcher Group, perhaps 1.7×10^9 years old, described by H. J. Hofmann and G. D. Jackson (1969) from stromatolitic cherts of the Belcher Islands of Hudson Bay.

Description of this assemblage has thus far been limited to the eight types of microorganisms detected in a single thin section of the dark gray cherts. Spheroidal microfossils (figs. 27, 34–36), compared to the Gunflint genus *Huroniospora* by Hofmann and Jackson (1969), are well-represented in the biota, as are double-walled spheroids (figs. 31–33) that seem somewhat similar to *Eosphaera* (note particularly the presence of a peripheral sporelike body in fig. 32). Also represented are colonial bacteria (fig. 28) referred to the genus *Biocatenoides*, originally described from the much younger Bitter Springs Formation of central Australia (Schopf 1968a), and nonseptate tubules (figs. 29–30) roughly comparable to *Eomycetopsis*, a second genus of the Bitter Springs assemblage. In some respects, therefore, the Belcher microflora may be roughly intermediate in evolutionary status between the Gunflint biota and a much younger assemblage; like the Gunflint, but in contrast to the younger biota, the Belcher assemblage contains no assured eucaryotic organisms.

Although Hofmann and Jackson (1969) do not interpret any of the Belcher fossils as being eucaryotes, they note the occurrence of a 'dark, circular or stellular object ... located on or near the periphery of some cells'; these rather irregular objects, shown in figs. 34–36, might be interpreted as possible evidence of intracellular organelles (and therefore of eucaryotic organization) as has been suggested for similar spots recently noted on several unicells from the Frustration Bay/Discovery Point locality of the Gunflint Iron-Formation (E. S. Barghoorn 1968, personal communication). Such an interpretation cannot be excluded entirely, particularly since demonstrably nucleated organisms from the Late Precambrian Bitter Springs Formation commonly contain somewhat similar structures (Schopf 1968a), as do probable eucaryotes from the approximately 1.3 billion year-old Beck Spring Dolomite of California (figs. 39, 40; Cloud et al. 1969). However, in view of the occurrence of intracellular organic bodies in some fossil blue-green algae (Croft and George 1959, Pls. 42, 44), resulting from the coalescence of cytoplasmic material during preservation, additional data are needed (e.g. optical

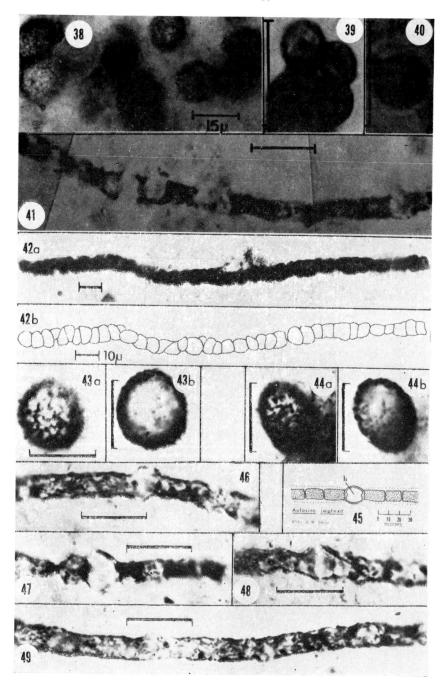

or electron microscopic evidence of the structural organization of such spots; statistical studies showing the consistent distribution of such spots in many cells; or an apparent mitotic sequence such as that shown in fig. 50) before such objects may be accepted as firm evidence of eucaryotic organization.

5.3. *Middle Precambrian microorganisms and atmospheric evolution*

Several lines of evidence indicate that oxygen-producing photosynthetic microorganisms dominated the Middle Precambrian biota. The predominantly photosynthetic nature of the Gunflint assemblage, for example, is amply attested to by the detailed morphological similarity of most of the Gunflint microfossils to extant blue-green algae; by the organization of these organisms in stromatolitic, biohermal communities reflecting a phototrophic growth habit; by the reported occurrence of isoprenoid hydrocarbons, possibly derived from chlorophyll, in the Gunflint cherts (Oró et al. 1965; Van Hoeven et al. 1969); and by carbon isotopic ratios, apparently indicative of photosynthetic fractionation, measured in both organic and inorganic carbon compounds of the Gunflint sediments (Hoering, according to Barghoorn and Tyler 1965; Smith et al. 1970). Similar morphological and geochemical evidence has been reported from several other Middle Precambrian deposits (text-fig. 1).

It seems evident, therefore, that at least locally, near sites of photosynthetic activity, the Middle Precambrian environment must have been oxygenic. Moreover, the wide geographic distribution of plant microfossils, algal stromatolites and oxidized sediments of similar geologic age suggests that

---

Figs. 38–49. Microfossils from the Late Precambrian of California and South Australia. Except as otherwise noted, line for scale in each figure represents 10 microns.

Figs. 38–42. Optical photomicrographs showing microorganisms in petrographic thin sections of stromatolitic black chert from the Beck Spring Dolomite (ca. 1.3×10^9 years old) of southern California (figs. 38–41 from Cloud et al. 1969; fig. 42 from Gutstadt and Schopf 1969): (38) 'Type b' spheroid of Cloud et al.; (39–40) 'Type d' spheroid of Cloud et al. exhibiting internal dark spots; (41–42) cyanophycean filaments.

Figs. 43–49. Optical photomicrographs showing microorganisms in petrographic thin sections of black chert (figs. 43, 44) and calcareous sediments (figs. 46–49) of the Skillogalee Dolomite (ca. 1.0×10^9 years old) of South Australia (Schopf and Barghoorn 1969): (43, 44) *Myxococcoides muricata* Schopf and Barghoorn; (45) line drawing of *Aulosira* (h = heterocyst), a living cyanophyte, for comparison with figs. 46–48; (46–48) heterocystous portion of *Archaeonema longicellularis* Schopf, a nostocalean cyanophyte; (49) septate portion of *A. longicellularis*.

oxygenic conditions were widespread, and that an oxygenic atmosphere may have been established at this point in earth history. Such conditions would be consistent with the presence of planktonic algae in the Gunflint cherts, and in other Precambrian iron-formations of comparable or even greater geologic age (LaBerge 1967; Cloud and Licari 1968), since an oxygenic atmosphere would result in the formation of an ozone layer that would absorb ultra-violet irradiation thereby permitting habitation of surface waters by phytoplankton. Finally, recent calculations seem to indicate that even in the absence of a photosynthetic biota, a highly oxygenic atmosphere (as much as 25% of the present atmospheric oxygen level) could have been maintained solely as the result of ultra-violet induced photodissociation of water vapor, and that such conditions may have prevailed much earlier than the Middle Precambrian (Brinkmann 1969). Although such evidence cannot be regarded as conclusive, and cogent arguments have been raised to support the more traditional view of an essentially anoxic Middle Precambrian environment, these new data suggest that oxygenic conditions may have been well-established as early as two billion years ago.

6. The Late Precambrian record

Perhaps the greatest advances in Precambrian Paleobiology of the past decade have accrued from the investigation of Late Precambrian sediments; several well-preserved microbiotas, exhibiting considerable biological diversity, recently have been described. These new discoveries appear to indicate that the Late Precambrian featured a series of important evolutionary innovations, including the origin, diversification and 'modernization' of many thallophytic stocks, that culminated in the development of multicellular organization near the beginning of the Paleozoic.

Since about 1960, and particularly during the past two or three years, microorganisms have been discovered in many Late Precambrian sedimentary units: the Beck Spring Dolomite of California (figs. 38–42; Cloud et al. 1969; Gutstadt and Schopf 1969), the Skillogalee Dolomite of South Australia (figs. 43, 44, 46–49; Schopf and Barghoorn 1969), the Bitter Springs Formation of central Australia (figs. 50–69; Schopf 1968a; Barghoorn and Schopf 1965), the Paradise Creek Formation of northern Australia (Licari et al. 1969), the Muhos Shale of Finland (Tynni and Siivola 1966), the Belt Supergroup of Montana (Pflug 1964, 1965, 1966b), the Nonesuch Shale of Michigan (Meinschein et al. 1964; Barghoorn et al. 1965), Brioverian cherts of

Normandy (Roblot 1963, 1964), and other sediments of Scotland, Sweden, Norway, Russia, Siberia, Poland and China (cited in Glaessner 1962, 1966). In addition, organic geochemical studies, demonstrating fundamental bio-chemical similarities between Late Precambrian and extant organisms (e.g. Barghoorn et al. 1965), have been carried out on several of these sediments (text-fig. 1). Moreover, the existence of Precambrian metazoans – tradi-tionally a point of controversy in paleontology – now seems generally accepted, although the oldest assured metazoans (Glaessner and Wade 1966), from the Pound Quartzite of the Ediacara Hills of South Australia, are apparently only of very latest Precambrian age (i.e. 570–680 million years).

6.1. The Bitter Springs microflora

Of the Late Precambrian microbiotas now known, the most diverse and best preserved is that reported from primary black cherts of the Bitter Springs Formation of Northern Territory, Australia, about 900 million years in age (Barghoorn and Schopf 1965; Schopf 1968a, 1970). This assemblage is composed of some 30 species of microorganisms including blue-green (figs. 52–55, 63–69), green (figs. 50, 51, 56) and possibly red algae (figs. 58, 59), probable fungi (figs. 60–62), possible dinophyceans (figs. 57), filamentous bacteria, and other organisms of less certain biological relationships (Schopf 1968a).

In several respects, the Bitter Springs microflora is comparable to the biota of the Middle Precambrian Gunflint Iron-Formation. Both assem-blages are primarily composed of procaryotic microorganisms that formed the undulating laminar mats of biohermal stromatolites; and both assem-blages were organically and structurally preserved in situ in relatively thin beds of carbonaceous chert, inorganically precipitated in shallow, extensive, elongate basins. The paleoecology of the two deposits appears to have been generally similar, although iron-bearing facies are not represented in the Bitter Springs Formation. In spite of these rather pronounced similarities in biotic organization, the assemblages differ markedly in biologic composi-tion; the Bitter Springs flora is considerably more 'modern' in appearance, including many species that are comparable to extant thallophytes and relatively few organisms for which modern morphological counterparts are unknown.

The dominant organisms of the Bitter Springs cherts are filamentous and coccoid blue-green algae. The filamentous taxa, referred to the Oscilla-toriaceae, Nostocaceae and Rivulariaceae, exhibit such typically cyanophy-

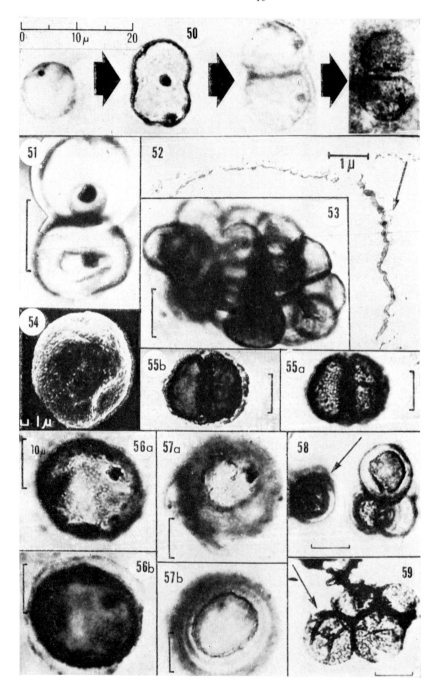

cean features as rounded (figs. 64, 66), attenuated (figs, 68, 69), or dilated terminal cells (fig. 65), disc- or barrel-shaped medial cells (figs 64–66, 68, 69), partial septations (fig. 66) and encompassing organic sheaths (figs. 66, 67). The coccoid (chroococcacean) algae occur singly (fig. 54), in sheath-enclosed pairs (figs. 55), or in colonies with the component cells embedded in an organic matrix (fig. 53); their multilayered cell walls (fig. 52) are similar to those of modern cyanophytes (Schopf 1970).

Of the 19 species of blue-green algae represented in the Bitter Springs microflora, 14 have been referred to modern families; of these, 6 or 7 seem comparable, at the specific or generic level, to living algae. For example, the colonial alga shown in fig. 53 (*Myxococcoides minor* Schopf) is comparable in morphology to members of the modern genus *Anacystis*; *Heliconema australiensis* Schopf (fig. 63) is essentially indistinguishable from modern *Spirulina*; the detailed morphology of *Oscillatoriopsis obtusa* Schopf (fig. 64) is found in extant algae of the genus *Oscillatoria*; *Cephalophytarion grande* Schopf (fig. 65) is quite similar to the living cyanophyte *Microcoleus vaginatus* (Vaucher) Gomont; the broad, sheath-enclosed filament shown in fig. 66 (*Paleolyngbya barghoorniana* Schopf) seems morphologically identical to living algae of the genus *Lyngbya*; and the prominent terminal hair exhibited by *Caudiculophycus rivularioides* Schopf (figs. 68, 69) is indicative of affinities to modern and fossil Rivulariaceae.

It seems evident, therefore, that blue-green algae were highly diversified in the Late Precambrian, and that certain cyanophytes have exhibited little or no evolution, at least in terms of organismal morphology, since Bitter Springs time; these organisms constitute an excellent example of evolutionary conservatism, perhaps the most striking such example now known from the geologic record.

Figs. 50–59. Spheroidal microfossils from the Late Precambrian Bitter Springs chert. Except as otherwise noted, line for scale in each figure represents 10 microns.

Figs. 50–59. Optical photomicrographs and transmission (fig. 52) and scanning (fig. 54) electron micrographs showing spheroidal algae in petrographic thin sections, an epoxy-embedded ultrathin section (fig. 52) and in hydrofluoric acid macerations (figs. 54, 59) of laminated black cherts from the Bitter Springs Formation (ca. 0.9×10^9 years old) of central Australia (Schopf 1968a, 1970): (50) *Glenobotrydion aenigmatis* Schopf apparently preserved at varying stages of a mitotic sequence; (51) partially plasmolyzed cells of *Caryosphaeroides pristina* Schopf containing probable nuclear residues; (52) multilayered (at arrow) cell wall of *Myxococcoides minor* Schopf; (53–55) *M. minor* Schopf; (56) *Globophycus rugosum* Schopf; (57) *Gloeodiniopsis lamellosa* Schopf; (58, 59) algal unicells exhibiting surficial triradiate markings (at arrows) of probable ceramialean affinities.

6.2. *Evolutionary trends in the Cyanophyta*

The Bitter Springs microflora contains the most diverse assemblage of blue-green algae known from a single geologic unit. It seems appropriate, therefore, to digress briefly at this point to consider the geologic distribution of the Cyanophyta, and the position of this assemblage in the evolutionary history of these procaryotic plants.

Although the primitive nature of the Cyanophyceae has long been assumed from studies of extant species, until quite recently the known fossil record offered little evidence of the antiquity and evolutionary development of the Class. As is shown in text-fig. 2, however, there has been a marked increase of interest in fossil cyanophytes in recent years. Of the 156 occurrences of cellularly preserved blue-green algae now known, nearly half have been reported during the past five years and more than 70 per cent since 1950; this surge of activity primarily reflects the large number of species recently reported from the Precambrian.

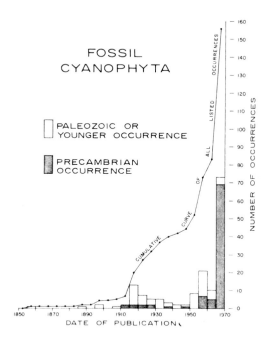

Text-fig. 2. Histogram showing date of publication and number of published occurrences of cyanophytes reported from Precambrian and Phanerozoic sediments since 1855.

Text-fig. 3 shows a histogram illustrating the distribution of reported cyanophytes (excluding stromatolites and similar organo-sedimentary structures) over geologic time; well over 50 per cent of these species have been reported from sediments of Precambrian age. With the exception of bacteria, whose geologic distribution is very incompletely documented, these primitive plants constitute the only known biologic group having an essentially unbroken fossil record dating back to the Early Precambrian.

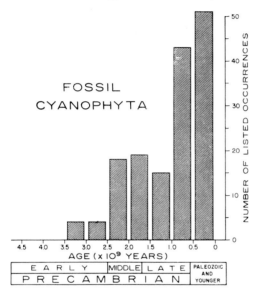

Text-fig. 3. Histogram showing known geologic distribution of cellularly preserved cyanophytes.

The geologic distribution of cyanophycean families is summarized in text-fig. 4. Coccoid (chroococcacean) unicellular blue-green algae are known from the Early Precambrian, and if the spheroidal Swaziland fossils are regarded as being of chroococcalean affinities, the evidence indicates that cyanophytes were extant more than 3.1 billion years ago. The earliest well-documented filamentous blue-greens are those reported from the Middle Precambrian Gunflint cherts; less established occurrences, particularly those reported by Gruner (1922, 1925) from the Pokegama Quartzite** and from

** Excellently preserved filaments from Gruner's locality, shown to the writer by G. R. Licari in Sept. 1969, are currently under study by P. E. Cloud, Jr. and Licari; note that the Pokegama is now thought to be only slightly older than the Gunflint Iron-Formation (G. B. Morey 1969, personal communication), rather than about 2.3 billion years in age as is indicated in text-figs. 1, 4 and 7.

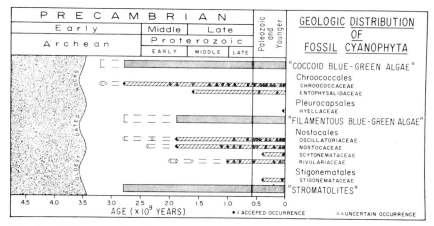

Text-fig. 4. Histogram showing known geologic distribution of cyanophycean families, based on cellularly preserved fossils.

the Soudan Iron-Formation extend the probable time-range of the filamentous Cyanophyceae (Oscillatoriaceae?) to about 2.7 billion years, approximately coinciding with the age of the oldest known stromatolites. Thus, it appears that blue-green algae originated and diversified early in earth history, with filamentous forms evolving from chroococcalean ancestors during the Early Precambrian and apparently concomitant with the appearance of biohermal, stromatolitic communities. Once established, certain of these organisms, characterized by wide ecological tolerance, inherent genetic stability and an unusually effective DNA-repair mechanism, exhibited little morphological evolution during subsequent geologic time.

6.3. *Eucaryotes from the Bitter Springs Formation*

Following the emergence and diversification of procaryotic microorganisms (bacteria and blue-green algae), the next major evolutionary advance involved the origin of the eucaryotic cell (containing a structurally defined nucleus and membrane-bound plastids, and exhibiting mitotic division). It frequently has been noted that the gap between these two distinct cellular types represents the greatest single discontinuity in the evolutionary hierarchy of living systems.

Although the *mode* of origin of eucaryotic organization may not be demonstrable by paleontologic means, decipherable evidence of its *time* of origin might be contained in the fossil record (for possible limitations, see section 2, above). This time of origin is of considerable interest, for it seems reasonably clear that the emergence of higher organisms (i.e. metazoans and metaphytes) must have been preceded by the evolution of primitive, mitotic and meiotic, eucaryotic microorganisms from their anucleate, asexual, procaryotic ancestors.

Returning to the microflora of the Late Precambrian Bitter Springs Formation, the coriaceous organic filaments shown in figs. 60–62 (*Eomycetopsis robusta* Schopf) are of probable fungal affinities, and are therefore presumably eucaryotic in organization. However, the reproductive structures of this organism are unknown and although a variety of criteria suggest a probable relationship to the Phycomycetes (Schopf 1968a), its phylogenetic position is somewhat uncertain. It may be noted that possible fungal spores have been reported from other Late Precambrian sediments (Pflug 1964; Schopf and Barghoorn 1969), but nowhere have spores and hyphae been observed in organic connection. Thus, although fungi are good candidates for early eucaryotic organisms, their presence in the Late Precambrian biota has not been demonstrated conclusively.

Nevertheless, the occurrence of eucaryotic plants in the Bitter Springs assemblage seems firmly established; this contention is supported by the following observations:

(1) Certain of the spheroidal members of the assemblage (e.g. *Caryosphaeroides* spp.) contain granular, subspherical, organic bodies (fig. 51) interpreted as representing remnants of degraded nuclei (Schopf 1968a). These unicells commonly appear to have been plasmolyzed during preservation, with a resulting separation of inner and outer wall layers (fig. 51), a feature not generally exhibited by procaryotic cells.

(2) Other members of the assemblage (*Glenobotrydion aenigmatis* Schopf) have been preserved in what appear to be stages of a mitotic sequence (fig. 50); that this represents true mitosis, rather than procaryotic fission, seems evidenced by the 'pinching in' of the cell walls near the early part of the sequence (second picture in fig. 50), by the central position of a spheroidal organic structure interpreted as representing nuclear material at this stage in the sequence, and by the symmetrical distribution of this presumed genetic material to the resulting daughter cells (third picture in fig. 50).

(3) Certain taxa of the assemblage (e.g. *Globophycus rugosum* Schopf, fig. 56; *Gloeodiniopsis lamellosa* Schopf, fig. 57) consistently contain peripheral, dark, organic structures, exhibiting definite and uniform morphology (stellate, circular or ring-shaped in outline depending on the species), interpreted as possible remnants of intracellular organelles (pyrenoids?).

(4) Still other spheroidal unicells (figs. 58, 59) exhibit surficial trichotom-

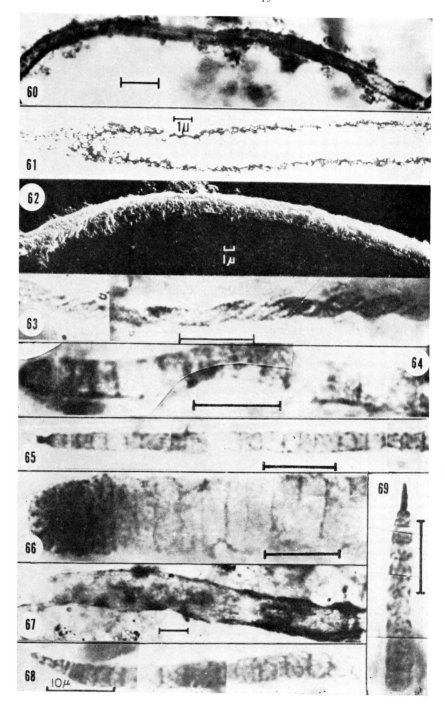

ous markings, presumably indicating that they were originally borne in a tetrad of four such cells, produced by a meiotic sequence. Tetrahedral spores of this type are relatively rare in thallophytes, but are characteristic of many Ceramiales (floridean red algae). Although apparently indicative of eucaryotic organization, such trilete marks should *not* be regarded as definite evidence of vascular plants as some workers have suggested.

(5) In size, shape, cell wall ultrastructure and general organization, several members of the biota bear striking resemblance to modern and fossil eucaryotic green algae (Pleurococcaceae, Chlorosphaeraceae, Chlorellaceae).

(6) *Eomycetopsis* (figs. 60–62) is probably a Phycomycete, and *Gloeodiniopsis* (fig. 57) is somewhat similar to primitive dinophyceans (*Gloeodinium montanum* Klebs); such affinities would be indicative of eucaryotic organization.

The evidence indicates, therefore, that the nucleated cell – with its potentialities for mitosis, meiosis, genetic exchange through sexuality and an alternation of diploid and haploid generations – had become established as early as 900 million years ago. Moreover, it seems probable that many, and perhaps all of these eucaryotic features were expressed in the Bitter Springs biota: mitosis seems evidenced by the sequence shown in fig. 50; the occurrence of meiosis seems indicated by the presence of the presumably haploid tetraspores shown in figs. 58 and 59; sexuality, and an alternation of diploid and haploid phases may be similarly inferred from these probable tetraspores, since their production in modern thallophytes is intimately associated with these features.

Figs. 60–69. Filamentous microfossils from the Late Precambrian Bitter Springs chert. Except as otherwise noted, line for scale in each figure represents 10 microns.

Figs. 60–69. Optical photomicrographs and transmission (fig. 61) and scanning (fig. 62) electron micrographs showing filamentous microorganisms in petrographic thin sections, an epoxy-embedded ultrathin section (fig. 61) and in hydrofluoric acid macerations (figs. 60, 62) of laminated black sherts from the Bitter Springs Formation (ca. 0.9×10^9 years old) of central Australia (Schopf 1968a, 1970): (60–62) *Eomycetopsis robusta* Schopf; (63) *Heliconema australiensis* Schopf; (64) *Oscillatoriopsis obtusa* Schopf; (65) *Cephalophytarion grande* Schopf; (66) *Paleolyngbya barghoorniana* Schopf; (67) *Siphonophycus kestron* Schopf; (68, 69) *Caudiculophycus rivularioides* Schopf.

6.4. The time of origin of eucaryotic organization

The presence of relatively advanced eucaryotes in the Bitter Springs assemblage, and their apparent absence from the ecologically comparable microbiota of the Middle Precambrian Gunflint Iron-Formation, has been interpreted to suggest that the eucaryotic cell may have first appeared at some time intermediate between the deposition of these two sediments (Schopf 1967, 1968a; Schopf and Barghoorn 1969; Cloud 1968; Cloud et al. 1969; Margulis 1968). Recent discoveries support this general concept and suggest that eucaryotic organisms were extant 1.3 billion years ago, perhaps first appearing near the beginning of the Late Precambrian.

One such discovery is that of a microbiota of limited diversity in siliceous and calcareous sediments of the Skillogalee Dolomite of South Australia, about 1000 million years in age (Schopf and Barghoorn 1969). In addition to heterocystous (figs. 46–48) cyanophycean (nostocalean) filaments and coccoid algal unicells (figs. 43, 44), the Skillogalee assemblage contains an unusual 'ascus-like microfossil of uncertain systematic position' (Schopf and Barghoorn 1969, Pl. II, fig. 1). Although this fossil bears little resemblance to any procaryotic structures, its probable eucaryotic relationships are uncertain. Similarly, certain of the relatively complex organisms (e.g. *Fibularix*) described by Pflug (1965) from the approximately 1.1 billion year-old Belt Supergroup of Montana are probably eucaryotes, although their precise affinities are unknown.

Recently, evidence of possible eucaryotic organization has been reported from cherty stromatolites of the Beck Spring Dolomite, perhaps 1.3 billion years in age, from southern California (Cloud et al. 1969). In addition to containing filamentous cyanophytes, perserved either as inorganic crystal chains (fig. 42; Gutstadt and Schopf 1969) or as organic filaments (fig. 41; Cloud et al. 1969), the Beck Spring assemblage includes spheroidal microfossils 'exhibiting dark spots suggesting the preservation of eucaryotic cell structures' (figs. 39, 40; Cloud et al. 1969). Although such evidence is not conclusive, the morphological regularity of these structures seems highly suggestive of a biologic (eucaryotic), rather than a diagenetic origin; this interpretation is further strengthened by the occurrence of tubular branching filaments, of probable chlorophycean affinities, in the Beck Spring cherts (G. R. Licari 1969, personal communication).

In summary, the presence of eucaryotic microorganisms in the Late Precambrian biota is firmly established. The oldest demonstrably nucleated cells are those of the Bitter Springs microflora (ca. 0.9×10^9 years), but presumptive eucaryotes are present in the Skillogalee Dolomite (ca. 1.0×10^9 years), the Belt Supergroup (ca. 1.1×10^9 years) and the Beck Spring Dolomite (ca. 1.3×10^9 years). Older microfossils, exhibiting morphology possibly suggestive of eucaryotic organization have been noted in the Belcher Group (ca. 1.7×10^9 years) and the Gunflint Iron-Formation (ca. $1.6–2.0 \times 10^9$ years); as is indicated in section 5.2., however, too few data are available to evaluate these latter two occurrences. It seems likely, therefore, that nucleated organisms were extant as early as 1.3 billion years ago, and it is possible that they first appeared near the beginning of the Late Precambrian (i.e. 1.7×10^9 years ago). This apparently late appearance of eucaryotic organization provides an evolutionary basis for the absence of metazoans and metaphytes from the Early and Middle Precambrian fossil record, and suggests that the 'biological species concept', which seems intimately tied to eucaryotic sexuality, is not strictly applicable to the earliest half of biological history. In addition, since eucaryotic organisms are fundamentally aerobic, the presence of such organisms during the early or middle part of the Late Precambrian seems incompatible with the Berkner-Marshall hypothesis (Berkner and Marshall 1965) that predicts an atmospheric oxygen concentration too low to support aerobic respiration until the beginning of the Paleozoic.

7. Summary

Text-fig. 5 shows a 'geologic clock' in billions of years that summarizes the general picture of early evolutionary development suggested by the recent studies discussed above.

Biological systems originated during the earliest third of geologic time, probably between four and one-quarter and three and one-half billion years ago. It is generally assumed that the primitive atmosphere was a highly reduced mixture, primarily composed of methane and ammonia, and that the earliest living systems were heterotrophic, using organic matter of abiotic origin as a carbon source; geologic evidence for these assumptions has yet to be detected, however, and by the beginning of the known fossil record, autotrophic organisms were apparently well established. The advent of algal photosynthesis, perhaps as early as three billion years ago, marked the beginning of a gradual transition from the presumably anoxic conditions of the primitive environment to the oxygenic atmosphere of later geologic time. Although relatively advanced in terms of biochemical complexity, the Early

Precambrian biota was phylogenetically quite primitive, consisting of non-colonial, procaryotic microorganisms, occurring earliest as isolated cells and somewhat later, with the evolution of the filamentous habit, as components of stromatolitic communities.

During the Middle Precambrian, algal bioherms flourished in shallow seas and littoral environments on a cosmopolitan scale, and planktonic algae were locally abundant. The Middle Precambrian flora was composed predominantly, or perhaps entirely, of procaryotic organisms including coccoid and filamentous blue-green algae and a variety of chemosynthetic bacteria. It remains to be demonstrated whether other organisms of this age were anucleate, eucaryotic, or perhaps of transitional organization. Photosynthetic microorganisms were widespread during the Middle Precambrian, and the evidence suggests that an oxygenic atmosphere may have become established during this segment of earth history.

The Late Precambrian biota contained a diverse assemblage of microscopic thallophytes, many of which are morphologically comparable to living species. Blue-green algae, highly diversified by this stage in evolutionary development, were the dominant components of extensive biohermal communities. Unlike earlier assemblages, however, Late Precambrian biocoenoses are known to have contained eucaryotic algae and probable fungi. Available evidence suggests that eucaryotes were probably extant as early as 1.3 billion years ago, and that this higher level of cellular organization first appeared near the beginning of the Late Precambrian. It seems likely, therefore, that the Late Precambrian featured a series of important evolutionary innovations including mitosis, meiosis, eucaryotic sexuality, and the origin and diversification of many thallophytic groups.

Although assured protozoans are as yet unreported from Precambrian sediments, eucaryotes probably gave rise to unicellular animals rather late in geologic time; the development of the metazoan grade of organization apparently occurred prior to 700 million years ago, near the close of the Precambrian. A chlorophycean stock served as the progenitor of vascular plants which had appeared by the close of the Silurian. The nature of the evolutionary transition from unicellular to multicellular organization, producing both metazoans and metaphytes, remains a fundamental unsolved problem in paleontology; this problem is perhaps best approached through a consideration of both morphological and chemical fossils, augmented by an understanding of comparative biochemistry and developmental biology.

This picture of gradually accelerating early evolutionary development, beginning rather slowly but markedly quickening with the emergence of eucaryotic organization, seems consistent with the fragmentary evidence currently available. Whether this represents an accurate description of the paleobiology of the Precambrian should become apparent with future investigations.

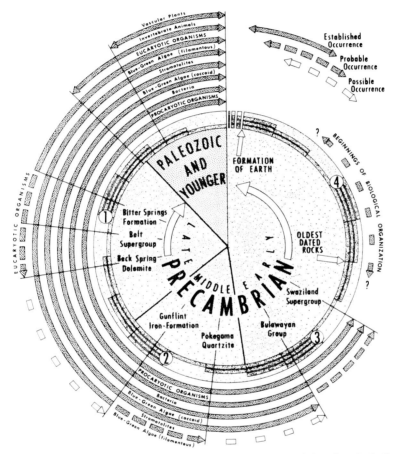

Text-fig. 5. 'Geologic clock' in billions of years showing known or inferred geologic distribution of various types of primitive organisms, based on morphological and chemical fossils discussed in text.

Acknowledgements

With minor differences (viz. the absence of a brief discussion of lower Paleozoic algae and the origin of vascular plants), this paper represents the combined, version of papers presented in the symposium on 'Major Evolutionary Events and the Geologic Record of Plants' at the XI International Botanical Congress in Seattle, Washington in August, 1969 (Schopf 1969a), and in the symposium on 'Ultra Microplankton (Part I)' at the North American Paleontological Convention in Chicago, Illinois in September, 1969 (Schopf 1969b).

The general trends of early morphological evolution here discussed are based on concepts briefly outlined previously (Schopf 1967) and recently updated (Schopf 1969c). The 'geologic clock' shown in text-fig. 5 is a revised version of a figure published earlier (Schopf 1967); this figure is modified after an idea first suggested to me by Elso Barghoorn in 1964.

The concepts here presented have been developed through discussions with many colleagues, and they have been particularly influenced by my association with E. S. Barghoorn, who initially encouraged my interest in these problems, by the studies of P. E. Cloud, Jr. and his coworkers, and by the writings of M. F. Glaessner. The section on organic geochemistry has benefitted greatly from discussions with I. R. Kaplan, W. G. Meinschein, J. Oró and K. A. Kvenvolden. I am grateful to the following individuals for providing photographs of Precambrian fossils: E. S. Barghoorn (figs. 13–25); H. J. Hofmann (figs. 27–36); K. A. Kvenvolden (fig. 12); G. R. Licari (figs. 38–41); B. Nagy (figs. 7–10); and A. A. Prashnowsky (figs. 11, 37). I am also grateful to H. T. Loeblich for keeping me abreast of the Russian literature, and I very much appreciate the excellent assistance of Mrs. Carol Lewis in compiling the data shown in text-figs. 2–4; the text-figures were drafted by Miss Julie Gunther. This work was supported by a faculty research grant from U.C.L.A., and by NASA Contract 9–9941.

References

ABELSON, P. H., 1954, Organic constituents of fossils; Carnegie Inst. Washington Year Book 53, 97.

ABELSON, P. H., 1959, Geochemistry of organic substances; Researches in Geochemistry, P. H. Abelson, ed. (Wiley, N.Y.) 79.

ABELSON, P. H., 1963, Geochemistry of amino acids; Organic geochemistry, I. A. Breger, ed. (Macmillan, N.Y.) 431.

ABELSON, P. H., 1966, Chemical events on the primitive earth; Proc. Natl. Acad. Sci. U.S. 55, 1365.

ABELSON, P. H., and P. E. HARE, 1968a, Recent origin of amino acids in the Gunflint chert; Program, Ann. Meeting, Geol. Soc. Amer., Mexico City, 2 (Abs.).

ABELSON, P. H. and P. E. HARE, 1968b. Recent amino acids in the Gunflint chert; Carnegie Inst. Washington Year Book 67, 208.

AI LSOPP, H. L., T. J. ULRYCH and L. O. NICOLAYSEN, 1968, Dating some significant events in the history of the Swaziland system by the Rb-Sr isochron method; Canadian J. Earth Sci. 5, 605.

ASHLEY, B. E., 1937, Fossil algae from the Kundelungu series of northern Rhodesia; J. Geol. 45, 332.

BARGHOORN, E. S. and S. A. TYLER, 1965, Microorganisms from the Gunflint chert; Science 147, 563.

BARGHOORN, E. S., W. G. MEINSCHEIN and J. W. SCHOPF, 1965, Paleobiology of a Precambrian shale; Science 148, 461.

BARGHOORN, E. S. and J. W. SCHOPF, 1965, Microorganisms from the Late Precambrian of central Australia; Science 150, 337.

BARGHOORN, E. S. and J. W. SCHOPF, 1966, Microorganisms three billion years old from the Precambrian of South Africa; Science 152, 758.

BERKNER, L. V. and L. C. MARSHALL, 1965, History of major atmospheric components; Proc. Natl. Acad. Sci. U.S. 53, 1215.

BONDENSEN, E., K. R. PEDERSEN and O. JØRGENSEN, 1967, Precambrian organisms and the isotopic composition of organic remains in the Ketilidian of south-west Greenland; Meddeleser om Grønland, Grønlands geologiske undersøgelse, Bd. 164, Nr. 4, 41 pp.

BRINKMANN, R. T., 1969, Dissociation of water vapor and evolution of oxygen in the terrestrial atmosphere; J. Geophys. Res. 74, 5355.

BURLINGAME, A. L., P. HAUG, T. BELSKY and M. CALVIN, 1965, Occurrence of biogenic steranes and pentacyclic triterpanes in an Eocene shale (52 million years) and in an Early Precambrian shale (2.7 billion years), a preliminary report; Proc. Natl. Acad. Sci. U.S. 54, 1406.

CAHEN, L., A. JAMOTTE and G. MORTELMANS, 1946, Sur l'existence de microfossiles dans l'horizon des cherts du Kundelungu supérieur; Ann. Soc. Géol. Belgique 70, B55.

CALVIN, M., 1969, Chemical Evolution (Oxford Univ. Press, N.Y.) 278.

CLOUD, P. E., JR., 1965, Significance of the Gunflint (Precambrian) microflora; Science 148, 27.

CLOUD, P. E., JR., 1968, Pre-Metazoan evolution and the origins of the Metazoa; evolution and environment, E. T. Drake, ed. (Yale Univ. Press, New Haven) 1.

CLOUD, P. E., JR. and H. HAGEN, 1965, Electron microscopy of the Gunflint microflora: preliminary results; Proc. Natl. Acad. Sci. U.S. 54, 1.

CLOUD, P. E., JR, J. W. GRUNER and H. HAGEN, 1965, Carbonaceous rocks of the Soudan iron formation (Early Precambrian); Science 148, 1713.

CLOUD, P. E., JR. and G. R. LICARI, 1968, Microbiotas of the banded iron formations; Proc. Natl. Acad. Sci. U.S. 61, 779.

CLOUD, P. E., JR., G. R. LICARI, L. A. WRIGHT and B. W. TROXEL, 1969, Proterozoic eucaryotes from eastern California; Proc. Natl. Acad. Sci. U.S. 62, 623.

CROFT, W. N. and E. A. GEORGE, 1959, Blue-green algae from the Middle Devonian of Rhynie, Aberdeenshire; Bull. British Mus. (Nat. Hist.) Geology 3, 339.

DEGENS, E. T., 1965, Geochemistry of sediments (Prentice-Hall, Englewood Cliffs, N.J.) 342 pp.

DEGENS, E. T., J. M. HUNT, J. H. REUTER and W. E. REED, 1964, Data on the distribution of amino acids and oxygen isotopes in petroleum brine waters of various geologic ages; Sedimentology 3, 199.

DONN, W. L., B. D. DONN and W. G. VALENTINE, 1965, On the early history of the earth; Bull. Geol. Soc. Amer. 76, 287.

ENGEL, A. E. J., B. NAGY, L. A. NAGY, C. G. ENGEL, G. O. W. KREMP and C. M. DREW, 1968, Alga-like forms in the Onverwacht series, South Africa: oldest recognized life-like forms on earth; Science 161, 1005.

FAURE, G. and J. KOVACH, 1969, The age of the Gunflint iron formation of the Animikie series in Ontario, Canada; Bull. Geol. Soc. Amer. 80, 1725.

FIESER, L. F. and M. FIESER, 1961, Advanced organic chemistry. (Reinhold, N.Y.) 1158.

GLAESSNER, M. F., 1962, Precambrian fossils; Biol. Rev. Cambridge Phil. Soc. 37, 467.

GLAESSNER, M. F., 1966, Precambrian paleontology; Earth-Sci. Rev. 1, 29.

GLAESSNER, M. F., 1968, Biological events and the Precambrian time scale; Canadian J. Earth Sci. 5, 585.

GLAESSNER, M. F. and M. WADE, 1966. The Late Precambrian fossils from ediacara, South Australia; Palaeontol. 9, 599.

GRUNER, J. W., 1923, The origin of sedimentary iron formations: the Biwabik formation of the Mesabi range; Econ. Geol. 17, 407.

GRUNER, J. W., 1924, Algae believed to be archaean; J. Geol. 31, 146.

GRUNER, J. W., 1924, Contributions to the geology of the Mesabi range; Minnesota Geol. Surv. Bull. No. 19, 71.

GRUNER, J. W., 1925, Discovery of life in the archaean; J. Geol. 33, 151.

GUTSTADT, A. M. and J. W. SCHOPF, 1969, Possible algal microfossils from the Late Precambrian of California; Nature 223, 165.

HARE, P. E. and R. M. MITTERER, 1967, Nonprotein amino acids in fossil shells; Carnegie Inst. Washington Year Book 65, 362.

HARRINGTON, J. W. and P. D. TOENS, 1963, Natural occurrence of amino-acids in dolomitic limestones containing algal growths; Nature 200, 947.

HAWLEY, J. E., 1926, An evaluation of the evidence of life in the archaean; J. Geol. 34, 441.

HOERING, T. C., 1967a, The organic geochemistry of Precambrian rocks; Researches in geochemistry, vol. 2, P. H. Abelson, ed. (Wiley, N.Y.) 87.

HOERING, T. C., 1967b, Criteria for suitable rocks in Precambrian organic geochemistry; Carnegie Inst. Washington Year Book 65, 365.

HOFMANN, H. J. and G. D. JACKSON, 1969, Precambrian (Aphebian) microfossils from Belcher Islands, Hudson Bay; Canadian J. Earth Sci. 6, 1137.

HOLMES, A., 1965, Principles of Physical Geology (Ronald, N.Y.) 1288.

HURLEY, P. M., H. W. FAIRBAIN and W. H. PINSON, JR., 1962, Unmetamorphosed minerals in the Gunflint formation used to test the age of the Animikie; J. Geol. 70, 489.

JOHNS, R. B., T. BELSKY, E. D. McCARTHY, A. L. BURLINGAME, P. HAUG, H. K. SCHNOES, W. RICHTER and M. CALVIN, 1966, The organic geochemistry of ancient sediments – part II; Geochim. Cosmochim. Acta 30, 1191.

KUZNETSOV, S. I., M. V. IVANOV and N. N. LYALIKOVA, 1963, Introduction to geological microbiology, translated from Vvedeniye v Geologicheskuyu Mikrobiologiyu (Acad. Sci. U.S.S.R. Press, Moscow, 1962) C. H. Oppenheimer, ed. of English edition. (McGraw-Hill, N.Y.) 252.

KVENVOLDEN, K. A., G. W. HODGSON, E. PETERSON and G. E. POLLOCK, 1968, Organic geochemistry of the Swaziland system, South Africa; Program, Ann. Meeting, Geol. Soc. Amer., Mexico City, 167 (Abs.).

KVENVOLDEN, K. A. and G. W. HODGSON, 1969, Evidence for porphyrins in Early Precambrian Swaziland System sediments; Geochim. Cosmochim. Acta 33, 1195.

KVENVOLDEN, K. A., E. PETERSON and G. E. POLLOCK, 1969, Optical configuration of amino-acids in Precambrian fig tree chert; Nature 221, 141.

LABERGE, G. L., 1967, Microfossils and Precambrian iron formations; Bull. Geol. Soc. Amer. 78, 331.

LICARI, G. R. and P. E. CLOUD, JR., 1968, Reproductive structures and taxonomic affinities of some nannofossils from the Gunflint iron formation; Proc. Natl. Acad. Sci. U.S. 59, 1053.

LICARI, G. R., P. E. CLOUD, JR. and W. D. SMITH, 1969, A new chroococcacean alga from the proterozoic of Queensland; Proc. Natl. Acad. Sci. U.S. 62, 56.

MACGREGOR, A. M., 1940, A Precambrian algal limestone in southern Rhodesia; Trans. Geol. Soc. South Africa 43, 9.

MACLEOD, W. D., JR., 1968, Combined gas chromatography-mass spectrometry of complex hydrocarbon trace residues in sediments; J. Gas Chromatogr. 6, 591.

MARGULIS, L., 1968, Evolutionary criteria in thallophytes: a radical alternative; Science 161, 1020.

MEINSCHEIN, W. G., 1965, Soudan formation: organic extracts of Early Precambrian rocks; Science 150, 601.

MEINSCHEIN, W. G., 1967, Paleobiochemistry; 1967 McGraw-Hill Yearbook of Science and Technology 283.

MEINSCHEIN, W. G., E. S. BARGHOORN and J. W. SCHOPF, 1964, Biological remnants in a Precambrian sediment; Science 145, 262.

MOORE, E. S., 1918, The iron formation on Belcher Islands, Hudson Bay, with special reference to its origin and its associated algal limestones; J. Geol. 26, 412.

NAGY, B. and H. C. UREY, 1969, Organic geochemical investigations in relation to the analyses of returned lunar rock samples; Life Sci. Space Res. 7, 31.

NAGY, B. and L. A. NAGY, 1969, Early Precambrian Onverwacht microstructures: possibly the oldest fossils on earth?; Nature 223, 1226.

NICOLAYSEN, L. O., 1962, Stratigraphic interpretation of age measurements in southern Africa; Petrologic Studies: a volume in honor of A. F. Buddington, A. E. J. Engel, et al., eds. (Geol. Soc. Amer., N.Y.) 569.

OBERLIES, F. and A. A. PRASHNOWSKY, 1968, Biogeochemische und elektronenmikroskopische Untersuchung präkambrischer Gesteine; Naturwiss. 55, 25.

ORÓ, J., D. W. NOONER, A. ZLATKIS, S. A. WIKSTRÖM and E. S. BARGHOORN, 1965, Hydrocarbons of biological origin in sediments about two billion years old; Science 148, 77.

ORÓ, J., D. W. NOONER, A. ZLATKIS and S. A. WIKSTRÖM, 1966, Paraffini c hydrocarbons in the Orgueil, Murray, Mokoia and other meteorites; Life Sci. Space Res. 4, 63.

ORÓ, J. and D. W. NOONER, 1967, Aliphatic hydrocarbons in Precambrian rocks; Nature 213, 1082.

PARK, R. and S. EPSTEIN, 1960, Carbon isotopic fractionation during photosynthesis; Geochim. Cosmochim. Acta 21, 110.

PARK, R. and S. EPSTEIN, 1961, Metabolic fractionation of C^{13} and C^{12} in plants; Plant Physiol. 36, no. 2, 133.

PFLUG, H. D., 1964, Niedere Algen und ähnliche Kleinformen aus dem Algonkium des

Belt Serie; Eer. Oberhess. Ges. Natur- u. Heilkunde Giessen Bd. 33, Heft 4, 403.

PFLUG, H. D., 1965, Organische Reste aus der Belt Serie (Algonkium) von Nordamerika; Paläont. Zeit. 39, 10.

PFLUG, H. D., 1966a, Structured organic remains from the Fig Tree series of the Barberton Mountain Land; Econ. Geol. Res. Unit, Univ. Witwatersrand, Johannesburg, Inform. Circ. 28, 1.

PFLUG, H. D., 1966b, Einige Reste niederer Pflanzen aus dem Algonkium; Palaeontographica Abt. B, Bd. 117, 59.

PFLUG, H. D., 1967, Structured organic remains from the Fig Tree series (Precambrian) of the Barberton Mountain Land (South Africa); Rev. Palaeobotan. Palynol. 5, 9.

PFLUG, H. D., W. MEINEL, K. H. NEUMANN and M. MEINEL, 1969, Entwicklungstendenzen des frühen Lebens auf der Erde; Naturwiss. 56, 10.

PONNAMPERUMA, C. A. and N. W. GABEL, 1968, Current status of chemical studies on the origin of life; Space Life Sci. 1, 64.

PRASHNOWSKY, A. A. and M. SCHIDLOWSKI, 1967, Investigation of Precambrian thucolite; Nature 216, 560.

RAMSAY, J. G., 1963, Structural investigations in the Barberton Mountain Land, eastern Transvaal; Trans. Geol. Soc. South Africa 66, 353.

ROBLOT, M. M., 1963, Découverte de sporomorphes dans des sédiments antérieurs à 550 M.A. (Briovérien); Compt. Rend. Acad. Sci. Paris 256, 1557.

ROBLOT, M. M., 1964, Sporomorphes du Précambrien Normand; Rev. de Micropaléont. 7, 153.

ROBLOT, M. M., M. CHAIGNEAU and L. GIRY, 1966, Étude au spectromètre de masse d'extraits d'un phtanite Précambrien; Compt. Rend. Acad. Sci. Paris 262, 544.

SCHIDLOWSKI, M., 1966, Zellular strukturierte Elemente aus dem Prakambrium des Witwaterrand-Systems (Süd Afrika); Z. Deutsch. Geol. Ges. Jahrgang 1963, Bd. 115, 783.

SCHOPF, J. W., 1967, Antiquity and evolution of Precambrian life; 1967 McGraw-Hill Yearbook of Science and Technology, 46.

SCHOPF, J. W., 1968a, Microflora of the Bitter Springs formation, Late Precambrian, central Australia; J. Paleontol. 42, 651.

SCHOPF, J. W., 1968b. Ultrathin section electron microscopy of organically preserved Precambrian microorganisms; Program, Ann. Meeting, Geol. Soc. Amer., Mexico City, 267 (Abs.).

SCHOPF, J. W., 1969a, Precambrian microorganisms and evolutionary events prior to the origin of vascular plants; Program, XI Internatl. Bot. Cong., Seattle, 192 (Abs.).

SCHOPF, J. W., 1969b, Organically preserved Precambrian microorganisms; J. Paleontol. 43, 898 (Abs.).

SCHOPF, J. W., 1969c, Recent advances in Precambrian paleobiology; Grana Palynologica 9, 147.

SCHOPF, J. W., 1970, Electron microscopy of organically preserved Precambrian microorganisms; J. Paleontol. 44, 1.

SCHOPF, J. W., E. S. BARGHOORN, M. D. MASER and R. O. GORDON, 1965, Electron microscopy of fossil bacteria two billion years old; Science 149, 1365.

SCHOPF, J. W. and E. S. BARGHOORN, 1967, Alga-like fossils from the Early Precambrian of South Africa; Science 156, 508.

SCHOPF, J. W. and E. S. BARGHOORN, 1969, Microorganisms from the Late Precambrian of South Australia; J. Paleontol. 43, 111.

SCHOPF, J. W., K. A. KVENVOLDEN and E. S. BARGHOORN, 1968, Amino acids in Precambrian sediments: an essay; Proc. Natl. Acad. Sci. U.S. 59, 639.

SIEGEL, S. M. and B. Z. SIEGEL, 1968. A living organism morphologically comparable to the Precambrian genus Kakabekia; Amer. J. Bot. 55, 684.

SMITH, J. W., J. W. SCHOPF and I. R. KAPLAN, 1970, Extractable organic matter in Precambrian cherts; Geochim. Cosmochim. Acta 34, 659.

SWAIN, F. M., G. V. PAKALNS and J. G. BRATT, 1966, Possible taxonomic interpretation of some Paleozoic and Precambrian carbohydrate residues; Program, Third Internatl. Meeting on Organic Geochem., London.

SWAIN, F. M., M. A. ROGERS, R. D. EVANS and R. W. WOLFE, 1967, Distribution of carbohydrate residues in some fossil specimens and associated sedimentary matrix and other geologic samples; J. Sed. Pet. 37, 12.

SWAIN, F. M., J. G. BRATT and S. KIRKWOOD, 1968, Possible biochemical evolution of carbohydrates of some Paleozoic plants; J. Paleontol. 42, 1078.

TYLER, S. A. and E. S. BARGHOORN, 1954, Occurrence of structurally preserved plants in Precambrian rocks of the Canadian Shield; Science 119, 606.

TYLER, S. A. and E. S. BARGHOORN, 1963, Ambient pyrite grains in Precambrian cherts; Amer. J. Sci. 261, 424.

TYNNI, R. and J. SIIVOLA, 1966, On the Precambrian microfossil flora in the siltstone of Muhos, Finland; Compt. Rend. Soc. Geol. Finlande 38, 127.

ULRYCH, T. J., A. BURGER and L. O. NICOLAYSEN, 1967, Least radiogenic terrestrial leads; Earth Planetary Sci. Letters 2, 179.

VALLENTYNE, J. R., 1963, Geochemistry of carbohydrates; Organic Geochemistry, I. A. Breger, ed. (Macmillan, N.Y.) 456.

VAN HOEVEN, W., 1969, Organic geochemistry. Ph.D. thesis, University of California, Berkeley (January, 1969).

VAN HOEVEN, W., J. R. MAXWELL and M. CALVIN, 1969, Fatty acids and hydrocarbons as evidence of life processes in ancient sediments and crude oils; Geochim. Cosmochim. Acta 33, 877.

WALCOTT, C. D., 1883, Precarboniferous strata in the Grand Canyon of the Colorado, Arizona; Amer. J. Sci. 26, 437.

WALCOTT, C. D., 1899, Precambrian fossiliferous formations; Bull. Geol. Soc. Amer. 10, 199.

WALCOTT, C. D., 1912, Notes on fossils from limestones of Steeprock Lake, Ontario; Geol. Surv. Canada Mem. 28, 16.

WALCOTT, C. D., 1914, Precambrian Algonkian algal flora; Smithsonian Inst., Misc. Coll. 64, no. 2, 77.

C. Ponnamperuma (ed.), Exobiology. © North-Holland Publishing Company

CHAPTER 3

The geology of juvenile carbon

P. C. SYLVESTER-BRADLEY

1. Introduction

Of the elements in the solar system, only hydrogen, helium and oxygen are more abundant than carbon. But carbon, in common with the other abundant elements of the solar system, is relatively volatile and forms highly volatile compounds. Consequently much of it has been lost from the Earth during its planetary and geological history. Nevertheless carbon is still twelfth in order of atomic abundance on the Earth (Johnson 1969). With the origin of life, carbon stepped onto the centre of the stage, and it is the element that dominates subsequent biological evolution. Previously, it had played a lesser but by no means unimportant role in another story, the story of chemical evolution, a story drawn on a much larger canvas than that of biological evolution. Chemical evolution is spread far more widely through the universe than is biological evolution: its time scale is much greater, its products are vast in comparison with life forms. Suns and planets are produced by chemical evolution in its planetary phase, continents and oceans during the geological stage.

The origin of life cannot, of course, be the product of biological evolution. It must have been the product of chemical evolution. To understand how it can have occurred we must trace the history of carbon through the planetary and geological phases involved.

2. Planetary evolution

Planetary evolution is a continuous, dynamic process. The Earth is one of the products of this process. It is a product that has by no means reached a final stage. As planetary evolution continues, the Earth is changing in composition and organization. There are many rival hypothetical models attempting to describe planetary evolution. All of them agree in postulating that the Earth has passed through a stage in which it was more homogeneous than it is at present. Many of them maintain that the Earth has been derived from a mixture of elements that was originally of the composition of the solar nebula and which in its earliest stages approximated to a random dispersal of elements with entropy at a maximum. Already by the time the Earth was differentiated from other planetary bodies a great change of composition must have taken place. Quantitatively the most important change must have been the loss of hydrogen, helium and other volatile elements. This process of loss can be properly described as an outgassing process. Coincidentally, there must have been some differentiation whereby the denser elements became concentrated at lower planetary levels. The present structure of concentric shells – atmosphere, hydrosphere, crust, mantle and core – must have been attained at the expense of the less-structured more-homogeneous primeval planetary material. Each shell differs from the primeval planet in that some elements have been lost and others gained. The extent of the change in composition could be quantitatively expressed by comparing tables of elemental abundance for each shell with that of the solar nebula. Even a qualitative comparison shows that the present mantle of the Earth must have changed far less in composition than have any of the other shells. The material of the mantle can therefore be regarded as more *primitive* than the material of core, crust, hydrosphere or atmosphere. It is known that the mantle has continued to contribute material throughout geological time to crust, hydrosphere and atmosphere. It is suspected by many that it has also contributed (at least during earlier geological time) a considerable amount of material to the core. Much of the volatile material lost from the mantle (or from the primeval planetary layers that were ancestral to the mantle) has not been retained in the atmosphere, but has been lost to outer space. Most if not all of the denser and more refractory material has been retained. The crust and the core have therefore grown during geological time at the expense of the mantle, which has shrunk.

This suggestion that the material of the mantle is more primitive than the material of the other envelopes of the Earth is consistent with the concept of what is regarded as primitive in terms of biological evolution. 'Primitive' means closer to the ancestor; 'advanced', further from the ancestor. In geological terms it is more usual, however, to speak of mantle-derived material as 'juvenile'. Juvenile water is that derived directly from the mantle;

meteoric-water, connate-water, ground-water are varieties of water distinguished by their present position in crust, atmosphere or hydrosphere; the steam which forms a major part of the volatile content of volcanic emanations is usually considered to be a mixture of juvenile and recycled water, recycled because it has previously passed through a meteoric stage.

In the same way we can regard carbon as juvenile when it is derived directly from the mantle, biological when it enters life-systems, recycled when, whatever its present position, it has once before in its history been through a biological stage. All three kinds of carbon are found to form both organic and inorganic compounds. Juvenile carbon may be found in abiogenic organic compounds; biological carbon may enter into the composition of inorganic carbonates within the body of an organism. Most sedimentary carbonates are biological but inorganic.

3. *Geological evolution*

The Earth seems to have existed in solid form for 4,600 million years. It will already have lost much of its primeval homogeneity by that time, and must already have had an atmosphere. The earliest rocks known are about 3,600 million years old. The first 1,000 million years can be regarded as 'Eogeological' time. Since that time geological processes have continued to modify the globe; we can speak of geological evolution as the continuation of planetary evolution. The most satisfactory division of geological time is perhaps that of Sutton (1968; see table 1). Recent discoveries have confirmed the supposition that during the course of geological evolution the mantle has contributed a considerable quantity of material to the crust, and that a certain amount of crustal material has been returned to the mantle. Likewise there is a continuous contribution of volatile material which escapes from the mantle and joins the atmosphere and hydrosphere, and a proportion of these volatiles are also returned to the mantle via the crust. No satisfactory method has yet been devised whereby we can distinguish whether material derived from the mantle is truly juvenile or is recycled. Nevertheless there is one very important evolutionary process which depends on the recycling of crustal material through the mantle and it will be necessary to examine this process in more detail (section 4).

Most models describing the progress of geological evolution suppose that the rate at which the mantle has outgassed volatiles and contributed differentiated material to crust and core has declined through geological time. Never-

TABLE 1

The geological time-scale; eons and chelogenic cycles (partly after Sutton 1968).

Millions of years before present	Eon	Chelogenic cycle
0		
	Phanerozoic	
600		Grenville
	Upper Proterozoic	
1000		
	Lower Proterozoic	Svecofennid
2000		
	Archaean	Shamvaian
2600		
	Katarchaean	Kola
3600		
	Eogeological	
4600		

theless a considerable volume of material continues to be derived from the mantle mainly through the activities of volcanoes. Probably volcanic activity has always been the main avenue for such contributions. Necessarily, then, we must suppose that if the rate of contribution was much higher in the past, the density of volcanic activity must also have been much greater. There is some geological evidence in favor of such a proposition. Although there is no evidence to suggest that volcanoes were much more widespread in Archaean or later time than they are today, it does seem that in Katarchaean time (2,600–3,600 million years ago) the whole area of continental crust may have been more mobile than it is today (Sutton 1969). Moreover, Sutton believes that each of the three succeeding eons were initiated by an event which left the crust traversed by a series of mobile belts more widespread than they are at the moment. Gradually these belts were sealed up and became rigid. Each eon was therefore characterized by a cycle of events (a 'chelogenic' cycle) which began with greater mobility and finished with the less mobile conditions that the globe is enjoying at the moment (table 1). Many of the volcanoes of today are concentrated along the mobile belts, and it seems certain that if mobile belts were more common during some periods of the past, then so were volcanoes. In Katarchaean time such volcanic remains are preserved in the greenstone belts that Sutton refers to as 'mini-geosynclines' (Sutton 1969), and of which the best-known and least metamorphosed is the Onverwacht Group of the Swaziland System of South Africa (see sections 9 and 11). The rate at which the mantle has contributed to the rest of the globe cannot therefore have declined steadily. We can postulate three stages:

(a) an Eogeological outgassing period, when vulcanicity was a process almost universally distributed over the face of the globe; first formation of the oceans; continental and oceanic crust first differentiated, each characterized by contrasting volcanic phenomena; 3,600–4,600 million years ago;

(b) a Katarchaean period of general crustal mobility; first development of intracontinental mobile belts; 2,600–3,600 million years ago;

(c) three cycles of rejuvenation and gradual decline; 0–2,600 million years ago.

It seems probable that an even earlier event occurred, when the whole primary atmosphere was lost (Sylvester-Bradley 1971b; Cloud 1968). This event seems to have also affected the moon and whatever body gave rise to the meteorites. It occurred 4,600 million years ago.

The Eogeological outgassing period will therefore have produced a secondary atmosphere. On the moon and on Mercury, this secondary atmosphere will have dispersed almost at once, but it has been retained on the other terrestrial planets and has evolved variously on all three. The effect of regional outgassing on a dry, solid, particulate surface has been studied experimentally by Mills (1969), who finds that when the volume of the escaping gas fluctuates, craters are formed, and these remain as surface scars even when the escape is not explosive (fig. 1). Many of the smaller craters on the moon may be fluidization scars recording a previous period when fluidization was widespread.

On the Earth, outgassing through fluidization has continued as a volcanic phenomenon characterized by surface craters (which usually seem to be explosive in nature) with necks that form diatremes cutting through the surrounding country rock (fig. 2). These diatremes are filled with a mixed breccia of country rock and volcanic debris, often with a carbonate cement that may contain both juvenile and recycled elements (e.g. the tuffisites described by Cloos 1941 and Coe 1966, see fig. 3; and diamond pipes filled with kimberlite, see fig. 1).

Other volcanic emanations in which fluidization does not play an important part are nevertheless always accompanied by gas-flow of some kind. Often the gases and vapors are carried up in solution in the molten rock; they may boil out on reaching the surface, or they may be trapped in a highly vesicular pumice. In general it seems that, in the earlier stages of volcanic activity, the gases involved are at very high temperatures. They are emitted in great volume and with great violence and provide much of the motive force of the eruption. Hydrogen is often abundant, and the gas mixture is both acid and reducing; iron is present in the ferrous state. Occasionally, some gases ignite; presumably these include hydrogen and hydrocarbons, but collection and analysis of such high-temperature gases is both difficult and dangerous. Later, the temperature and the volume of gas emitted declines; water vapor and carbon dioxide become dominant, and hydrochloric acid and sulphur compounds are important constituents. Gas emission continues at even lower temperatures long after the eruption has ceased; hot liquid water is dominant and from it are deposited a characteristic suite of hydrothermal minerals. These include carbon compounds. Although the water involved must include much that is recycled, there is strong evidence that juvenile elements contribute to the contents of many geothermal waters. For example, the famous fumaroles of Larderello (Rittmann 1962) are developed in a geological setting suggesting that they could be entirely produced by ground-water reaching still-hot magma chambers at depth. Nevertheless they have a vapor content with a chemical and isotopic composition that can only be ascribed to juvenile contribution. Helium and argon are among these volatiles, the total content of which includes over 4% carbon dioxide, and 0.015% methane.

4. Global tectonic theory

If the mantle is regarded as the residuum of a primordial homogeneous protoplanet, then its subsequent differentiation has been in a vertical direction; upwards to form the outer layers, downwards to form the core. The mantle itself is evidently layered, but it is legitimate to suppose that it still possesses some lateral homogeneity. Moreover, both hydrosphere and atmosphere,

P. C. Sylvester-Bradley

(a)

(b)

though also structured in layers, approximate to lateral homogeneity. Nor is there any evidence to suggest that the core possesses any form of lateral variation. The crust of the Earth on the other hand has a complicated lateral structure that has evidently been acquired as the result of a long-continued dynamic process of differentiation. Recently it has been possible to construct a very convincing hypothesis that accounts for crustal evolution by postulating a model in which the crust of the earth is divided up into a finite number of rigid plates, all of which are moving in relation to each other (McKenzie 1970; Dewey and Horsfield 1970; Oxburgh and Turcotte 1970). The plates are separated from each other by three kinds of fracture, all three of which are marked as the site of earthquake foci.

First, there are fractures along which the plates shear against each other; the movement along the fault planes is lateral and earthquakes initiated along them have relatively shallow foci. They are known as transform faults. Volcanoes are seldom located along these lines.

Secondly, there are fractures which mark lines along which the plates are being driven towards each other, and indeed in every case one plate is found to be over-riding the other. The crust in such regions is in a mobile condition. These are, in fact, the 'mobile belts' of the present globe. Earthquakes along them are crowded and range from shallow, through medium, to deep focus, and follow planes inclined to the surface of the globe so that they slope down through the crust from the surface and into the mantle. They are believed to mark the shear plane along which the lower plate is being forced down under the upper. Above these planes there is a belt of active volcanoes whose products are dominantly andesitic. The upper of the two plates is characterized by orogenic and metamorphic phenomena which include the emplacement of huge volumes of granite.

The third kind of fracture along which plates join is a region in which the plates are moving away from each other. As they move away, new crustal material of basaltic composition wells up from the mantle, and freezes onto the trailing edges of the plates. These fractures therefore form regions in which new crust is created. They are invariably found in oceanic regions, and can be called 'spreading ridges'. Volcanic islands are frequent and suboceanic fissure eruptions continuous. Earthquakes are much rarer than

---

Fig. 1a and b. Vertical and oblique views of experimental fluidization scars made under the direction of A. A. Mills in the University of Leicester; the diameter of the large crater in fig. 1b is 10 cm.

(a)

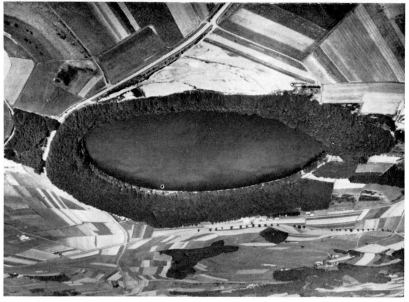

(b)

those associated with the other two kinds of fracture zone, and they are confined to those caused by volcanic eruption.

The rigid plates themselves are constructed out of three different kinds of material. They are divided both by vertical and by horizontal discontinuities. The plates are not composed only of crustal material. The horizontal discontinuity (the Mohorovičic Discontinuity) is in fact the plane separating a crustal layer from a mantle layer. This layer of upper-mantle is welded on to the overlying crust and is continuously shearing across an underlying more plastic layer of the mantle (the asthenosphere). The vertical discontinuities found crossing these rigid plates separate two kinds of crust, one composed of the basaltic oceanic crust (which, as we have seen, is formed along the spreading belts); the other composed of the very varied but dominantly granitic continental crust, three times as thick as the oceanic crust, and much older, different parts of it having been formed during earlier eons along mobile belts long since extinct (Sutton 1969).

Although all the plates that cover the globe are composed of two layers, an upper crustal layer and a lower mantle layer, they do not all have both continental and oceanic crust. Some have oceanic crust only. However, no plates are known in which the crustal layer is entirely continental.

According to plate tectonic theory, new material is being contributed by the mantle to the crust in two regions: along the spreading ridges (where it is basaltic) and along the mobile belts. But the mobile belts are not only the scene of new additions to the crust; they are also the site along which old oceanic crust is being consumed by the mantle. The whole of the oceanic crust along these regions is being consumed, but part of it is being regurgitated. This recycling process presents the petrologist with an extremely interesting and still not fully understood problem. During the course of it, differentiation leads to the gradual acidification of the original basaltic crust. Presumably, the more basic differentiate is returned to the mantle. New continental crust produced along the mobile belts therefore consists partly of material recycled from the oceans, and partly of juvenile material derived from the mantle.

It will be seen that the majority of volcanoes are situated along the spreading ridges and mobile belts that bound the rigid tectonic plates. However,

Fig. 2a and b. Vertical and oblique views of a terrestrial volcanic fluidization crater (the Pulvermaar, in the Eifel district of Germany). Early activity was probably explosive; later activity seems to have involved quiescent fluidization. Photos by Landesbildstelle Rheinland-Pfalz (by permission).

there are other regions in which volcanoes are found. These are along deep fractures found within the plates. Although little movement appears to be taking place along these fractures, they do seem to reach right down to the mantle, and they therefore form intracontinental volcanic zones. Rittmann (1962) has suggested that they are tensional fractures caused by the tendency of the high, continental part of a plate to flow outwards as a response to lateral hydrostatic disequilibrium. He suggests that the fissures which result open from above downwards in contrast to those characteristic of mobile belts, which open from below upwards. He goes on to suggest that the tensional fissures penetrate deeper into the mantle. Certainly the products of vulcanicity are quite different, and these intracontinental igneous provinces are prominent sources of juvenile carbon (section 10).

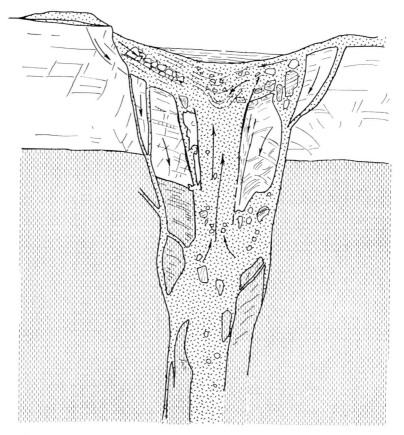

Fig. 3. A terrestrial fluidization diatreme (redrawn from Cloos 1941).

5.   *The energetics of evolution*

Evolutionary processes are always time-directed. An evolving system always changes in the direction of greater organization and towards a decrease in ᵌntropy. According to the second law of thermodynamics, such systems must necessarily operate under conditions that lie far from equilibrium. They must always be powered by an energy flow, and be situated between an energy source and an energy sink (Sylvester-Bradley 1967; 1971a; Morowitz 1968). Energy is usually dissipated in the form of heat; an evolving system must in such cases be at a higher temperature than its surroundings. The Earth as a whole is powered by both external and internal sources of energy. The external sources include gravitational and magnetic energy, but by far and away the major contribution is in the form of radiant solar energy. Internal energy is derived from the gradual decay of radio-active elements in the mantle and the crust. Both solar energy and internal energy are ultimately of course derived from the galactic energy that was stored in the elements as they were formed in an earlier stage of cosmic evolution. Evolution within the two outer layers of the globe, the atmosphere and the hydrosphere, is largely powered by solar energy. Evolution within the mantle and core is almost exclusively powered by internal energy. Evolution within the crust is mainly powered by internal energy, but partly powered by solar energy. For example, the organization achieved as a result of sedimentary processes is effected by the interaction of the weather and sea on the crust; erosion, denudation and sedimentation are all processes powered mainly by the sun. All life processes are ultimately powered from the same source. But volcanic and metamorphic processes are powered with energy derived from the mantle.

Plate tectonic theory makes considerable demands on mantle-derived energy. At one time it was supposed that all this energy was transferred from the mantle to the crust by convection currents which rose under the spreading ridges and descended under the mobile belts. Now that it has been possible to plot the course of these belts with considerable accuracy, however, the shape and size of the convection cells that it would be necessary to postulate have become too varied for credibility. It now seems that, though convection currents in the mantle may well be responsible for the transfer of energy, they are not related in any simple way in size or shape with either the rigid plates or with the global fractures that form their boundaries. The direction of movement must be a resultant of all the forces acting on the plates.

In common with all other evolving parts of the crust, the evolution of carbon compounds demands an energy source, and the source will depend on the environment of evolution; in volcanic regions it is likely to be mantle derived; on the surface, it may well be solar. The evolution of carbon compounds in atmosphere or hydrosphere is more probably solar-powered; that leading to the origin of carbon compounds in the mantle will certainly be powered by internal energy.

6. *The steady state*

All terrestrial environments are diabatic. Indeed, the adiabatic system of classical thermodynamics is entirely imaginary and hypothetical. No such system exists in nature. But diabatic systems vary in the degree to which they approach equilibrium, and sometimes near-to equilibrium conditions are maintained for relatively long periods of time. This is especially the case when the system incorporates buffers that keep the composition in a steady state. This necessarily means that the system itself is evolving too slowly during these periods for any change to be evident. Both the present atmosphere and the present hydrosphere approximate to a steady-state, and are close enough to equilibrium to make a comparison with true equilibrium meaningful. In its simplest terms, a steady-state is achieved if the contributions from the mantle to atmosphere and hydrosphere (in terms of igneous rocks and volatiles) are exactly balanced by a subtraction of sediments, water and air to form metamorphic rocks (Sillén 1967). This simple equation implies an intimate exchange of volatile elements not only between the ocean and its sediments, but also between continental crust and the atmosphere. Sillén postulates a simplified model with five reservoirs, continental basement, continental sediments, oceanic basement, oceanic sediments, and the ocean. There is continuous exchange taking place between all five reservoirs, and indeed with other reservoirs not considered in the simplified model (atmosphere and mantle are of course both involved). The main buffering of both ocean and atmosphere is achieved by the large reservoirs of silicate and carbonate which contribute to the oceanic sediments. The way that the approach to a steady-state has been achieved has been explored in a series of fascinating papers by Sillén (1965, 1966, 1967) and Siever (1968a, b). But from the point of view of the history of carbon it is more important to investigate in what respects atmosphere and hydrosphere *depart* from the steady-state as we trace them back in time, for it is in this departure that we have evidence of evolutionary change.

One of the most important factors in the geochemical cycle which keeps the present composition of the atmosphere stable is the release of free oxygen as a result of photosynthesis. There is therefore no doubt that in prebiological times the atmosphere will have had a composition quite different to that of today. A number of plausible models have been proposed. For example, Rutten (1966, 1969a, 1970) postulates two previous periods in the geological history during which the oxygen level remained constant. One period was maintained at about 0.001 × present atmospheric level; this is termed the 'Urey level' and lasted from the origin of the Earth through Eogeological time into Katarchaean time. During Katarchaean time the oxygen level rose rapidly until by the beginning of Archaean time it had reached a second constant level (the 'Pasteur level') at about 0.01 × present level. During the Lower Proterozoiceon, this level was broken and the oxygen level fluctuated until it reached the present level during the early Phanerozoic. Cloud (1968) postulates a similar time scale on rather different premises. With the increase of oxygen there is a decrease of carbon dioxide accounted for by the development of a carbonate sink in oceanic sediments, as well as by the quantitatively much less significant fixation of carbon in living organisms. Both Rutten and Cloud agree that the Eogeological atmosphere held its carbon in the form of carbon dioxide together with (during the earlier stages) carbon monoxide – for a significant proportion of hydrogen was probably also present at that time. Methane is not postulated as a likely component in more than trace amounts, though it will certainly have existed in the primary atmosphere, which can be assumed to have had a composition close to that still retained by Jupiter and Saturn.

Many other models of atmospheric evolution have been proposed and Brinkmann (1969) shows that on theoretical grounds the oxygen concentration could have reached levels as high as 0.25 × the present under prebiological conditions. Siever (1968b) insists that the only possible tests that can be applied to such models must depend on a critical analysis of the kinds and amounts of sediments produced during the periods under study. Siever considers specially the deposition of limestone, salt, chert and reconstituted silicates; Cloud (1968) and Rutten (1962) draw special attention to peculiarities in the Precambrian sedimentary record, in particular to the occurrence of pyrite sands, pitch blende and the banded iron formations, and to the absence or rarity in the earlier rocks of red beds, carbonates and pure quartz sands.

7. Isotopic ratios

According to our definition, all carbon in prebiological time must be regarded as juvenile. During Katarchaean time, however, it will have rapidly passed through the biological and geochemical cycles and so will have lost its juvenile status. Most of it will have passed into the sedimentary carbonate reservoir. From that time onwards much of it will have been recycled many times through the atmosphere under near to steady-state conditions and it will have approached isotopic equilibrium in each phase. The average $_{12}C:_{13}C$ ratio in the carbon dioxide of the present atmosphere is about 89.2:1; or, expressed in $\delta_{13}C$ values, about $-7.0\permille$. The carbon in sedimentary limestones is heavier (enriched in $_{13}C$), $_{12}C:_{13}C = 88.6:1$ ($\delta_{13}C =$ $0.0\permille$). The organic carbon in the biosphere is much lighter (depleted in $_{13}C$) and rather variable, averaging 91.0 (or $-13.0\permille$), with marine plants less depleted in $_{13}C$ than terrestrial plants. The inorganic carbon in the biosphere (e.g. as found in the shells of marine animals) is not significantly different from that in limestone sediments, which is not surprising, as limestone sediments are almost entirely biogenic. Organic carbon in sedimentary rocks (e.g. kerogen, crude oil and coal) has a range rather wider than that of the organic carbon of the biosphere, with an average not significantly different (Rankama 1963). Biogenic methane has a very wide range. For example, Nakai (1961) reports $\delta_{13}C$ values of -69 to $-84\permille$ for methane of biochemical origin; and Colombo et al. (1970) report -16 to $-70\permille$ for methane derived from German coal gases. Thucholites are radio-active bitumens found both in igneous associations (e.g. pegmatites) and in sedimentary rocks. The former have often been claimed as magmatic and juvenile (Mueller 1963), the latter as recycled and biogenic (Schidlowski 1966, 1968). Isotope ratios of all of them are much depleted in $_{13}C$ with δ values ranging from -22 to $-45\permille$ (Berger 1966; Hoefs and Schidlowski 1967; Dubrova and Nesmelova 1968; Krouse and Modzeleski 1970).

   The isotope ratios of juvenile carbon also range widely, but it is not always easy to determine whether a particular sample analysed is truly juvenile. Diamonds are almost certainly of magmatic origin and have $\delta_{13}C$ values ranging from -4 to $-8\permille$, close to those of atmospheric carbon dioxide (Rankama 1963; Vinogradov et al. 1965, 1967). Juvenile carbonates (carbonatites, hydrothermal minerals) have similar values. They are therefore depleted in $_{13}C$ when compared to marine biogenic carbonates. The carbon dioxide collected from hot springs (which may be juvenile, or more probably a mixture of juvenile and recycled) give results which are somewhat variable. They may be close to those characteristic of marine carbonates (Cheminée et al. 1969), or close to atmospheric CO_2. But the associated methane is much lighter and with a much wider range than is present in biogenic organic carbon. Craig (1953) gives some values almost as heavy as atmospheric CO_2, and others as light as in terrestrial plants; and Ferrara et al. (1963) quotes a range for methane from the Larderello fumaroles even more depleted in $_{13}C$ ($\delta_{13}C = -23.0$ to $-28.8‰$), and this is consistent with the results of Hulston and McCabe (1962), who give values of -16.1 to $-28.5‰$ for the methane in New Zealand fumaroles, and -1 to $-6.6‰$ for the associated carbon dioxide. The methane collected from the associated hot pools is more variable and reaches a $_{13}C$ depletion as high as $-63‰$.

Inorganic carbon with ratios in the organic range have been described by a number of authors with $\delta_{13}C$ values sinking to $-56‰$. (Thode et al. 1954; Feely and Kulp 1957; Spotts and Silverman 1966; Hodgson 1966). These are explained by supposing they represent oxidized organic carbon. Conversely Galimov and Petersil'ye (1967, 1968) describe organic compounds enriched in $_{13}C$ ($\delta_{13}C = -3$ to $-13‰$), and thus overlapping the usual range of carbonates. They explain this result by postulating that the hydrocarbons are synthesized from juvenile carbon in the mantle without further fractionation.

Summarizing, we may say that the isotope ratios fall into two groups; inorganic carbon is relatively enriched in $_{13}C$, organic carbon relatively impoverished. Terrestrial inorganic carbonates can reach a value as high as $+24.8‰$ (probably juvenile; from a mica peridotite dyke; see Deines 1968). Biogenic organic carbon can be depleted to a figure as low as $-89.3‰$ (in a Pleistocene algal mat, see Kaplan and Nissenbaum 1966). Most inorganic carbon is relatively enriched in $_{13}C$ in comparison with atmospheric CO_2, most organic carbon is relatively depleted in $_{13}C$ when compared with CO_2. Diamond and graphite have values which fall close to CO_2 ($\delta_{13}C$ about $-6‰$).

The carbon isotope ratios in meteorites show the same differentiation between organic and inorganic carbon. In carbonaceous chondrites the carbonates are much more enriched in $_{13}C$ than most of those on the Earth; values of up to $+61.8‰$ have been quoted by Krouse and Modzeleski (1970), and only in one specimen was a reading under $+25‰$ obtained. The organic compounds show a depletion ranging from $+6$ to $-28‰$ (Krouse and Modzeleski 1970 and Kvenvolden et al. 1970), which may therefore also be richer in $_{13}C$ than are organic compounds in terrestrial samples. Diamond ($-5.7‰$) and graphite ($-6.3‰$) from the Novo Urei ureilite (Vinogradov et al. 1967a) fall within the terrestrial range. Extensive fractionation has therefore taken place in meteorites, and has probably been achieved by kinetic rather than equilibrium effects.

At one time it was suggested that the isotope ratios of terrestrial carbons would distinguish juvenile from biogenic compounds. But the problem is not a simple one. Though the atmosphere and hydrosphere approach equilibrium between carbon dioxide and organic compounds on the one hand and sedimentary carbonates on the other, kinetic effects seem to modify the isotopic ratios in all compounds containing juvenile carbon. It is clear that there is no single isotopic ratio characteristic of all juvenile carbon, though most seems to fall between -6 and $-7\%_0$. Some fractionation must have already taken place in the mantle and subsequent migration through the crust may lead to further fractionation (Galimov 1967; Colombo et al. 1970), as does metamorphism (Deines and Gold 1969). It is true that the juvenile carbon of carbonatites differs in $\delta_{13}C$ values from the carbon of biogenic limestone (Taylor et al. 1967; Conway and Taylor 1969) and the difference certainly indicates a different history of fractionation. But similar ratios can be achieved by many routes, and give no certain indication of a similar history. Often a particular deposit contains both juvenile and recycled carbon. Rye (1966) has investigated the origin of hydrothermal fluids. He has analysed the isotopic ratios of hydrogen, oxygen and carbon, and concludes that the majority of the carbon is juvenile; the contribution of recycled carbon (derived from limestone) increased from about 20% at the beginning of the deposition of hydrothermal calcite, and reached about 50% in late stages.

8.   Biological markers

The discovery that many inorganic crustal accumulations contain both juvenile and recycled carbon suggests that the same may apply to accumulations of organic compounds. Geological deposits of this kind often contain direct fossil evidence of their origin. Thus fossiliferous inorganic limestones have their counterpart in fossiliferous organic coals; neither is suspected to have any juvenile carbon component. But there are other deposits which give evidence of a mixed parenthood. The evidence of the biological contribution is often easier to distinguish than is the contribution of juvenile carbon. The most controversial accumulations are those composed of hydrocarbons and associated bitumens. These are found abundantly in four geological situations: first, dispersed in sedimentary rocks, particularly in bituminous shales;

secondly, concentrated in oil pools; thirdly, dispersed in igneous rocks; and fourthly as gangue minerals in hydrothermal veins. In all four situations diagenetic effects have produced a degree of 'maturation'; the more stable organic compounds survive for longer than the less stable. Even so, relatively unstable organic molecules can survive for a surprising length of time. Florkin (1969) and Grégoire (1966) have shown that some proteins have survived through most of Phanerozoic time. The older fossil proteins are less varied chemically than more recent proteins; this is ascribed to maturation. Similarly there are other organic molecules which are stable enough to survive through the whole of geological time. Notable among these are isoprenoid hydrocarbons and porphyrins. Though protein molecules can be regarded as certain biological markers, isoprenoids and porphyrins can in certain circumstances be synthesized abiologically. Other less diagnostic hydrocarbons can readily be synthesized in both biological and abiological environments. Optical activity seems to be certain evidence for biological activity and, in hydrocarbon mixtures, odd-carbon predominance is with some certainty regarded as evidence of biogenesis.

9. *Crustal accumulations of organic carbon*

It is instructive to examine the distribution of these biological markers in each of the four crustal environments mentioned above. First, dispersed organic matter in sedimentary rocks has been found throughout the geological column; the best preserved sediments of Katarchaean age are found in the Swaziland System of the Barberton Mountains in South Africa (a typical 'greenstone belt', dominated by volcanic rocks of the ophiolitic suite). At the base of the System, the thick Onverwacht Series contains dispersed organic material of considerable variety. The Onverwacht Series is overlain by the Fig Tree Series which contains a considerably greater quantity of organic material of a rather different composition. Both the Onverwacht and the Fig Tree Series are being actively studied by organic geochemists at the moment. Biological markers, including microfossils, have been found in both. There is some evidence to suggest that there is a considerable difference between the two. It is even possible that the Onverwacht remains are pre-biological, so that the so-called fossils are the remains of prebiological 'organized elements', and that other biological markers are later contaminants (Nagy 1970; Sylvester-Bradley 1971b). It is more probable that they are truly biological, and that the origin of life antedated the Swaziland

System. Carbonaceous rocks of earlier Katarchaean time (see section 11) have been insufficiently studied.

Secondly, oil pools are crustal reservoirs of trapped hydrocarbon, floating on groundwater. It is universally agreed that all the oil has been concentrated in its present position after migration through the surrounding sedimentary rocks. The migration of oil through a rock follows the same pattern as the migration of groundwater. If it passes through rocks containing dispersed organic matter, it will pick up the soluble fraction and add this to its bulk. The total oil reservoir may therefore contain oil of more than one source. The groundwater reservoir on which it floats is usually saline. Apparently it always contains live bacteria which use the oil-water surface as a feeding ground. All crude oils investigated contain biological markers. Isoprenoids, porphyrins, and optical activity are all common, and odd-carbon predominance is evident in many of the younger reservoirs. There are three opinions about its source. By far the majority believe that all oil is biogenic. I am not one of them. I believe that the oil in oil pools is always a mixture of polygenetic oils, most of them biogenetic but some of them abiogenetic (Sylvester-Bradley and King 1963; Sylvester-Bradley 1964; Robinson 1966). A small but vocal minority (mainly of Russian workers) believe that most oil is abiogenetic (Kudryavtsev 1959; Kropotkin 1959; Kravtsov 1959, 1967; Florovskaya 1964, 1967). It is important to realize before dismissing this last view that the biological markers which support the first view may all be contaminants, picked up either during the course of migration, or from the activities of the bacteria in the underlying brines.

Thirdly, hydrocarbons and bitumens are found both dispersed and in small concentrations in igneous rocks. The most thoroughly studied are those in the alkaline provinces of the Kola Peninsula in Russia. Petersil'ye after detailed study over many years is firmly of the conclusion that the carbon is juvenile (Petersil'ye 1961, 1962, 1964, 1966, 1968; Petersil'ye et al. 1961a, b, 1965: see also Ikorskii 1964; Ikorskii and Romanikhin 1964; Zakrzhevskaya 1964). The discovery that in Greenland a similar but geologically much older igneous province also contains hydrocarbons and bitumens of the same kind greatly strengthens Petersil'ye's conclusions (Pedersen and Lam 1970; Petersil'ye and Sørensen 1970; Sobolev et al. 1970). The isotope ratios, as we have already seen (section 7) are much enriched in $_{13}C$ compared with those found in most terrestrial organic compounds (Lebedev and Petersil'ye 1964; Zezin et al. 1967; Galimov and Petersil'ye 1967, 1968a). This makes the hypothesis of Goldberg and Chernikov (1968), who believe that the hydrocarbons are biogenic oils which have seeped in from 'the surrounding rocks', unlikely, even if the 'surrounding rocks' rich in bitumen could be found. Most geologists would probably now agree that the carbon in both bitumens and gases is juvenile, syngenetic with the igneous rocks with which they are associated, and derived from the mantle.

Hydrocarbons and bitumens are also found incorporated in kimberlites (Beskrovnyy 1958, 1968) and carbonatites (Beskrovnyy and Baranova 1963; Heinrich and Anderson 1965). Methane and other hydrocarbon gases and solid and liquid bitumens occur fairly frequently in granites and other igneous rocks (Kalyuzhnyy and Kovalishin 1967; Mogarovskiy et al. 1968), though they are often associated with pegmatites or mineral veins and should perhaps be considered hydrothermal in origin. Igneous thucholites have already been mentioned (section 7) and are probably juvenile (Spence 1930; Mueller 1963; Berger 1966; Krouse and Modzeleski 1970).

Fourthly, hydrocarbons and bitumens often occur in hydrothermal veins. They are frequently associated with low-temperature mineralization, more particularly with fluorspar and with mercury ores (Kashkay and Nasibov 1968). As mineral veins normally cut through sedimentary strata, it is likely that the ascending hot solutions will pick up recycled carbon, and it will be rare to find organic compounds with carbon of unmixed parentage (Bocharova 1964). It is not surprising, therefore, that their analysis has given ambiguous results (Mueller 1954, 1970; Sylvester-Bradley and King 1963; Aucott and Clarke 1966; Eglinton et al. 1966; Mogarovskiy and Markov 1966; Ponnamperuma and Pering 1966; Mayskiy and Trufanov 1967; Kashkay and Nasibov 1968; Pering and Ponnamperuma 1969). Methane occurs commonly in hydrothermal gases (Elinson and Polykovskii 1963; Ferrara et al. 1963; Shcherbak 1964; Gunter and Musgrave 1966; see section 7). As we have seen, its isotopic composition favors the hypothesis that it contains a mixture of juvenile and recycled carbon (Hulston and McCabe 1962). Hydrothermal bitumens can often be seen to have formed simultaneously with other minerals, particularly with fluorspar and with calcite, when they can occur in primary fluid inclusions. Hydrocarbons may also be involved in secondary fluid inclusions, and may secondarily replace other minerals, particularly calcite (Dymkov et al. 1967).

10. Crustal accumulations of inorganic carbon

In the previous section evidence has been marshalled to show that methane and organic compounds, though dominantly the result of biogenic processes, are also produced in small quantity abiogenetically from juvenile carbon. Most juvenile carbon derived directly from the mantle is, however, oxidized and is found as carbon dioxide, carbon monoxide and as carbonates. This is in direct contrast to the extraterrestrial occurrence of carbon in carbonaceous chondrites, in which the bulk of it forms organic compounds, and only a very minor percentage occurs as carbonate (Sylvester-Bradley 1971a).

There is some evidence to suggest that outgassing events involving carbon on the Earth show a succession in which methane, always much less abundant than carbon dioxide, is nevertheless present in early stages, but gradually declines until it finally disappears altogether (Elinson and Polykovskii 1963).

In section 4 we have shown that there are three main regions of the globe along which the mantle makes contributions to the crust: along spreading ridges, along mobile belts, and along intracontinental rifts. All three regions are characterized by volcanoes, and the main contribution of juvenile carbon is undoubtedly in the form of volcanic gases, most of which escapes straight into the atmosphere, only to join the crust in solid form after it has passed through the biological cycle. However, probably all igneous rocks retain a relatively small amount of carbon dioxide entrapped in minerals as fluid inclusions. For example, Roedder (1965) has shown that it is almost ubiquitous in crystals of olivine within basalts. But by far the most spectacular contribution made by juvenile carbon is in the form of carbonate, and carbonates (particularly calcite) are abundant as gangue minerals in hydrothermal veins. These are universally distributed as late stage igneous phenomena in all mobile belts. Undoubtedly some of this carbon is recycled, but just as certainly, some of it is juvenile. An isotopic analysis of the carbon and oxygen compositions of calcite in mineral veins from Mexico, for example, 'strongly point to a magmatic source for the hydrothermal water, and to a deep-seated source for at least 50% of the carbon in the calcites' (Rye 1966). Calcite, dolomite, strontianite and witherite are all relatively common carbonates occurring as gangue minerals. Carbonatization is one of the most widespread processes responsible for hydrothermal alteration, and carbon dioxide one of the most abundant fluids found in primary fluid inclusions. It seems certain that the first hydrothermal fluid to separate from a felsic magma will be greatly enriched in carbon dioxide (Burnham 1967). The abundance of carbon in carbonates in these hydrothermal situations, and the probability that a considerable proportion of it is juvenile renders it likely that the same applies to the methane and bitumen found in hydrothermal situations as discussed in section 9.

But the bulk of juvenile carbonates are igneous. They are specially associated with the anorogenic intracontinental volcanic regions mentioned in section 4, and with rift valleys in particular. Four kinds of igneous phenomena are important: (1) intrusive carbonatites and carbonatitic volcanoes; (2) kimberlites and diamond pipes; (3) explosion craters and maars; (4) fluidized diatremes and intrusive tuffisites.

To this we may add a fifth phenomenon, not certainly igneous: crypto-explosion structures.

10.1. Carbonatites

Intrusive carbonatites are best known from their occurrence in association with the great rift valleys of Africa; they probably occur in all continents, but have in the past often been misidentified as metamorphic marbles. They have been identified over a wide area of Africa, and from Europe, Asia, North America and South America (Heinrich 1966; Tuttle and Gittins 1966). They are known from Greenland (Stewart 1970) Australia (Crohn and Gellatly 1969) and the Cape Verde Islands and will no doubt eventually be discovered in Antarctica. They are often associated with alkaline igneous provinces on the one hand and with kimberlites on the other (Heinrich 1966; Frantsesson

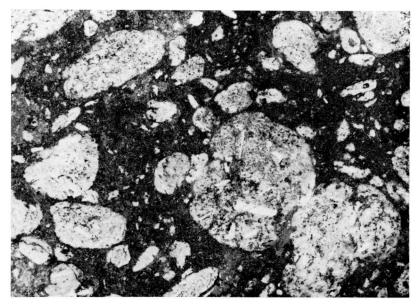

Fig. 4. The petrological texture within a fluidized carbonatite dyke; photograph from a specimen collected by D. S. Sutherland in Uganda.

(a)

(b)

1968). Although many carbonatite complexes are associated with rift valleys (as in Africa and in the Rhine graben) they are also found towards the margins of continental platforms.

The intrusion of carbonatites is clearly dominated by carbon dioxide, and carbonatite dykes often exhibit a texture that suggests that the plutonic fragments were brought up from below in streams of gas, sorted and suspended in a fluidized system, and simultaneously rounded by attrition (King and Sutherland 1966; fig. 4).

There is one active carbonatite volcano in Africa (Guest 1956; Dawson 1966; fig. 5). The ash and lava erupted is dominantly sodium carbonate; almost all the insoluble matter is calcium carbonate; silicates are insignificant. The surrounding country is dominated by the alkalinity of the run-off, and of the surrounding lakes (e.g. Lake Natron). Eruptions destroy much of the fauna and flora of the immediate vicinity, and past catastrophes can be recognised in the local geological record (Bishop 1968). An extinct carbonatite volcano has been described from Southwest Africa (Janse 1969). This volcano is characterized by a ring-shaped structure, as are most of the volcanoes in East Africa. Holmes (1965) equates the vulcanicity with the escape of immense volumes of gas, and widespread fluidization.

(c)

Fig. 5a, b and c. The most spectacular source of juvenile carbon in the world. The volcano Oldoinyo Lengai (Africa). The white ash is composed of almost pure carbonates, mainly sodium carbonate. Photos by M. A. Khan (a and b) and N. J. Guest (c).

10.2. Kimberlites

Kimberlites are associated with carbonatites as a cratonic phenomenon. They are likewise best-known from African and Russian shield areas but are known also from the other continents (e.g. Shoemaker et al. 1962; Hearn 1968). They are found within diatremes – volcanic pipes cutting vertically through the crustal rocks and widening as they ascend (Davidson 1964; Holmes 1965; Dawson 1967). The pipes are filled with a breccia composed of a hydrothermally altered mixture of country-rock and kimberlite. The kimberlite itself is believed to be derived from the mantle, and includes diamond as an accessory mineral; it is often associated with blocks of peridotite and (more rarely) eclogite. The whole complex shows abundant evidence of emplacement as a result of fluidization involving an immense proportion of carbon dioxide and water (Harris et al. 1970). Holmes (1965) speaks of the unweathered rock containing 10% H_2O and CO_2. Kennedy and Nordlie (1968) believe that the pipes narrow down at depth to a system of dykes. They envisage an explosive gas drive from a depth of not less than 200 km which drilled its way upwards and 'blew its top' to form the pipe on reaching the surface. But kimberlites may not always have been emplaced explosively, and Edwards and Howkins (1966) show that in some cases no thermal alteration is involved, no new joint pattern is imposed, and the kimberlite never reached the surface. They infer that the preservation of the diamonds, which were formed at depth, is due to the release of pressure at low temperatures. The isotopic composition of the carbon in diamonds is near that of mantle-derived graphite (Kropotova et al. 1967), although it might be expected that the diamond should be enriched in $_{13}C$ (Bottinga 1969). Kimberlite deposits contain both organic compounds (section 9) and carbonates; both are believed to have been acquired as the result of secondary hydrothermal effects; whether the carbon involved is juvenile, recycled, or a mixture of the two is disputable. There is no dispute that the carbon of the diamonds is juvenile.

10.3. Maars

Maars are ring-craters occupied by lakes. The type-locality is the Eifel district of Germany (Rutten 1969b). A fine example is the Pulvermaar (fig. 2). They result from a phenomenon found in all continents (Shoemaker 1962; Ollier 1967). Holmes (1965) refers to them as *fluidization craters*, and classes with them the ring craters of Ruwenzori in Africa. Mueller and

Veyl (1957) have described the recent eruption of a new maar in South America. Some of the craters have clearly been formed explosively; others, however, seem to have resulted from the non-explosive eruption of a diatreme in a state of active fluidization, just as in the experiments of Mills (fig. 1) (1969) and as envisaged by Holmes (1965, page 314). The German Maar fields are still the scene of the copious ebullition of mineral waters charged with carbon dioxide, and the carbonates of the Laacher See have isotopic ratios which show the carbon is juvenile (Taylor et al. 1967).

10.4. Intrusive tuffisites

In south Germany the Swabian tuffisite pipes so graphically described by Cloos (1941) are clearly fluidization phenomena involving carbon dioxide. Just as with the maars, the surrounding country is rich in carbon dioxide springs. The pipes do not now reach the surface in the form of craters. They have been eroded to expose the tuff-filled conduits and dykes (fig. 3). Intrusive tuffs are presumably always the result of fluidization. Coe (1966) has described carbonate-rich tuffisites from Ireland. There is no necessity for the carbon dioxide involved to be juvenile. It may well have resulted from remobilization if the ascending magma reached limestone. Isotopic evidence may help to decide whether juvenile gases are involved, but even then it must be remembered that metamorphism can effect the ratio (Deines and Gold 1969).

10.5. Cryptoexplosion structures

One of the most puzzling of present-day geological controversies concerns the origin of craters which show evident signs of violent explosive activity. The literature has been admirably classified and summarized (Freeberg 1966, 1969). One of their characteristic features (though not always present) is the development of shatter cones. Bucher (1936), in a classic paper, described a number of occurrences from the United States, and concluded that they resulted from the sudden liberation of pent-up volcanic gases, a phenomenon recalling the origin postulated for kimberlite pipes (Kennedy and Nordlie 1968) and implying that the crater overlies a diatreme. Subsequently Dietz suggested that shatter cones indicated that the craters were formed by meteoritic impact, and so were 'astroblemes'. Dietz and Bucher contributed to a lively discussion in the American Journal of Science (Bucher 1963; Dietz 1963) and discussed particularly the Ries and Steinheim structures of Germany. Although Bucher has since died, and majority opinion has come down strongly on the side of the Ries being due to impact (Preuss and Schmidt-Kaler 1969), the controversy is far from dead. The Sudbury Basin (Ontario) and the Vredefort Dome (South Africa) are just two of the structures in dispute (see Freeberg 1966, 1969 for references). The puzzle is exacerbated by the discovery in several of the disputed areas both of evidence pointing to the existence of a diatreme and of features regarded as indicative of meteoritic impact. The particular point of interest for our present enquiry is that the diatremes always seem to involve carbon dioxide, and the deposition of carbonates, and thus (by implication) juvenile carbon. It is even possible that in some cases meteoritic impact has triggered off the eruption of a carbon-rich diatreme. For example, the Steinheim Basin is a cryptoexplosion structure lying some 40 km to the south-west of the Ries Basin. It is interpreted by Groschopf and Reiff (1969) as very probably of impact origin. It may or may not be underlain by a diatreme; it certainly is associated with carbon dioxide springs (Holmes 1965).

11. Prebiological sources of food

It is now clear that the origin of life took place either in Pregeological time or shortly after the beginning of Katarchaean time. It is legitimate to suppose that during Pregeological time all the processes which are still contributing juvenile carbon to crust, atmosphere and hydrosphere were already in operation. Indeed it is highly probable that they were doing so at that time at a much enhanced rate. It is almost certain that when life originated the primary atmosphere had already been lost, and had been replaced with the secondary atmosphere derived directly from the outgassing processes we have been considering in this chapter. Although some chemists favor an early secondary atmosphere dominated by methane, most geologists agree that methane will have been much subsidiary to carbon dioxide and carbon monoxide (Galimov et al. 1968b).

Although the many 'prebiological experiments' that followed Miller's famous synthesis of amino acids (Miller 1953) have shown that abiogenic organic compounds must have formed in a non-oxidizing atmosphere, it is difficult to see how such compounds could have sustained within the atmosphere any form of life unable to employ photosynthesis. It has therefore been usual to postulate that life originated in the ocean which (in the words of Haldane 1929) formed a 'hot, dilute soup'. But Sillén (1965) has shown that thermodynamic considerations require the soup to be so thin that it is difficult to see how it could be regarded as a nutrient broth; he refers to 'the myth of the probiotic soup'.

On the other hand, accumulations of organic compounds are certain to have occurred in the Eogeological crust as they do today, and it is likely that these accumulations formed the first food for the first organisms (Sylvester-Bradley 1971a). Today there is evidence, as we have seen, that the concentration of organic substances is greater in continental crust than it is in oceanic crust. In Eogeological times it is likely that the oceanic crust had much the same structure and composition as it has today. But the same does not apply to the continental crust. The earliest rocks of Katarchaean time, which form the continental nuclei, have been shown by Windley and Bridgwater (1971) to fall into two groups. The rocks of the better-known group from the high-level cratons are dominated by granites and traversed by greenstone belts; they are characterized by low-grade metamorphism (see sections 3 and 9). In this group the relations of the greenstone belts to the basement are obscured by the emplacement of later granites. Geologically older rocks are found in the other group, which are high-grade metamorphic rocks of deeper level, though their radiometric ages have been updated during subsequent periods of recrystallization. These high-grade rocks are dominated by granulites which are intruded by calcic anorthosites. Similar anorthosites are postulated to form the highland areas of the Moon (Wood et al. 1970) and Windley (1970) has suggested that both on the Moon and on the Earth the development of anorthosites was a feature of the early crust. The granulite basement is also traversed by belts of metapelite. These are silli-manite-mica-schists exhibiting the same high-grade metamorphism as the surrounding basement but representing supracrustal rocks that must once have consisted of aluminous sediments and basic lavas. Dr. Windley tells me that many of these schists are rich in graphite. They are clearly analogous to the greenstone belts which characterize the lower-grade metamorphic rocks of the first group. They probably represent the earliest crustal carbonaceous deposits of the geological record. Windley and Bridgwater maintain that they represent the cover-deposits of a basement which was much thinner and much more mobile than it later became. It seems probable that it was the carbon in these schists of the cratonic nuclei that supported the origin of life. The hydrothermal ebullition of hot water and the volcanic emanation of carbon-rich gases may well have provided not only the food which formed the first nutrient medium, but also the background energy which gave rise to life.

Acknowledgements

Many of the ideas I express in this paper have been formulated in conversation with my colleagues in the University of Leicester, and I would particularly like to thank Drs. T. D. Ford, J. D. Hudson, M. J. Le Bas, A. J. Meadows, A. A. Mills, D. S. Sutherland and B. F. Windley.

References

AUCOTT, J. W. and R. H. CLARKE, 1966, Nature, Lond. 212 (5057) 61.

BERGER, R., 1966, Trans. Am. Geophys. Union. 47, 495.

BESKROVNYY, N. S., 1958, Dokl. Akad. Nauk. SSSR 122, 119.

BESKROVNYY, N. S., 1968, Dokl. Acad. Sci. USSR, Earth Sci. Sect. 178, 145.

BESKROVNYY, N. S. and T. E. BARANOVA, 1963, Dokl. Acad. Sci. USSR, 149, 62–65. (Transl. from Dokl. Akad. Nauk. SSSR 149 (4), 918–921.)

BISHOP, W. W., 1968, Trans. Leicester Lit. Phil. Soc. 62, 22.

BOCHAROVA, G. I., 1964, Dokl. Acad. Sci. USSR (Earth Sci. Sect.) 156, 140–141. (Dokl. Akad. Nauk. SSSR 156, 590–591.)

BOTTINGA, Y., 1969, Earth Planet. Sci. Letters 5, 301.

BURNHAM, C. W., 1967, Hydrothermal fluids at the magmatic stage, In: Geochemistry of hydrothermal ore deposits, ed. H. L. Barnes. (Holt, Rinehart and Winston, New York) 34.

CHEMINÉE, J., R. LETOLLE and P. OLIVE, 1969, Bull. Volcanologique 32, 469.

CLOOS, H., 1941, Geol. Rundschau 32, 703.

CLOUD, P. E., JR., 1968, Pre-metazoan evolution and the origins of the metazoa; In: Evolution and environment, ed. E. T. Drake. (Yale Univ. Press, New Haven) 1.

COE, K., 1966, Quart. J. Geol. Soc. Lond. 122, 1.

COLOMBO, U., F. GAZZARRINI, R. GONFIANTINI, G. KNEUPER, M. TEICHMÜLLER and R. TEICHMÜLLER, 1970, Advan. Org. Geochem. 577.

CRAIG, H., 1953, Geochim. Cosmochim. Acta 3, 53.

CROHN, P. W. and D. C. GELLATLY, 1969, Austral. J. Sci. 31 (9), 335.

DAVIDSON, C. F., 1964, Econ. Geol. 59, 1368.

DAWSON, J. B., 1966, Oldoinyou Lengai – an active volcano with sodium carbonatite lava flows; In: Tuttle and Gittins, 155.

DAWSON, J. B., 1967, A review of the geology of kimberlite; In: Ultramafic and related rocks, ed. F. J. Wyllie. (Wiley, New York) 241.

DEINES, P., 1968, Geochim. Cosmochim. Acta 32 (6), 613.

DEINES, P. and D. P. GOLD, 1969, Geochim. Cosmochim. Acta 33, 421.

DEWEY, J. F. and B. HORSFIELD, 1970, Nature, Lond. 225, 521.

DUBROVA, N. V. and Z. N. NESMELOVA, 1968, Geokhimiya 9, 1066.

DYMKOV, YU. M., V. A. USPENSKIY and B. V. BRODIN et al., 1967, Vop. Prikl. Radiogeol. 2, 122.

EDWARDS, C. B. and J. B. HOWKINS, 1966, Econ. Geol. 61 (3), 537.

EGLINTON, G., P. M. SCOTT, T. BELSKY, A. L. BURLINGAME, W. RICHTER and M. CALVIN, 1966, Advan. Org. Geochem. 41.

ELINSON, M. M. and V. S. POLYKOVSKII, 1963, Geochemistry 1963, 799.

FERRARA, G. C., G. FERRARA and R. GONFIANTINI, 1967, Carbon isotopic composition of carbon dioxide and methane from steam jets of Tuscany; In: Nuclear geology on geothermal areas. Spoleto, 1963, ed., Tongiorgi. (Cons. Naz. Ric. Lab. Geol. Nucl., Pisa).

FEELY, H. W. and J. L. KULP, 1957, Bull. Am. Ass. Petroleum Geol. 41, 1802.

FLORKIN, M., 1969, Fossil shell 'conchiolin' and other preserved biopolymers; In: Organic Geochemistry. (Springer-Verlag, Berlin) 498.

FLOROVSKAYA, V. N., 1964, Vestn. Mosk. Univ., Ser. Geol. 1964 (2), 3.

FLOROVSKAYA, V. N., 1967, Vestn., Mosk. Univ., Ser. Geol. 22, 49.

FRANTSESSON, Y. V., 1968, Dokl. Akad. Nauk. SSSR 183 (6), 1404.

FREEBERG, J. H., 1966, Geol. Surv. Bull. 1220, 91.

FREEBERG, J. H., 1969, Geol. Surv. Bull. 1320, 39.

GALIMOV, E. M., 1967, Geokhimiya 12, 1504.

GALIMOV, E. M., 1968, Izv., Akad. Nauk. SSSR Ser. Geol. 5, 29.

GALIMOV, E. M. and I. A. PETERSIL'YE, 1967, Dokl. Akad. Nauk. SSSR. 176, 914.

GALIMOV, E. M. and I. A. PETERSIL'YE, 1968, Dokl. Akad. Nauk. SSSR 182, 182.

GALIMOV, E. M., N. G. KUZNETSOVA and V. S. PROKHOROV, 1968. Geokhimiya 11, 1376.

GOLDBERG, I. S. and K. A. CHERNIKOV, 1968, Geochem. Intern. 5, 402.

GRÉGOIRE, C., 1966, Bull. Inst. Roy. Soc. Natl. Belg. 42, 1.

GUEST, N. J., 1956, Tanganyika Geo. Surv. Rec. 4, 56.

GUNTER, B. D. and B. C. MUSGRAVE, 1966, Geochim. Cosmochim. Acta 30, 1175.

HALDANE, J. B. S., 1929, The origin of life. The Rationalist Annual, 1929, 148-153, reprinted in The Origin of Life by J. D. Bernal, 1967. (Weidenfeld and Nicolson, London) 242.

HARRIS, P. G., W. Q. KENNEDY and C. M. SCARFE, 1970, Volcanism versus plutonism – the effect of chemical composition; In: Mechanism of Igneous Intrusion, ed. G. Newall and N. Rast. (Gallery Press, Liverpool) 187.

HEARN, B. C., 1968, Science 159, 622.

HEINRICH, E. W., 1966, The Geology of Carbonatites. (Rand McNally, Chicago) 555.

HEINRICH, E. W. and R. J. ANDERSON, 1965, Am. Mineral. 50, 1914.

HODGSON, W. A., 1966, Geochim. Cosmochim. Acta 30, 1223.

HOLMES, A., 1965, Principles of Physical Geology. 2nd edition. (Nelson, London) 1288.

HOEFS, J. and M. SCHIDLOWSKI, 1967, Science 155, 1096.

HULSTON, J. R. and W. M. MCCABE, 1962, Geochim. Cosmochim. Acta 26, 399.

IKORSKII, S. V., 1964, Dokl. Akad. Sci. USSR Earth Sci. Sect. 157, 90.

IKORSKII, S. V. and A. M. ROMANIKHIN, 1964, Geochem. Intern. 2, 245.

JANSE, A. J. A., 1969, Bull. Geol. Soc. Am. 80, 573.

JOHNSON, F. S., 1969, Space Sci. Rev. 9, 303.

KALYUZHNYY, V. A. and Z. I. KOVALISHIN, 1967, Geol. Geokhim. Goryuch. Iskop. Akad. Nauk. Ukr. SSR 9, 5.

KAPLAN, I. R. and A. NISSENBAUM, 1966, Science 153, 744.

KASHKAY, M. A. and T. N. NASIBOV, 1968, Geochem. Intern. 5, 932; 934. (Transl. from Geokhimiya, 1968, 1132–1134.)

KENNEDY, G. C. and B. E. NORDLIE, 1968, Econ. Geol. 63, 495.

KENVOLDEN, K., J. LAWLESS, K. PERING, E. PETERSON, J. FLORES, C. PONNAMPERUMA, I. R. KAPLAN and C. MOORE, 1970, Nature, Lond. 228, 923.

KING, B. C. and D. S. SUTHERLAND, 1966, The carbonatite complexes of eastern Uganda; In: Tuttle and Gittins, 1966, 73–126.

KRAVTSOV, A., 1959, The non-organic origin of methane and its homologues in igneous formations; In: The Origin of Life on the Earth. (Pergamon, London) 118.

KRAVTSOV, A. I., 1967, Proc. USSR All-Union Oil Gas Genesis Symposium, Moscow, 314. (English Transl. Express Transl. Serv. 0094/6.)

KROPOTKIN, P. N., 1959, The geological conditions for the appearance of life on the earth, and the problems of petroleum genesis; In: The Origin of Life on the Earth. (Pergamon, London) 84.

KROPOTOVA, O. I., V. A. GRINENKO and G. N. BEZRUKOV, 1967, Geokhimiya 8, 1003.

KROUSE, H. R. and V. E. MODZELESKI, 1970, Geochim. Cosmochim. Acta 34, 459.

KUDRYAVSTEV, N. A., 1959, Trudy Vnigri, All-Union Geol. Prospecting Res. Inst. 142.

LEBEDEV, V. S. and I. A. PETERSIL'YE, 1964, Dokl. Acad. Sci. USSR Earth Sci. Sect. 158, 153. (Transl. from Dokl. Akad. Nauk. SSSR 158, 1102.)

McKENZIE, D. P., 1970, Endeavour 29, 39.

MAYSKIY, YU. G. and V. N. TRUFANOV, 1967, Geol. Geokhim. Gorynch. Iskop. Akad. Nauk. Ukr. SSR 9, 30.

MILLER, S. L., 1953, Science 117, 528.

MILLS, A. A., 1969, Nature, Lond. 224, 836.

MOGAROVSKIY, V. V. and A. B. MARKOV, 1966, Geochem. Intern. 1966, 347. (Transl. from Geokhimiya 4, 459–463, 1966.)

MOGAROVSKIY, V. V., A. K. MEL'NICHENKO and G. I. KARAPETOVA et al., 1968, Dokl. Akad. Nauk. Tadzh. SSR 11 (1), 50.

MOROWITZ, H. J., 1968, Energy Flow in Biology. (Academic Press, New York-London).

MUELLER, G., 1954, Compt. Rend. Cong. Géol. Int. 19, Alger. 12, 279.

MUELLER, G., 1963, Nature 198, 731.

MUELLER, G., 1970, Indications of high-temperature processes in organic geochemistry; In: Adv. Org. Geochem. 1966. (Pergamon, Oxford) 443.

MUELLER, G. and G. VEYL, 1957, The birth of Nilahue, a new maar type volcano at Rininahue; In: Chile. Congr. Geol. Internal. XX, Mexico. Sec. 1, Vulcanologia del Cenozoico, Vol. 2, 375.

NAGY, B., 1970, Geochim. Cosmochim. Acta 34, 525.

NAKAI, N., 1960, J. Earth Sci. Nagoya Univ. 8, 174.

NAKAI, N., 1961, J. Earth Sci. Nagoya Univ. 9, 59.

OXBURGH, E. R. and D. L. TURCOTTE, 1970, Bull. Geol. Soc. Amer. 81 (No. 6).

PEDERSEN, K. R. and J. LAM, 1970, Grön. Geol. Undersög. Bull. 82.

PERING, K. and C. PONNAMPERUMA, 1969, Geol. Soc. Amer. 7, 172.

PETERSIL'YE, I. A., 1958, In Russian. Dokl. Acad. Sci. USSR 122, 1086.

PETERSIL'YE, I. A., 1961, English Transl., Izvest. Akad. Nauk. SSSR, Ser. Geol. 12, 19.

PETERSIL'YE, I. A., 1962, Geochemistry 14.

PETERSIL'YE, I. A., 1964, The gas constituent and trace bitumens of igneous and metamorphic rocks of the Kola peninsula; In: Int. Geol. Cong., 22 d., India. Rep. Pt. 1, 19.

PETERSIL'YE, I. A., 1966, Organic matter in igneous and metamorphic rocks of the Kola Peninsula; In: Chemistry of the Earth's Crust. Vol. 1, 47–61. Edited by A. P. Vinogradov. Translated from the Russian. Israel Program for Sci. Transl., Jerusalem.

PETERSIL'YE, I. A., S. V. IKORSKII, L. I. SMIRNOVA, A. M. ROMANIKHIN and E. B. PROS-KURYAKOVA, 1961a, Geochemistry 945.

PETERSIL'YE, I. A. and Y. B. PROSKURYAKOVA, 1961b, English transl. Izvest. Akad. Nauk. SSSR. Ser. Geol. 4, 55.

PETERSIL'YE, I. A., Y. D. ANDREYEVA and Y. V. SVESHNIKOVA, 1965, Dokl. Acad. Sci. USSR. Earth Sci. Sect. 161, 65.

PETERSIL'YE, I. A. and H. SØRENSEN, 1970, Lithos 3, 59.

PONNAMPERUMA, C. and K. PERING, 1966, Nature 209, 979.

PREUSS, E. and H. SCHMIDT-KALER (editors), 1969, Geol. Bavarica 61, 9.

RANKAMA, K., 1963, Progress in Isotope Geology. Interscience, New York. 705 pp.

RITTMANN, A., 1962, Volcanoes and their activity. (Wiley, New York) 305 pp.

ROBINSON, SIR ROBERT, 1966, Nature 212, 1291

ROEDDER, E., 1965, Amer. Mineral. 50, 1746.

RUTTEN, M. G., 1962, The Geological Aspects of the Origin of Life on Earth. (Elsevier, Amsterdam).

RUTTEN, M. G., 1966, Palaeogeog. Palaeoclimatol. Palaeoecol. 2, 47.

RUTTEN, M. G., 1969a, Sedimentary ores of the early and middle Precambrian and the history of atmospheric oxygen; In: Sedimentary Ores, Ancient and Modern, (Revised), ed. C. H. James, pp. 187–195. Dept. of Geology, University of Leicester. Sp. Publ. No. 1.

RUTTEN, M. G., 1969b, The Geology of Western Europe. (Elsevier, Amsterdam) 520 pp.

RUTTEN, M. G., 1970, Space Life Sci. 2.

RYE, R. O., 1966, Econ. Geol. 61 (8), 1399, 5.

SCHIDLOWSKI, M., 1968, Wissenschaft Technik, 18/68, 566.

SCHIDLOWSKI, M., 1966, Contr. Mineral. Petrol. 12, 365.

SHCHERBAK, V. P., 1964, Dokl. Acad. Sci. USSR. Earth Sci. Sect. 157, 151.

SHOEMAKER, E. M., C. H. ROACH and F. M. BYERS, 1962, Diatremes and uranium deposits in the Hopi Buttes, Arizona; In: Petrologic Studies: A volume in honour of A. F. Buddington. (Ed.: A. E. J. Engel. Geol. Soc. Amer. 327.)

SIEVER, R., 1968a, Earth Planet. Sci. Letters 5, 106.

SIEVER, R., 1968b. Sedimentology 11, 5.

SILLÉN, L. G., 1965, Arkiv. Kemi 24, 431.

SILLÉN, L. G., 1966, Tellus 18, 198.

SILLÉN, L. G., 1967, Science 156, 1189.

SOBOLEV, U. S., T. Y. BAZAROVA, N. A. SHUGUROVA, L. S. BAZAROV, Y. A. DOLGOV and H. SØRENSEN, 1970, Grön. Geol. Undersög. Bull. 81.

SPENCE, H. S., 1930 Amer. Mineral. 15, 499.

SPOTTS, J. H. and S. R. SILVERMAN, 1966, Amer. Mineral. 51, 1144.

STEWART, J. H., 1970. Grön. Geol. Undersög. Bull. 84.

SUTTON, J., 1968, Proc. Geol. Ass., 78, 493.

SUTTON, J., 1969, Transl. Leicester Lit. Phil. Soc. 63, 26.

SYLVESTER-BRADLEY, P. C., 1964, Discovery 25, 37.

SYLVESTER-BRADLEY, P. C., 1967, Proc. Geol. Ass. 78, 137.

SYLVESTER-BRADLEY, P. C., 1971a, Carbonaceous Chondrites and the Prebiological Origin of Food. In: Molecular Evolution, eds. R. Buvet and C. Ponnamperuma (North-Holland, Amsterdam).

SYLVESTER-BRADLEY, P. C., 1971b, Environmental Parameters for the Origin of Life. Proc. Geol. Ass. 82, 87.

SYLVESTER-BRADLEY, P. C. and R. J. KING, 1963, Nature 198, 728.

TAYLOR, JR., H. P., J. FRECHEN and E. T. DEGENS, 1967, Geochim. Cosmochim. Acta 31, 407.

THODE, H. G., R. K. WANLESS and R. WALLOUCH, 1954, Geochim. Cosmochim. Acta 5, 286.

TUTTLE, O. F. and J. GITTINS (editors), 1966, Carbonatites. (Interscience, New York).

VINOGRADOV, A. P., O. I. KROPOTOVA and V. I. USTINOV, 1965, Geoch. Intern. 495. Transl. from Geokhimiya 6, 643.

VINOGRADOV, A. P., O. I. KROPOTOVA, G. P. VDOVYKIN and V. A. GRINENKO, 1967a, Geochem. Intern. 229. Transl. from Geokhimiya 3, 267.

VINOGRADOV, A. P. and O. I. KROPOTOVA, 1967b, Izv. Akad. Nauk. SSSR. Ser. Geol. 11,3.

WINDLEY, B. F., 1969, Amer. Ass. Petrol. Geol. Mem. 12, 899.

WINDLEY, B. F., 1970, Nature 226, 333.

WOOD, J. A., J. S. DICKEY, JR., U. B. MARVIN and B. N. POWELL, 1970, Science 167, 602.

ZAKRZHEVSKAYA, N. G., 1964, Dokl. Acad. Sci. USSR. Earth Sci. Sect. 154, 128.

ZEZIN, R. B., V. S. LEBEDEV and Y. D. SYNGAYEVSKIY, 1967, Dokl. Akad. Nauk. SSSR 177, 236.

CHAPTER 4

Primordial organic chemistry

NORMAN W. GABEL and CYRIL PONNAMPERUMA

1. Introduction

When faced with the question of origins, we are faced with the problem of what may have happened. There is no single description of what may have happened just as there is no single description of what will happen. Since evaluations of forthcoming occurrences are based upon present-day trends of high probability, it is just as valid to examine the past through present-day knowledge with sets of information having a high probability of primordial occurrence.

Thus, when we address ourselves to the question of the origin of organic molecules, we are faced not only with Wheland's dilemma of delimiting organic chemistry (Wheland 1949), but with attempting to discern *what might have been* from *what is* and *what may be*. Scientific objectivity can not proceed without operational terminology. Therefore, from *what is* it is necessary to define systems (groups of particles; sets of information) and to study the interrelationships within these systems. Fortunately, the word *system* has been defined in the natural sciences in a manner which lends itself to study and observation (Chemical Bond Approach Project 1964). The task which is left to us is to examine the systems, place them in perspective, and assess their probability. Although these definitions of systems may be arbitrary, they are made for the same reasons that scientific disciplines are arbitrarily delimited. *Life* has been arbitrarily delimited from *non-life* for the purpose of study and observation. Perhaps it would be better in attempting to understand the origin of life to either set no limits or to set new limits.

In a systematic approach to observational phenomena it is useful to apply the idea of sets to our systems (Selby and Sweet 1969). The systems which

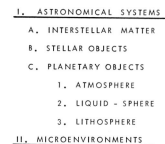

I. ASTRONOMICAL SYSTEMS

   A. INTERSTELLAR MATTER

   B. STELLAR OBJECTS

   C. PLANETARY OBJECTS

      1. ATMOSPHERE

      2. LIQUID - SPHERE

      3. LITHOSPHERE

II. MICROENVIRONMENTS

Fig. 1. Systems under consideration for the primordial origin of organic molecules.

will be delimited can be depicted schematically in outline form (fig. 1). Of the two main headings, only microenvironments has not been previously defined. A microenvironment is a system in which a mass or energy parameter becomes limiting or dominating which then results in a predominant set of chemical reactions or physical states (including a repeating sequence of change of physical state). Examples of this set are reaction conditions which simulate vulcanism, saline ponds, brackish tidal areas, thermal springs, and mineral and hydrocarbon deposits.

A question of system boundaries arises when the effect of energies impinging upon a system is considered. For convenience of discussion, radiant energy impinging upon a system is not considered to be part of the system until it is incorporated into the system. The radiant energy impinging upon the chemical systems under consideration is extended over the entire electromagnetic spectrum (fig. 2).

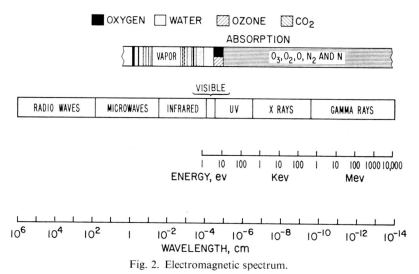

Fig. 2. Electromagnetic spectrum.

An estimate of the intensity of electromagnetic radiation which would be available for chemical reactions has been made only for solar-type planetary objects. High intensities of solar radiation (table 1) and a long path-length through planetary atmospheres could have resulted in significant amounts of photochemical reactions even if absorption coefficients were very small. As indicated by table 1 the intensity of the solar flux also decreases as the wavelength of the radiation decreases. It must be remembered, however, that the energy of a photon of electromagnetic radiation increases with decreasing wavelength (fig. 2). To initiate bond-breaking of simpler molecules and

TABLE 1

Solar flux at different wavelengths.

Å	Solar flux (cm^{-2} sec^{-1})
2900	7×10^{14}
2600	4×10^{14}
2400	9×10^{13}
2000	2×10^{13}

concomitant synthesis of more complex materials the simpler molecules must be capable of absorbing the energy of a given wavelength of radiation which is impinging upon them.

It would be instructive at this point to review what is known about the chemical composition of the systems under consideration and to catagorize some of the energies which would be part of or would be impinging upon these systems.

2. *Chemical composition*

Russell's (1929) discovery that hydrogen is the most abundant element in the universe indicates that the other elements of the universe and all atmospheres must be or have been present at some time in their reduced states. Oxygen, nitrogen, and carbon account for only 0.05% of the total mass of the sun, whereas hydrogen constitutes 87%; helium follows next in abundance to hydrogen (Brown 1949). The larger solar-type planetary objects, such as Jupiter and Saturn, contain large amounts of methane, ammonia, and water (Opik 1962). If the evidence of Tilton and Steiger (1965), that the age of the

Earth is 4.75 billion years, is accepted, then meteorites which have been dated as being 4.5 billion years old would be possible examples of the material from which planetary objects were formed. In these meteorites the metals are generally found in their reduced states (Mason 1962).

TABLE 2

Chemical equilibria.

	$K_{25° C}$
$C + 2H_2 \rightarrow CH_4$	8×10^8
$N_2 + 3H_2 \rightarrow 2NH_3$	7×10^5
$H_2 + \frac{1}{2}O_2 \rightarrow H_2O$	4×10^{41}
$CO_2 + 4H_2 \rightarrow CH_4 + 2H_2O$	7×10^{21}
$S + H_2 \rightarrow H_2S$	6×10^{15}

The thermodynamic properties of hydrogen, carbon, carbon dioxide, methane, nitrogen, ammonia, oxygen, and water are known (table 2) and the equilibrium composition of mixtures of these materials has been determined (Urey 1952a, 1956; Miller and Urey 1959). If the rate at which hydrogen escapes from the Earth is assumed to be constant, it is possible to calculate the pressure of hydrogen on the Earth 4.5 billion years ago. By taking into account the amount of hydrogen necessary for the reduction of iron, carbon, oxygen, nitrogen, and sulfur, Miller and Urey (1959) suggested the figure of 1.5×10^{-3} atmosphere as a reasonable concentration of hydrogen in the Earth's primordial atmosphere. The Urey equilibrium for the formation of calcium carbonate from calcium silicate precludes a carbon dioxide pressure greater than 10^{-8} atmospheres.

$$CaSiO_3 + CO_2 \rightarrow CaCO_3 + SiO_2$$

If 1.5×10^{-3} atmosphere of hydrogen is assumed to be a reasonable figure, the corresponding pressure of methane would be 4×10^3 atmospheres. Miller and Urey (1959) concluded that since ammonia is very rapidly converted to nitrogen and hydrogen by any energy source, a reducing atmosphere consisting of small amounts of hydrogen, ammonia, and water and a moderate pressure of methane and nitrogen would constitute a reasonable atmosphere for the primordial Earth.

Cadle and Allen (1970) have underscored the complexity of the photochemically initiated reactions which can occur in a structured planetary atmosphere. Any comprehensive treatment of this subject should incorporate meteorological considerations. There still exists some room for argument about the exact nature of primordial planetary atmospheres. Rasool and McGovern (1966) have pointed out that hydrogen is a prerequisite for a methane-ammonia atmosphere. The gravitational escape velocity of hydrogen based on the current exospheric temperature maximum (ca. 2000 K°) indicates that a methane-ammonia atmosphere could only have persisted on the primordial Earth for 10^5–10^6 years after accretion. Holland's (1961) model of the Earth's atmosphere suggests that this exospheric temperature maximum is higher than what actually existed and that 500–1000 K° is more realistic. In McGovern's (1969) thermal model of the primitive atmosphere, ammonia would have shielded water from photodissociation and would have created a zone similar in some respects to the present-day ozone layer. The primitive atmosphere would eventually evolve after 10^8–10^9 years into an environment dominated by nitrogen with small amounts of carbon dioxide and water.

Rubey (1955) and Abelson (1966) have contended that the majority of the carbon in the primitive atmospheres of the inner planets was in an oxidized state. Anders (1968) and his associates have postulated the existence of relatively non-volatile carbon compounds in a cooling solar nebula. It would have been these carbon compounds which would have accreted during the formation of the inner planets if methane had been too volatile. The suggestion that carbon was incorporated into the inner planets as complex organic matter was first made by Urey (1952a). Nevertheless, the high temperatures which would have been produced during accretion (Latimer 1950) would have insured the pyrolysis of this complex organic matter (with the possible exception of porphyrins) into a secondary primordial atmosphere (Brown 1949; Urey 1952a) which was still relatively reducing compared to the present-day oxidizing atmospheres of the inner planets (Brandt and McElroy 1968).

In one of the most exhaustive investigations on the products obtained by subjecting mixtures of gases to an energy source, Abelson (1953–1954) studied 20 different combinations of hydrogen, methane, carbon monoxide, carbon dioxide, ammonia, nitrogen, water, and oxygen. Organic molecules, having two or more carbon atoms, formed only when the mixture was non-oxidizing with respect to methane. The corollary implication of these experiments is that reducing conditions are necessary for the synthesis of organic molecules in significant quantities on primitive planetary objects. The escape velocity of hydrogen from any gravitational field is less than that of any other element. Therefore, during the process of chemical evolution, all astronomical systems are continually being depleted of hydrogen even when nuclear processes are not operative. Since the production of complex organic molecules from the primordial gases requires an increase in the oxidation state of some of the atoms of the synthesized molecules, it is just as important that the environment of the system gradually becomes relatively more oxidizing as it is to begin with a reducing environment for the accumulation of diverse organic molecules. If the environment did not gradually become relatively more oxidizing, there would be no chemical evolution. No system which is in equilibrium ever evolves.

3. Energies

The energy which would be available to this reducing liquid and gaseous mixture for the initiation of chemical reactions would be impinging electromagnetic radiation, electric discharges, ionizing radiation from radioactive decay, and heat in the form of volcanoes and hot springs (table 3; Miller and Urey 1959). An examination of table 3 shows that ultraviolet light is by far the most intense source of electromagnetic radiation available for the synthesis of more complex materials from the simple primordial molecules. Thompson et al. (1963) and Watanabe et al. (1953) measured the absorption coefficients of a number of gases with argon in the reference cell. Two of the main constituents of the primordial atmosphere, methane and nitrogen, have absorption coefficients of less than 10^{-4} cm^{-1} between 1850 and 2000 Å. In contrast, ammonia and water have an appreciable absorption in this region

TABLE 3

Energy available for synthesis of organic compounds
on earth-like planetary objects.

Source	Energy (cal cm^{-2} yr^{-1})
Ultraviolet light (2500 Å)	570
Electric discharges	4
Radioactivity	0.8
Heat	0.13

(figs. 3 and 4). Although the absorption of methane extends to 1450 Å (Terenin 1959), it is highly probable that ammonia and water served to a large extent both as initiators and reactants of any initial homolytic photo-

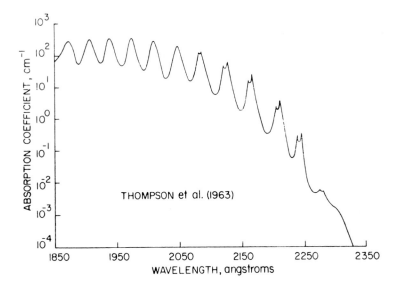

Fig. 3. Absorption coefficient of NH_3 as a function of wavelength.

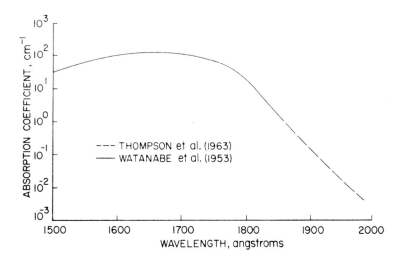

Fig. 4. Absorption coefficient of H_2O as a function of wavelength.

chemical reactions. The dissociation products of methane also absorb at higher wavelengths. For example, the CH_3, CH_2 and CH radicals absorb electromagnetic radiation up to 2800 Å (Urey 1952a).

The ionizing radiation of the solar wind (Mackin and Neugebauer 1966; Buhler et al. 1969) would play an important role in the heterolytic photodissociation of methane and nitrogen in the upper atmosphere. The intensity and composition of the solar wind and the effects of its albedo from planetary objects has not as yet been assessed.

Next in importance as a source of energy are cathodic electrical pulses such as lightening and corona discharges from pointed objects (Schonland 1953). These are heterogeneous phenomena which would occur at or near a planetary surface and therefore could have promoted the transfer of the reaction products to primordial oceans and microenvironments.

The principle sources of ionizing radiation at the surface of Earth-type planetary objects are potassium-40, uranium-238, uranium-235 and thorium-232 (Swallow 1960). The energy of potassium-40 seems to be quantitatively more important and is furthermore in the form of penetrating beta and gamma rays. In contrast almost 90% of the energy from uranium-238 and thorium-232 is emitted as alpha particles, which may not be penetrating enough to have a significant effect. Calculations show that the decay of potassium-40 in the Earth's crust today gives rise to 3×10^{19} calories per year; 2.6×10^9 years ago this would have been 12×10^{19} calories per year.

Volcanic heat is another form of energy which may have been effective in microenvironments (Rittmann 1962). The numerical value in table 3 has been calculated on the assumption of 1 km^3 of lava emission per year at 1000°C. Heat would also be provided to planetary objects by meteoritic impact (Hochstim 1963). The generalized system which is under consideration for the primordial origin of organic molecules is a mixture of simple molecules composed primarily but not exclusively of carbon, hydrogen, nitrogen, phosphorus, oxygen, and sulfur in a reducing environment which gradually changes to a relatively more oxidizing environment while being acted upon by various energy sources (fig. 5).

4. The effect of energies on the simple molecules

4.1. Methane

Methane and nitrogen are capable of undergoing photochemical dissociation at wavelengths of ionizing radiation. Walker and Back (1963) investigated the photolysis of methane with light from a microwave discharge through helium, which consisted mostly of the strongly self-reversed 584 Å resonance line. The methane pressure was varied from 0.001 to 0.3 mm and the identi-

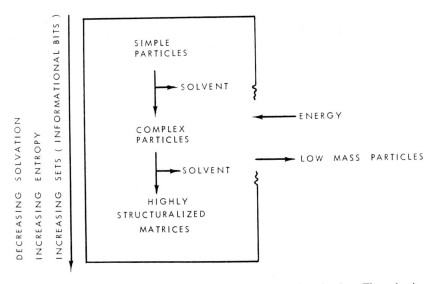

Fig. 5. The set under consideration is defined as atoms and molecules. The solvation sphere particles can be considered to be quasi-ordered environmental parameters. In addition to hydrogen, other low mass particles with an energy greater than but opposite in sign to the gravitational field will leave the system.

fiable products were hydrogen, ethane, ethylene, and acetylene. Walker and Back presented the following mechanisms as an explanation for the observed products.

$$CH_4 + h\nu \rightarrow CH_4^+ + e^-$$
$$CH_4^+ + CH_4 \rightarrow CH_5^+ + CH_3$$

$$CH_4 + h\nu \rightarrow CH_3^+ + H + e^-$$
$$CH_3^+ + CH_4 \rightarrow C_2H_5^+ + H_2$$

$$CH_5^+ + e^- \rightarrow CH_5^* \begin{array}{l} \nearrow CH_4 + H \\ \rightarrow CH_3 + H_2 \\ \searrow CH_2 + H_2 + H \\ CH + 2H_2 \end{array}$$

$$C_2H_5^+ + e^- \rightarrow C_2H_5^* \begin{array}{l} \nearrow C_2H_4 + H \\ \rightarrow C_2H_3 + H_2 \\ \searrow C_2H_2 + H_2 + H \end{array}$$

$$2\,CH_3 \rightarrow C_2H_6$$
$$2\,CH_2 \rightarrow C_2H_4$$
$$2\,CH \rightarrow C_2H_2$$

Mass spectrometric studies by Meisels et al. (1966) of the radiolysis of methane indicated the occurrence of similar processes including the production of small amounts of larger alkanes at higher energies. Ausloos et al. (1964) investigated the photolysis and radiolysis of equimolar mixtures of methane and completely deuterated methane as a function of pressure. Their results showed that CH_2 (carbene) and CD_2 (deuterocarbene) are produced in sufficient quantities to account for the presence of polymerization products by an insertion mechanism. Knox and Trotman–Dickenson (1957) concur with these results. The failure of Munson et al. (1965) to detect the following reaction in the high pressure mass spectroscopy of methane supports this viewpoint by Ausloos et al. (1964).

$$CH_5^+ + CH_4 \rightarrow C_2H_5^+ + 2H_2$$

Enright (1959) and Dolle (1958) found that when methane is irradiated with gamma rays from a Co^{60} source for three weeks there is produced a 10–24% conversion to heavy unsaturated hydrocarbons. Although the unsaturation could have resulted from dehydrogenation after polymerization, a preponderance of vinyl end groups would indicate that a radical chain mechanism must have played some part in the formation of these polymers.

Methane has also been subjected to high and low intensity electron beams. Lampe (1957) irradiated methane at both 50 and 150 mm of pressure with a beam of 2 meV electrons. The products were hydrogen, ethane, propane, butane, isobutane and ethylene. The product distribution seemed to be independent of pressure and intensity of radiation. A small amount of higher polymeric material was also formed. Lampe believed the reaction to be completely free radical in nature. Decomposition by an electron beam having energies between 15 and 100 eV was studied by Manton and Tickner (1960). The products, which were immediately frozen out, consisted mainly of ethane, ethylene, and acetylene. There were also small quantities of propane, propylene, butane, butene, butadiene, and pentene. Williams (1959) studied the chemical decomposition of several gases in a low energy electron beam. The behavior of the gas systems and the yields of products suggested that an important primary process was nonionizing excitation by electron impact. The mass spectrogram of methane 10^{-5} to 10^{-6} sec after electron bombardment was recorded by Tunitskii and Kupriyanov (1957). Their investigations revealed the presence of the following ions: C^+, CH^+, CH_2^+, CH_3^+, CH_4^+, CH_5^+, C_2H^+, $C_2H_2^+$, $C_2H_3^+$, $C_2H_4^+$, $C_2H_5^+$ and $C_2H_6^+$; CH_5^+ and $C_2H_5^+$ were present in the largest quantities.

The electric arc discharge through methane is one of the known commercial methods of producing acetylene (Yamamoto 1954; Eremin 1958; Pevera and Hess 1958). Other hydrocarbon products in addition to acetylene are also formed. Starodubtsev et al. (1961) exposed methane to a high frequency electric discharge and examined the species produced in the reaction chamber by infrared spectroscopy. They explained the products they observed by the following sets of reactions:

$$CH_4^+ + CH_4 \rightarrow CH_5^* + CH_3^+$$
$$CH_5^* \rightarrow CH_3 + H_2$$
$$CH_4^+ + e^- \rightarrow CH_4^* \rightarrow CH_3 + H$$
$$CH_3^+ + CH_4 \rightarrow C_2H_5^+ + H_2$$
$$C_2H_5^+ + e^- \rightarrow CH_3 + CH_2$$
$$2\ CH \rightarrow C_2H_2$$
$$2\ CH_2 \rightarrow CH_2{=}CH_2$$
$$2\ CH_3 \rightarrow CH_2{=}CH_2 + H_2$$

The ultraviolet spectrum emitted from methane when submitted to a high–frequency electric discharge was studied by Badareu and Popovici (1960). Analysis of the spectrum disclosed the presence of CH, CH^+, acetylene and benzene within the zone of the discharge. Badareu et al. (1959) had previously reported the deposition of aromatic hydrocarbons near the electrodes and suggested that they resulted from the polymerization of CH radicals.

Methane has also been submitted to pyrolysis in order to obtain acetylene (Sherwood 1960). This thermal cracking of methane is believed to be a short-chain free radical reaction (Germain and Vaniscotte 1958). A somewhat different pyrolytic treatment of methane, which may be more pertinent to chemical evolution, is its treatment at 350°–550° in the presence of a transition metal oxide and 5–10% water (Elian and Le Pingle 1953). Higher hydrocarbons are produced. An extensive discussion, including theoretical models, of chemical reactions in electrical discharges through hydrocarbons has been published recently as the proceedings of a symposium on this subject (Blaustein 1969).

4.2. Nitrogen

Between 1570 and 430 Å nitrogen also undergoes photoionization. Weissler et al. (1959) measured the ion intensity as a function of photon energy, but did not determine the types of ionic species present. Nitrogen in the upper atmosphere is dissociated, to some extent, by absorption of solar radiation, and its recombination reactions may give rise in part to the aurora borealis phenomenon (Herzberg and Herzberg 1948; Barth 1961).

When nitrogen is subjected to an electric discharge there is produced what has been called active nitrogen. According to Evans and Winkler (1956), active nitrogen consists chiefly of atomic nitrogen and excited N_2 molecules that may result from their recombination. Since nitrogen does not oligomerize to any great extent, the interesting reactions of active nitrogen within the context of chemical evolution are its reactions with methane. One hundred years ago Berthelot (1869) found that hydrogen cyanide was produced by sparking a mixture of nitrogen and hydrocarbons. The current status of research on active nitrogen has been recently reviewed by Wright and Winkler (1968).

Gartaganis and Winkler (1956) found that the main products from methane and active nitrogen are hydrogen cyanide and hydrogen. At low temperatures a small amount of dark brown polymer is deposited. They tentatively concluded that the reaction is carried substantially by hydrogen atom reactions since the characteristic induction period disappeared when atomic hydrogen was generated in the flow system. At 45° only 1% of the methane reacted, but above 280° the reaction was substantially complete.

$$N + CH_4 \rightarrow HCN + H_2 + H$$
$$H + CH_4 \rightarrow CH_3 + H_2$$
$$CH_3 + N \rightarrow HCN + 2H$$

The reactions of atomic nitrogen with unsaturated hydrocarbons are highly exothermic as can be seen from the measurements of Evans et al. (1956).

$$CH_2=CH_2 + N \rightarrow HCN + CH_3 + 62 \text{ kcal.}$$
$$CH_3 + N \rightarrow HCN + 2H + 10 \text{ kcal.}$$

The above reactions are of consequence since methane is partially polymerized to unsaturated high molecular weight hydrocarbons by the previously discussed types of energies. Weininger (1960) studied the reactions of polyethylene with atomic nitrogen. The terminal ends of the polyethylene used in this study were unsaturated. The gaseous products consisted of 80% hydrogen cyanide and 20% cyanogen. Weininger proposed the following mechanism for the formation of the gaseous products:
formation of a nitrogen complex:

$$CH_2=CH-CH_2R + N \rightarrow CH_2-CH-CH_2R$$
$$\underset{\cdot}{\overset{\diagdown \quad \diagup}{N}}$$

Chain unzipping:

$$\text{complex} \rightarrow HCN + H \text{ (or } CN + 2H) + CH_2=CH-R$$

Free-radical chain termination:

$$\text{complex} \rightarrow \text{HCN (or CN} + \text{H)} + \cdot \text{CH}_2\text{—CH}_2\text{—R}$$

Because 10% of the polyethylene was cross-linked, secondary hydrogen abstraction was apparently an important process.

4.3. Ammonia

Although the NH radical is produced via irradiation only below 1550 Å, the NH_2 radical is produced throughout the lower end of the absorption spectrum of ammonia (Schnepp and Dressler 1960). The NH radical is stable up to 36° K but NH_2 becomes unstable at 20° K. Therefore, in the interstellar dust and gases and on the large outer planets these radical species would be expected to have an appreciable lifetime. When ammonia is irradiated at 1470 Å, it decomposes to nitrogen and hydrogen (Jucker and Rideal 1957). Passage of an electric discharge through ammonia has been shown by electron paramagnetic measurements to produce active nitrogen as well as hydrogen atoms (Cole and Harding 1958). Gamma irradiation of ammonia gives stoichiometric quantities of nitrogen and hydrogen (Dolle 1958). The production of ammonium ion is an important process in the radiolysis of gaseous ammonia. The investigations of Dorfman and Noble (1959) indicate that the following process occurs with high efficiency.

$$\text{NH}_3^+ + \text{NH}_3 \rightarrow \text{NH}_4^+ + \text{NH}_2$$

Liquid ammonia can be expected to be found on some of the large outer planets. Glow discharge electrolysis of liquid ammonia yields hydrazine as its chief product (Hickling and Newns 1959). Therefore, the existence of liquid pools of hydrazine and ammonia would not be unlikely. A unique characteristic of liquid ammonia, as apposed to liquid water, is its ability to solvate the metallic state of alkali and alkaline Earth metals (Symons 1959). If these metals were in their reduced states and escaped from reaction with water, they might be found in solution in the liquid ammonia. Physical chemical studies of dilute metal-ammonia solutions indicate the principal solution species as the solvated metal cation $M^+(NH_3)_n$, the solvated electron $e^-(NH_3)n$, the monomeric metal atom, the dimeric metal atom and the negative metal anion M^- (Jolly 1965). In view of the reducing efficacy of hydrazine and liquid ammonia solutions of metals, it is doubtful whether chemical evolution would proceed to any great extent if these compounds were present in excess, since the production of more complex organic materials from the primodial molecules requires an increase in the oxidation state of some of the atoms of the synthesized molecules.

4.4. Water

The 'liquid'-sphere of the earth is a hydrosphere and as such water has always preoccupied man in his philosophical thinkings.

Since the energies available for chemically changing the water are radiant energies, it becomes germane to examine the effect of radiolysis on atmospheric water, aqueous solutions and frozen water with and without other molecules in the matrix. Anbar (1965) and Moorthy and Weiss (1965) have contributed particularly illuminating chapters to a symposium on this subject.

Irradiation by solar flux of atmospheric gas-phase H_2O presumably results in photodissociation and the eventual formation of hydrogen and oxygen. Radiation-induced chemical changes in water have yielded substantive evidence for the primary production of the species H, OH, e^-(aq), H_3O^+, H_2 and H_2O_2 (Matheson 1962). The hydrogen atoms, hydroxyl radicals and hydrated electrons have only a transitory existence in the liquid state.

Hydrated electrons are unusually efficacious nucleophilic agents. Paraffinic hydrocarbons, alcohols, amines and ethers dissolved in water are relatively unreactive toward hydrated electrons (Anbar and Hart 1965; Hart et al. 1964a) Olefins are polarizable and are, therefore, subject to attack. Carbonyl compounds dissolved in water react very rapidly with hydrated electrons (Hart et al. 1964a, b). In general all solubilized molecular dipoles which can accomodate an additional electron will react with hydrated electrons. The reactivity of aromatic compounds with hydrated electrons is a function of the electron density of the pi-orbitals and closely follows the relationships symbolized by Hammett's sigma-function (Anbar 1965).

Studies on the irradiation of ice and frozen aqueous solutions have substantiated the existence of the transient species produced in water (Moorthy and Weiss 1965). In addition, gamma-irradiation of ice produces positive holes in the crystalline matrix. Secondary reactions of other molecules in the matrix involve not only the radiation-produced species in water but also the delocalized positive holes.

The irradiation of water produces species which are reducing (e^-(aq), H, H_2) and species which are oxidizing (H_3O^+, OH, H_2O_2). Since the reducing species have the least mass and consequently the greatest mobility, reducing reactions might be favored in non-equilibrium irradiated solutions. In irradiated ice, however, the existence of delocalized positive holes would indicate that oxidations and reductions of foreign molecules in the matrix should be equally facilitated.

5. *The synthesis of complex organic molecules*

It is a tribute to Oparin (1924, 1964) and Haldane (1928) that, before the overwhelming abundance of hydrogen in the universe was determined astrophysically (Russell 1929), they foresaw the need for the primordial atmosphere to be reducing in order for complex organic materials to form. The Oparin–Haldane hypothesis presupposed a long chemical evolution before the appearance of life. During this period there would be an accumulation of complex molecules and the eventual generation of replicating molecules that would evolve into recognizable life-forms.

Horowitz (1945) claims that metabolic pathways should also be regarded as a recapitulation of prebiological syntheses. He postulates that the first living entity was a heterotroph which used for its structural network and energy metabolism those prefabricated molecules that were present in its environment. A depletion of the supply limited further multiplication until the living system adapted itself to the use of other substrates which were then synthesized into the ones it lacked. Horowitz maintains that the final step in a modern biological synthesis of a particular molecule must have evolved first.

Astronomical systems

(a) Interstellar matter. Recent radioastronomic observations at the National Radioastronomy Observatory in Green Bank, West Virginia have disclosed the presence of formaldehyde in the interstellar spaces of the 15 galactic regions which were examined (Snyder et al. 1969). Other organic molecules will probably be detected through the use of this technique. Radioastronomers now postulate that polyatomic organic molecules are abundant throughout interstellar space. It is conceivable that this complex organic matter (especially porphyrins) could have contributed to the organic milieu of a primitive planet if a comet which had been formed from such interstellar matter was captured by the planetary object. The thermal outgassing which follows accretion may have pyrolized most complex organic material. However, some porphyrins have unusual thermal stability and may have survived pyrolysis. The most salient feature of the discovery of polyatomic organic molecules in interstellar space is the ubiquity of organic matter.

(b) Stellar objects. The only parts of stellar objects which have been observed by spectrochemical methods are the radiating particles of the photosphere and the absorbing particles of the upper stellar atmospheres. Additional information on the composition of type-G stars may be extrapolated from the examination of sun spots. Thus far, the only diatomic species of carbon which have been recorded are C_2, CH, CN, and CO (McKellar 1960).

(c) Planetary objects: (1) Atmosphere. (2) 'Liquid'-sphere. (3) Lithosphere.

Microenvironments
What follows is a brief compaction of the material previously presented in a review article by Ponnamperuma and Gabel (1968), plus more recent experiments on the abiogenic evolution of organic molecules of biological significance. The experiments which are discussed were selected for the purpose of illustration and to support significant points in the discussion.

5.1. Amino acids

By far the greatest amount of experimentation in the field of chemical evolution has been concerned with the origin of amino acids. The fact that amino acids are readily formed and detected was probably the main impetus for the avalanche of experiments performed on the prebiological synthesis of amino acids following the success of the classical experiment of Miller (1953). A mixture of methane, ammonia, water and hydrogen was exposed to an electric discharge from tesla coils for approximately one week. Many organic compounds were formed and among those that could be identified were glycine, alanine, β-alanine, aspartic acid and glutamic acid. Abelson (1953–1954) studied the effect of an electric discharge on 20 different combinations of hydrogen, methane, carbon monoxide, carbon dioxide, ammonia, nitrogen, water and oxygen. Amino acids were formed only when the mixture was non-oxidizing with respect to methane. When the oxides of carbon predominated in the mixtures, no amino acids could be detected. Pavloskaya and Pasnyskii (1959) found that an apparatus which removed formed hydrogen from the system enhanced the synthesis of amino acids. This is understandable in view of the fact that a carbon atom from methane has its oxidation number increased when it is incorporated into an amino acid. In all of these experi-

ments glycine and alanine were the predominant products. The previously
described electric discharge experiments on methane alone also have produced predominantly two-carbon compounds.

Miller (1957) had suggested that a Strecker synthesis is involved in the
formation of amino acids by electric discharge in the heterogenous phase
system of methane, ammonia and water (fig. 6). The reaction of methane and
ammonia in the absence of water to yield a-aminonitriles indicates that an
alternative mechanism could be valid (Ponnamperuma and Woeller 1967).
This direct synthesis of a-aminonitriles could then give rise to amino acids.
Since nitriles are also products of this reaction system, they may be the
precursors of the a-amino derivatives. Cyano groups are known to stabilize
unpaired electrons on adjacent carbon atoms through electron delocalization (Lewis and Matheson 1949). These cyano

$$
\cdot \underset{\overset{|}{R}}{\overset{\overset{\displaystyle R}{|}}{C}} - C \equiv N \longleftrightarrow \underset{\overset{|}{R}}{\overset{\overset{\displaystyle R}{|}}{C}} = C = N \cdot
$$

substituted radicals could then react with amide radicals NH_2, or azene
radicals NH, to produce the a-aminonitriles. The experimental results of
Taube et al. (1967) on the formation of amino acids in the heterogeneous
gas-solid system of methane, ammonia, water and nitrogen at 900–1100°
passed through a silica flow reactor indicates that the Strecker synthesis is
not a main reaction pathway.

$$RCHO + NH_3 + HCN \rightleftharpoons RCH(NH_2)CN + H_2O$$

$$RCH(NH_2)CN + 2H_2O \longrightarrow RCH(NH_2)COOH + NH_3$$

$$RCHO + HCN \rightleftharpoons RCH(OH)CN$$

$$RCH(OH)CN + 2H_2O \longrightarrow RCH(OH)COOH + NH_3$$

Fig. 6. Strecker synthesis of amino acids.

One of the first photochemical experiments to determine whether amino
acids could be synthesized by ultraviolet irradiation of simple molecules was
performed by Abelson (1953–1954). A solution of ammonia, ammonium
formate, sodium cyanide and ferrous sulfate was irradiated at 2536 Å.

Glycinonitrile was formed which upon hydrolysis yields glycine. Groth (1957) reported that when mixtures of ethane or methane plus ammonia and water were photolyzed at 1470 or 1295 Å at relatively high pressures, glycine, alanine, sarcosine and several unidentified amino acids and polymers resulted. Groth and Von Weyssenhoff (1960) irradiated the same mixtures at 4070 Å and 2196 Å and claimed that glycine, alanine, and aminobutyric acid were formed. Ammonia and water probably serve both as initiators and reactants in the formation of amino acids by photolytic reactions in the ultraviolet region. Ionizing radiation has been used as the energy source for several syntheses of amino acids. Dose and Rajewsky (1957) irradiated a mixture of methane, ammonia, water, hydrogen, carbon dioxide and nitrogen with X-rays and gamma rays and demonstrated that some of the organic products were amino acids. Hasselstrom et al. (1957) irradiated a solution of ammonium acetate with β-particles and found glycine, aspartic acid, and diaminosuccinic acid in the products. Palm and Calvin (1962) exposed a mixture of methane, ammonia and water to 5 meV electrons from a linear accelerator and were able to identify glycine and alanine among the end products. Using N-acetylglycine as a starting material Dose and Ponnamperuma (1967) produced a large number of amino acids by irradiating this amino acid derivative with gamma rays.

Harada and Fox (1964) have extensively studied the effect of high temperatures on a mixture of methane, ammonia and water. The mixture was passed over quartz or alumina in a glass reaction tube which was maintained at 900° to 1000° C. The effluent material was absorbed in aqueous ammonia, subjected to acid hydrolysis, and examined by an amino acid analyzer. These investigators claimed that 14 of the amino acids commonly found in protein could be synthesized: glycine, alanine, aspartic acid, threonine, serine, glutamic acid, proline, valine, leucine, isoleucine, tyrosine and phenylalanine.

The formation of so many amino acids under conditions which presumably simulate various environmental niches of a primordial planet again testifies to the ubiquity of polyatomic organic molecules. However, many of the amino acids formed in these experiments do not occur naturally in biological material. Furthermore, a large number of the naturally occurring amino acids have not been found in these experiments. Most of the identifications have been based upon only one analytical technique. Very often more than one compound in the same class will have the same retention time, Rf value, or elution time on ion exchange columns. With one exception, where the identification has been corroborated by ion-exchange chromatography, gas chromatography and mass spectrometry (Ponnamperuma and Flores 1966;

Ponnamperuma and Woeller 1967), the identifications have been based on either paper chromatography or ion-exchange chromatography alone. The preponderance of glycine and alanine amongst the products of these experiments is indicative of the high concentration of starting materials and relatively short reaction times compared to geophysical reality. It is very likely that high dilution, long reaction times, and a gradual removal of hydrogen would result in the detectable syntheses of many more amino acids. Since the products are the results of collisions of energetically excited particles, a high ratio of energy to particles is required with the further proviso that the relatively less thermodynamically stable aggregates (which in this case would be polyatomic carbon compounds), do not extensively collide with each other when energetically excited.

5.2. Purines and pyrimidines

The first demonstration of the synthesis of a purine from a readily formed primordial chemical was Oró's (1960) synthesis of adenine from ammonium cyanide. Although the concentrations of ammonium cyanide were much too high in his experiments to be considered reasonable for a prebiotic Earth, they

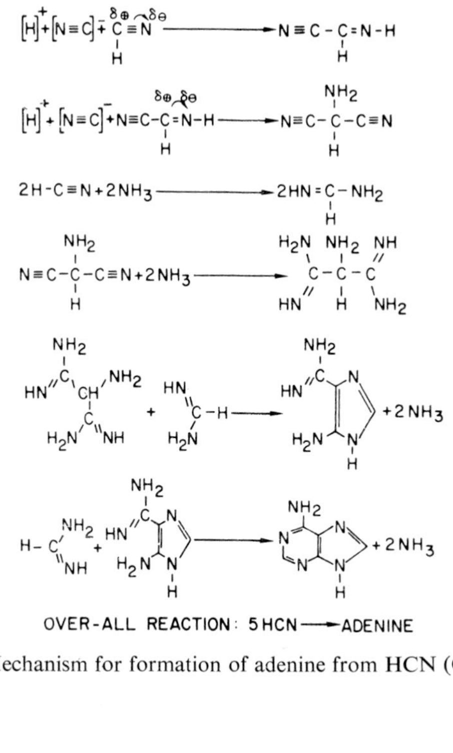

OVER-ALL REACTION: 5 HCN ⟶ ADENINE

Fig. 7. Mechanism for formation of adenine from HCN (Oró 1961).

are not at all unreasonable for the large outer planets. Oró's laboratory experiments may very likely have duplicated one of the fundamental chemical processes occurring on Jupiter and Saturn. A detailed study of the reaction has indicated that formamidine and 4-aminoimidazole-5-carboxamidine were the immediate precursors of adenine. Oró's suggested mechanism is illustrated by fig. 7. This synthesis has been confirmed by Lowe et al. (1963).

Of perhaps greater significance to the origin of purines on the primoridal Earth is the synthesis of adenine by the β-irradiation of methane, ammonia and water. Since Ponnamperuma et al. (1963) detected hydrogen cyanide in their reaction mixture, the synthetic pathway may be the same as in the experiments carried out by Oró. The conversion to adenine was enhanced by the removal of hydrogen which is not surprising since hydrogen is the other product formed when hydrogen cyanide is produced from methane with nitrogen or ammonia. Ultraviolet irradiation of dilute solutions of hydrogen cyanide also produce adenine and guanine (Ponnamperuma 1965). The investigations by Sanchez et al. (1966a) have shown that aminomalononitrile is an important intermediate in the synthesis of purines. The aminomalono-nitrile can follow the pathway formulated by Oró (fig. 7) or can form the hydrogen cyanide tetramer which is capable of undergoing photochemical rearrangement to 4-aminoimidazole-5-carbonitrile (fig. 8). This compound via hydrolysis and reaction with hydrogen cyanide, formamide, or cyanogen can produce a variety of purines.

The search for pyrimidines in mixtures of methane, ammonia and water that have been subjected to various energy sources has thus far been un-successful. The experiments of Ferris et al. (1969) suggest that the rigorous exclusion and removal or formed oxygen might be necessary. A possible precursor of the pyrimidine ring structure, cyanoacetylene, has been shown by Sanchez et al. (1966b) to be one of the major nitrogen-containing products from an electric discharge on a mixture of methane and ammonia. When cyanoacetylene and urea or cyanate salts are heated together at 100°, cytosine is produced (Ferris et al. 1968).

5.3. Carbohydrates

Neither monosaccharides nor polysaccharides have been detected as products of the reactions of hypothetical primordial gas mixtures. Formaldehyde, however, has been reported in the products obtained when methane, ammo-nia, and water mixtures are subjected to an electric discharge (Miller and

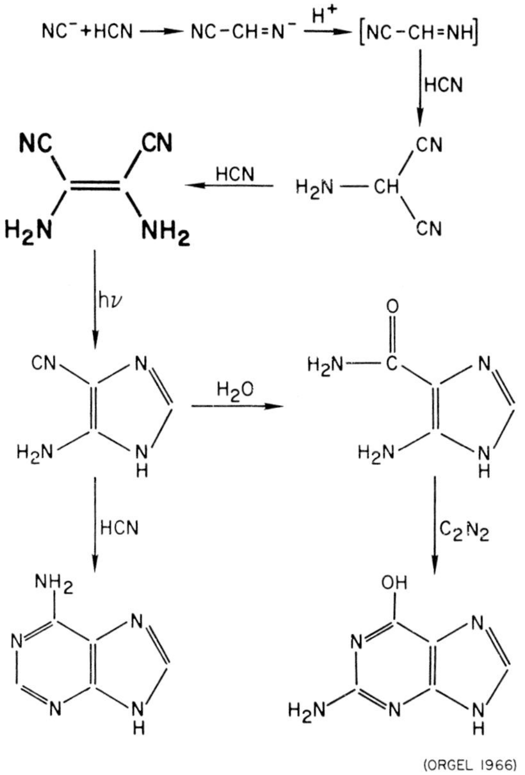

Fig. 8. Adenine and guanine from HCN tetramer.

Urey 1959) or ionizing radiation (Palm and Calvin 1962). Radiation-produced hydrated electrons have been demonstrated to be unreactive with aqueous formaldehyde because of the inhibition to nucleophilic attack which is facilitated by adjacent electron donors bonded by sigma-orbitals (Gordon et al. 1963).

Butlerow (1861) demonstrated that formaldehyde could be converted into a mixture of monosaccharides when heated in an aqueous alkaline solution. Loew (1886, 1889), Euler and Euler (1906), Langenbeck (1942) and Mayer and Jaschke (1960a,b) have studied the base-induced condensations of formaldehyde extensively. All of these experiments have been conducted in strongly alkaline media using relatively high concentrations of formaldehyde. Glycolaldehyde has been isolated from this mixture and is apparently the primary condensation product of two molecules of formaldehyde. Other condensation products that have been identified are glyceraldehyde, dihydroxyacetone, erythrose, pentoses and hexoses (Orthner and Gerisch 1933; Akerlof and Mitchell 1963). Loew's experiments indicated that alkaline Earth hydroxides and the weakly basic hydroxide of lead and tin are the best catalysts for sugar formation. Akerlof and Mitchell (1963) observed that condensation of formaldehyde to a sugar mixture does not take place in the presence of tetramethylammonium hydroxide, a strong base, unless the salt of an alkaline Earth metal is introduced. It has also been observed that an induction period is necessary for the appearance of the monosaccharides but that this induction period could be eliminated by the addition of glycolaldehyde, dihydroxyacetone, or higher sugars (Mayer and Jaschke 1960).

$$2\ CH_2O \longrightarrow CH_2OH \cdot CHO$$

$$CH_2OH \cdot CHO + CH_2O \longrightarrow CH_2OH \cdot CO \cdot CH_2OH$$

$$CH_2OH \cdot CHO + CH_2OH \cdot CO \cdot CH_2OH \longrightarrow CH_2OH \cdot CHOH \cdot CHOH \cdot CO \cdot CH_2OH$$

$$2\ CH_2OH \cdot CHO \longrightarrow CH_2OH \cdot CHOH \cdot CHOH \cdot CHO$$
$$OR\ CH_2OH \cdot CO \cdot CHOH \cdot CH_2OH$$

$$2\ C_4 \longrightarrow \left[C_8 \right] ? \longrightarrow C_3 + C_5$$

Fig. 9. Synthesis of sugars from formaldehyde.

The condensation of two molecules of formaldehyde thus appears to be a slow process. Since aldol condensations, which would proceed after this initial step, are known to be generally base catalyzed, the specific catalytic activity of divalent metal cations is probably involved in the formation of glycolaldehyde from two molecules of formaldehyde. This reaction implies that the carbon atom of one molecule of formaldehyde must become a nucleophilic center in order to attack the carbon atom of a second molecule of formaldehyde. Formaldehyde exists primarily as its hydrate in aqueous solutions (Walker 1964). Coordination of the oxygen atoms with divalent cations would greatly enhance their inductive effect on the carbon atom decreasing its electron density and thereby stabilizing the anion which would result from the ionization of a proton attached to carbon.

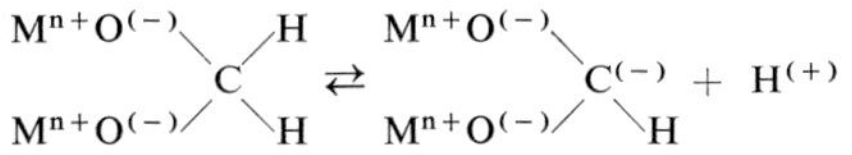

$$
\begin{array}{c}
M^{n+}O^{(-)} \diagdown \quad \diagup H \\
\hspace{2em} C \\
M^{n+}O^{(-)} \diagup \quad \diagdown H
\end{array}
\rightleftarrows
\begin{array}{c}
M^{n+}O^{(-)} \diagdown \\
\hspace{2em} C^{(-)} + H^{(+)} \\
M^{n+}O^{(-)} \diagup \quad \diagdown H
\end{array}
$$

The polyvalent cations, being Lewis acids, undoubtedly also coordinate with the carbonyl oxygen atoms of nonhydrated formaldehyde making the carbon atom much more easily attacked by a nucleophilic agent.

Some opposition has been encountered in proposing formaldehyde as the primordial precursor of monosaccharides. Horowitz and Miller (1962) pointed out that the high concentrations of formaldehyde used in most of the experiments were unrealistic for a primordial environment. Objections have also been raised to the use of very basic solutions. Abelson (1966) maintains that the concentration of free ammonia in the seas and atmosphere was never very large and that a strongly alkaline oceon never existed.

In an effort to circumvent these objections, Gabel and Ponnamperuma (1967) used a simulated hydrothermal spring as a reaction medium. Aqueous solutions of formaldehyde were refluxed over kaolinite. Such a situation may more faithfully reproduce actual prebiological conditions. At a formaldehyde concentration of 0.3 M only trioses, tetroses and pentoses could be detected. When the concentration of formaldehyde was reduced to 10^{-2} M, hexoses were formed as well (table 4). One possible explanation of why hexoses were formed in the more dilute solution is that these condensations were heterogeneous surface reactions. In a more dilute solution, a reactive intermediate would remain on the surface for a longer period of time, whereas in a concentrated solution, it would be more quickly displaced by molecules in

TABLE 4

Synthesis of sugars from formaldehyde.
Per cent conversion.

	(With alumina)		(With kaolinite)	
Concentration of formaldehyde	0.33 M	10^{-2} M	0.33 M	10^{-2} M
Hexoses	—	4.0	—	3.3
Pentoses	22	2.3	3.0	3.0
Tetroses	17	2.4	3.9	4.4
Trioses	38	—		

solution. The longer any given reactant remains on the catalytic surface the greater are its chances of undergoing further reactions to form higher molecular weight products.

Since the metabolism of carbohydrates proceeds by way of phosphorylated sugar intermediates, Halmann et al. (1969) have studied the phosphorylation of monosaccharides in dilute aqueous conditions through the mediation of various condensing agents. D-ribose and orthophosphate in the presence of cyanogen or cyanamide undergo condensation to produce β-ribofuranose-1-phosphate. Halmann has proposed a cyclic intermediate involving the condensing agent to explain the formation of the product.

5.4. Nucleosides and nucleotides

When a dilute aqueous solution (10^{-3} M) of adenine, ribose and phosphate is exposed to ultraviolet light, adenosine has been reported to form (Ponnamperuma et al. 1963). Deoxyadenosine has also been synthesized in a similar manner by exposing adenine and deoxyribose to ultraviolet light in the presence of hydrogen cyanide (Ponnamperuma and Kirk 1963). Recently, Reid et al. (1967) showed that several isomers are formed in this reaction. Most of them are hydrolytically unstable with the exception of deoxyadenosine which yields an approximate conversion of 0.3%. The mechanism of the reaction is probably based upon the formation of cyanamide from the aqueous hydrogen cyanide under the conditions of the experiment (Schimpl et al. 1965). Cyanamide has been shown to act as a dehydrating agent even in the presence of water when irradiated with ultraviolet light (Ponnamperuma and Peterson 1965).

In studies on the synthesis of nucleotides, the heterogeneous reactions that may have taken place in partially dried up tidal areas has also been examined. In simulating these conditions, an intimate mixture of the nucleosides was heated with an inorganic phosphate salt (Ponnamperuma and Mack 1965). Mononucleotides were identified in the end products. Although the highest conversions were obtained at 160°, a small amount of the nucleotides was obtained at temperatures as low as 50°. Since water is not incompatible with these reactions and does not hinder them unless present in large excess, these conditions can be described as hypohydrous. Rabinowitz et al. (1968) have demonstrated that the phosphorylation which occurs in these reactions is the result of the prior thermal polymerization of orthophosphate to polyphosphate. Schwartz and Ponnamperuma (1968) phosphorylated adenosine with tripolyphosphate and higher polyphosphates in aqueous solutions which were ca. 0.5 M in phosphorus. The 2′, 3′ and 5′ isomers were obtained. These polyphosphates were found to be effective phosphorylating reagents for adenosine over a wide pH range.

5.5. Lipids

Although there is a considerable amount of literature on the production of hydrocarbons from methane, few if any of the experiments were carried out for the purpose of studying the prebiological origin of hydrocarbons.

Ponnamperuma and Woeller (1964) compared the effects of various kinds of electric discharges on methane. From a high intensity arc, the product was a clear fluid that gave distinct peaks on a gas chromatography column. Aromatics predominated with benzene being the most abundant product followed next by toluene. In contrast the semicorona discharge yielded a material that was not easily resolved by gas chromatography. Benzene and toluene were virtually absent. The most prominent peaks on the gas chromatogram were identified by mass spectrometry as 2,2-dimethylbutane 2-methylpentane, 3-methylpentane, 2,4-dimenthylhexane and 3,4-dimethylhexane. In the unresolved fraction, nuclear magnetic resonance and mass spectrometry data indicate the predominance of cyclohexyl derivatives. The use of a molecular sieve demonstrated that there were no normal or isoprenoid hydrocarbons in this system. These results are of interest since some naturally occurring hydrocarbon deposits have been found to have this composition (Ponnamperuma and Pering 1966, 1967). Gelpi et al. (1969) have found isoprenoids and other hydrocarbons in terrestrial graphite. Munday et al. (1969) have synthesized acyclic isoprenoids by the gamma-irradiation of isoprene.

When a mixture of methane and water is exposed to a semicorona electric discharge, several monocarboxylic acids from C_2 to C_{12} are produced. Those above C_6 were characterized through mass spectrometry as having branched chains (Allen and Ponnamperuma 1967). The results of Meisels et al. (1966) and Ausloos et al. (1964) indicate that large unbranched chains of hydrocarbons result from the radiolysis of methane. Fatty acids above C_6 may result from oxidation of these hydrocarbon chains.

5.6. Porphyrins

The Oparin–Haldane hypothesis assumes that the first organisms or proto-organisms were heterotrophs which utilized the organic material of their reducing environment. As the organic material became depleted and the environment gradually changed to an oxidizing one, an autotrophic organism would have had the best chance for survival at the earth's surface. What the latter statement implies is that some organisms had already incorporated porphyrins or proto-porphyrins within their metabolic systems before the reducing to oxidizing transition occurred. It is an open question whether the escape of hydrogen from the atmosphere or the evolvement of photosynthetic organisms had the greatest effect on the change from a reducing atmosphere to an oxidizing one. Nevertheless, photochemically oriented autotrophs must have existed during the transition. Cloud (1968) has well documented the geological evidence for this transition and presents a plausible argument for the evolutionary development of autotrophs.

Until very recently, all attempts to detect pyrroles or porphyrins in the end products of primitive atmospheres subjected to the various energy sources had been unsuccessful. Hodgson and Baker (1967) obtained evidence for the formation of porphyrins from a dilute aqueous solution of formaldehyde and pyrrole under conditions simulating geochemical abiogenesis. A more biochemically oriented approach used σ-aminolevulinic acid as the starting material. This amino acid is the biogenic precursor of pyrroles and it was found that ultraviolet irradiation could induce its condensation to porphyrin pigments (Szutka 1966).

Finally, Hodgson and Ponnamperuma (1968) successfully demonstrated that porphyrins are present in the residue formed by passing an electric discharge through methane, ammonia and water vapor. The identification was made on the basis of absorption spectra, solubility, chromatographic behavior, metal complexing and fluorescence spectra. When pyrrole was included in a discharge experiment, the yield of porphyrin material increased markedly.

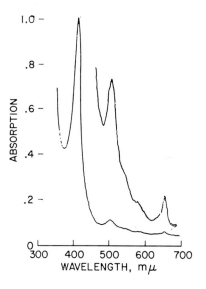

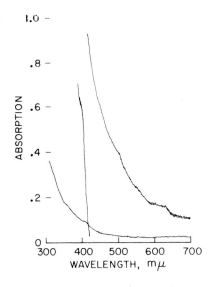

Fig. 10. Electric discharge porphyrin from methane, ammonia, water and pyrrole.

Fig. 11. Pigment from electric discharge in methane, ammonia and water.

Fig. 10 is the spectrum of the pigments obtained when pyrrole was included in the discharge mixture. Fig. 11 shows a corresponding spectrum of the product of a typical discharge experiment. A set of peaks similar to those of the pyrrole-augmented experiment is evident although somewhat distorted and obscured by dense background absorption.

5.7. Polypeptides

Simulated primordial syntheses of polypeptides have received more attention than the syntheses of other biological polymers. This is probably due to the great emphasis that has been placed on the primordial origin of amino acids. Fox (1965) has long been an exponent for the anhydrous or hypohydrous thermal origin of all primordial organic compounds. Fox and Harada (1958) have shown that in the presence of a proportionally large amount of glutamic acid or aspartic acid, an intimate mixture of all 18 amino acids normally present in proteins can be thermally polymerized at a temperature of 180°–200° C. These polymers have been described as proteinoids. The molecular weights increased from 3600 in a proteinoid made at 160° to 8600 in one made at 190°. The maximum molecular weight which has been obtained is 80,000. The polymers give a positive biuret test, can be hydrolyzed back to amino acids and are attacked by various proteolytic enzymes (Fox and Harada 1960).

It has been argued that the energy available from volcanic activity is quite small compared to the other available sources of energy (Horwitz and Miller 1962). Nevertheless, it cannot be denied that organic reactions, including polymerization of amino acids with the formation of peptide bonds, would occur under conditions simulating a volcano. A glaring weakness in this type of pathway for the origin of polypeptides is the necessity of postulating highly concentrated, intimate mixtures of the amino acids which are to be polymerized.

Oró and Guidry (1961) reported the interesting observation that glycinamide or glycine can be thermally polymerized in concentrated ammonia. Although their experiments demonstrate that anhydrous conditions are not necessary for thermal polymerization of amino acids, it is doubtful whether such large concentrations of ammonia ever existed on the primitive Earth. Their experiments, however, may be significant in the chemical evolution of the larger planets of the solar system.

The synthesis of peptides directly from dilute solutions of amino acids has also been accomplished through the use of such condensing agents as cyanamide and dicyanamide. Both compounds are known to be formed upon ultraviolet irradiation of aqueous solutions of hydrogen cyanide or through the irradiation of a mixture of methane, ammonia and water (Schimpl et al. 1965). Ponnamperuma and Peterson (1965) have reported that glycylleucine and leucylglycine were formed from a dilute aqueous solution of glycine and leucine subjected to ultraviolet irradiation in the presence of cyanamide at *p*H 5. Mechanistically, ultraviolet radiation probably increased the concentration of carbodiimide, a tautomer of cyanamide which can act as a condensing agent for amino acids.

$$H_2N-CN \rightarrow H\dot{N}-C\equiv N \leftrightarrow HN=C=\dot{N} \rightarrow HN=C=NH$$

Exposing carboxylic acids to far ultraviolet irradiation is known to promote homolytic scission of the C-H bonds on the *a*-methylene carbon atoms followed by dimerization of the resulting radicals to substituted succinic acids (Pfordte and Leuschner 1959). Irradiation of ammonia produces NH and NH_2 radicals. Since none of these processes appeared to be operating and dipeptides and tripeptides were the resultant products, it would appear that the ultraviolet light did not act directly on the amino acids.

Carbodiimides react with carboxyl groups to form an adduct with an extremely good leaving-group on the carboxyl carbon atom. Since amines are much better nucleophiles than water, the amino group of a second amino acid could then attack the carbon atom of the aforementioned adduct forming a tetrahedral intermediate. Disruption of the unstable intermediate would produce the dipeptide and urea.

$$H_2NCH_2\overset{O}{\overset{\|}{C}}-OH + HN=C=NH \rightarrow H_2NCH_2\overset{O}{\overset{\|}{C}}-O-C\overset{NH_2}{\underset{NH}{<}}$$

$$H_2NCH_2\overset{O}{\overset{\|}{C}}-O-C\overset{NH_2}{\underset{NH}{<}}$$

→ Glycylleucine + urea

$$H_2N-CH-COOH$$
$$|$$
$$CH_2$$
$$|$$
$$CH$$
$$CH_3 \quad CH_3$$

In a similar experiment Steinman et al. (1965a) reported the formation of the dipeptides and tripeptides of alanine when a dilute solution of the amino acid was treated with dicyanamide. This reaction, which is facilitated by acidic conditions, also yields a small amount of peptides at a neutral *p*H.

Steinman et al. (1965b) proposed the formation of an intermediate in the reaction similar to the one in the preceding experiment.

Rabinowitz and Ponnamperuma (1969) have reported that aqueous solutions which contained amino acids and polyphosphates (both 0.01 to 0.1 M) produced peptides at 70° or ambient room temperature at various *p*H's. The *p*H optimum is near neutrality (*p*H 7–8) and the reaction still proceeds in alkaline but not in acid medium. The condensation reaction occurs even with pyrophosphate and at room temperature, with small yields (0.4–0.5%) (Rabinowitz et al. 1969). The yields increased with the length of the chain of the linear polyphosphate (up to 13.6%) as had originally been suggested by Gabel (1965) for polymerizations of biochemicals in dilute solutions. It is conceivable that detectable polymerizations would occur at much greater dilution if the reactants were absorbed at an interface or incorporated into a solubilized or suspended matrix. When radioactively labeled glycine was left undisturbed for several days with an aqueous mixture of low molecular weight polyphosphates (10^{-3} M in phosphorus), and inorganic salts of sea water composition at *p*H 7.4, a substantial amount of radioactively labeled material was no longer dialyzable (Gabel 1966).

The formation of peptides from a mixture of methane, ammonia and water was first reported by Ellenbogen (1958). The mixture was subjected to ultraviolet irradiation in the presence of suspended particles of inorganic minerals. According to Ellenbogen, a large amount of organic material including polypeptides could have been synthesized during the accretion process which is believed to have occurred during the formation of the primordial Earth. The existence of some high molecular weight material is postulated even before the crust of the planet was actually formed.

Ponnamperuma and Flores (1966) reported the formation of peptides from a methane, ammonia and water mixture that had been subjected to an electric discharge. The peptides appeared to contain three to four amino acid residues. Upon hydrolysis, nine amino acids were identified. A hypothesis for the formation of polypeptides that would circumvent the primary formation of amino acids has been given by Kliss and Matthews (1962). They have proposed that aminocyanomethylene $H_2N—\dot{C}=C=\dot{N}$, a 1,3 biradical, would form from ammonium cyanide and hydrogen cyanide via diaminoacetonitrile. Addition-type polymerization would then lead to polymeric peptide precursors having a repeating unit which contained

$$NH_4CN \rightarrow \overset{\overset{\displaystyle NH}{\|}}{HC}{—}NH_2 \xrightarrow{\;HCN\;} H{—}\overset{\overset{\displaystyle NH_2}{|}}{\underset{\underset{\displaystyle CN}{|}}{C}}{—}NH_2 \rightarrow H_2N—\dot{C}=C=\dot{N} \;+ NH_3$$

amidine groups rather than amide groups. The addition of hydrogen cyanide, mild hydrolysis, and loss of carbon dioxide would lead to polyglycine. The side

$$H_2N-\dot{C}=C=\dot{N} \rightarrow \left(\begin{array}{c} -C=C=N- \\ | \\ NH_2 \end{array}\right)_n \rightarrow \left(\begin{array}{c} -C-CH=N- \\ | \\ NH_2 \end{array}\right)_n$$

$$\xrightarrow{HCN} \left(\begin{array}{c} CH \\ | \\ -C-CH-NH- \\ \| \\ NH \end{array}\right)_n \xrightarrow[-CO_2]{H_2O} \left(\begin{array}{c} -C-CH_2-N- \\ \| \\ O \end{array}\right)_n$$

chains of other amino acids could be constructed by reactions at the tertiary methine carbon atom before or after hydrolysis and elimination of of the cyano group. In support of this hypothesis it was found that when methane and ammonia were subjected to an electric discharge and the residue heated with dilute acid, the resulting material appeared to be a polypeptide (Matthews and Moser 1966). The limitation of this reaction scheme is that mild hydrolytic conditions would not affect the cyano group but would succeed only in converting the amidine groups to amide groups. The alleged polypeptide should have a cyano group on every alpha carbon atom. Furthermore, Huntress et al. (1969) have concluded after an intensive study by ioncyclotron resonance of the gas phase reactions of hydrogen cyanide that polymerization processes were not of major importance. Akabori (1955) proposed a hypothesis concerning the formation of what he called the 'foreprotein' from aminoacetonitrile (Ponnamperuma and Woeller 1967) which would give rise to a polyglycine. Again the repeating groups of this 'foreprotein' would be amidine groups rather than amide groups. This type of condensation reaction was first reported by Krewson and Couch (1943).

$$H_2NCH_2CN \rightarrow \left(\begin{array}{c} -NHCH_2C- \\ \| \\ NH \end{array}\right)_n$$

$$\xrightarrow{H_2O} \left(\begin{array}{c} -NHCH_2C- \\ \| \\ O \end{array}\right)_n$$

Hanafusa and Akabori (1959) successfully used this reaction to produce diglycine and triglycine. Akabori et al. (1956) demonstrated that *a*-methylene groups of polyglycine can condense with aldehydes under mildly basic conditions to produce a variety of side chains.

5.8. Polynucleotides

Beadle (1960) once remarked that primordial polynucleotides could have consisted of an adenine-thymine type of polymer which subsequently developed into the present day polynucleotides. Thermal phosphorylation of nucleosides by inorganic phosphate salts produces some oligonucleotides (Ponnamperuma and Mack 1965). The temperature required for this reaction is ca. 150° and the yields are quite small. Just as in the synthesis of nucleotises via the same reaction conditions, Rabinowitz et al. (1968) have shown that the actual phosphorylating agent is a mixture of polyphosphate salts formed from the various orthophosphate salts used in the experiment.

6. Phosphate Polymers

The first organisms or protolife forms are assumed by most investigators to have been obligate heterotrophs which used the materials of their environment for energy to reproduce and maintain their structure. Although it is conceivable that some simple chemicals of the environment could have been utilized directly, the universal occurrence of inorganic and organic phosphates in the metabolic pathways of every arbitrarily designated living thing indicates the fundamental importance of phosphates and especially the phosphoric anhydride bond for the maintenance of life processes.

Since the phosphoric anhydride bond is important as the repository of metabolic energy in biological systems, considerable attention has been focused by those interested in the origin of life on the abiotic formation of pyrophosphate from orthophosphate. Miller and Parris (1964) obtained a 1–2 percent conversion of hydroxyapatite to calcium pyrophosphate in the presence of a dilute solution of cyanate salts.

There is ample reason for believing that cyanate ion may have been present in primordial waters. Urea, which is produced in abundance in experiments on so-called primordial gas mixtures, thermally decomposes in aqueous solution to ammonium cyanate. Orthophosphoric acid and cyanate salts in water readily form carbamyl phosphate, a mixed anhydride. It is

$$H_3PO_4 + NCO^{(-)} + H^{(+)}_{aq} \rightarrow H_2N-\overset{\overset{\displaystyle O}{\|}}{C}-O-\overset{\overset{\displaystyle O}{\|}}{\underset{\underset{\displaystyle OH}{|}}{P}}-OH$$

probably through this intermediate that the pyrophosphate is formed.

Steinman et al. (1964) reported a 1–½ percent conversion of orthophosphate to pyrophosphate when dicyandiamide was used as a condensing agent at *pH* 1.

Beck and Orgel (1965) and Lohrmann and Orgel (1968) surveyed the efficacy of a number of condensing agents in water. They were able to confirm the results of Miller and Parris (1964) but they could not repeat the experiments of Steinman et al. (1964).

The purine and pyrimidine nucleotide diphosphates and triphosphates, although universally occurring in all arbitrarily designated life forms, are not the only compounds with phosphoric anhydride bonds to be found in living tissue.   Linear inorganic polyphosphate chains were first reported as a constituent of yeast cells by Wiame (1947a, b).   Since then a large number of microorganisms have been found to contain inorganic polyphosphate chains (Harold 1966). Recently, inorganic polyphosphates have been detected in spinach (Miyachi 1961), the dodder (Tewari and Singh 1964), corn roots (Vagabov and Kulaev 1964) and the wax moth *Galleria* (Nimierko 1962) The highly interesting observation has been made by Griffin et al. (1965) that inorganic polyphosphates of high molecular weight are formed in rat liver nuclei. Josan et al. (1965) also reported the presence of high molecular weight inorganic polyphosphates in rabbit liver.

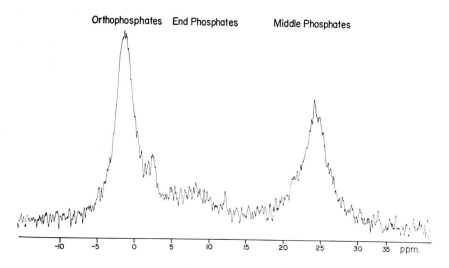

Fig. 12. The ^{31}P NMR spectrum of an extract prepared from 4 kg of sheep brains. The spectrum and its interpretation were provided through the courtesy of Drs. T. Glonek and T. C. Myers.

Gabel and Thomas (1971) have studied the occurrence and distribution of polyphosphates in vertebrate tissue. Polyphosphates having apparent chain-lengths ranging from less than ten to over 5,000 orthophosphate units have been isolated from the adult rat brain as well as every other tissue which was examined. There appears to be three times as much polyphosphate in the mammalian brain as there is in the liver. Adult rat brain contains at least 15 μg of phosphorus as polyphosphate per g of fresh tissue. The identity of the polyphosphate material was inferred from its isolation procedure, its chromatographic behavior and color development and its metachromatic effect. The technique of ^{31}P nuclear magnetic resonance spectroscopy (Crutchfield et al. 1967) was applied to an extract prepared from 4 kg of sheep brains. The spectrum disclosed absorption in the middle phosphate region (fig. 12; Glonek et al. 1971). The chemical shift of the signals and their respective areas preclude the possibility that these signals arose solely from a mixture of ortho-, pyro- and tripolyphosphates. Glonek et al. (1971) have also unequivocally identified the phosphate material from the electron-dense particulate structures of *Micrococcus lysodeikticus* as long-chain, linear, inorganic polyphosphate. Gabel (1965) has based his model for precellular organization on long-chain, linear, inorganic polyphosphate.

7. Discussion

The successful experiments of Miller (1953) and Abelson (1953–1954) were a challenge to vitalistic egocentricity. Just as organic compounds were once thought to be synthesized only by 'living' organisms, resulting in the mis-nomer 'organic chemistry' (Wheland 1949), it appears that man's bias toward the idea that the hand of man was necessary for the laboratory synthesis of complex 'biological' compounds hampered investigation into the products which resulted when so-called primordial molecules were acted upon by geological and astronomical energy sources. During the past decade there again appears to have been a bias on the part of natural science that all of the important molecules for life processes were 'organic'. Vitalism dies slowly. It may be misleading to assume that all of the important parameters which govern life processes and human behavior have been discovered. Ego-centricity dies more slowly. The experimental observations of the biophysi-cist Petersen (1943) which correlate meteorological phenomena with physio-logical variability in genetically matched identical human triplets coupled with the lucid description of the chemical unity and diversity of the universe by the astrophysicist Merrill (1963) is reason enough to be cautious in our conclusions on the nature of life.

Since the solar system is relatively young with respect to the age of the universe (Tilton and Steiger 1965), it is somewhat difficult to assign definition to the idea of a primordial molecule. The energies simulating those arising from astronomical and geological sources very effectively alter the composition of the mixtures of simple molecules to which they are applied. If matter does not exist without energy, then matter does not exist without some degree of complexity. For the development of complex carbon compounds on a planetary object, it was necessary for a reducing environment (relative to carbon) to gradually change to a more oxidizing one.

The stability and accumulation of organic materials in an Earth-type primordial environment are problems of considerable scope and importance. Although these problems can be studied in the laboratory under simulated environments (Vallentyne 1964; Hare and Mitterer 1966), much more meaningful data can be obtained through the systematic study of ancient sediments. For example, Schopf et al. (1968) examined three fossiliferous Precambrian cherts, which were approximately one, two and three billion years old, for the presence of amino acids. Twenty-two amino acids were present at a concentration of approximately 10^{-8} moles per g of sediment. All of them were either naturally occurring ones or were known degradation products of biological amino acids. These investigators believe that the distribution of the amino acids which were detected are indicative of a biotic, rather than abiotic, origin. Kvenvolden et al. (1969) have determined the optical activity of these amino acids by gas chromatography (Pollock and Oyama 1966). Most of the material consisted of L-isomers. These findings may be consistent with earlier studies which inferred that biochemically complex organisms may have been in existence more than 3.1 billion years ago (Schopf and Barghoorn 1967).

Optical activity has been viewed by many investigators as the most general characteristic of biologically derived chemicals (Wald 1957). Van't Hoff (1908) postulated that circularly polarized radiation could be used to synthesize optically active compounds. Experimentation confirmed this hypothesis (Karagunis and Drikos 1934; Davis and Ackerman 1945). Lee and Yang (1956) have shown that the electrons emitted in beta-decay are polarized, and that if they slow down and lose some of their energy the resulting gamma rays are left rather than right circularly polarized. Vester et al. (1959) and Ulbricht (1959) suggested that the optical asymmetry of terrestrial molecules is related to the structure of matter itself and is a reflection of the asymmetry of our part of the universe. The only experimental evidence which has been offered for this viewpoint has been presented by Garay (1968). Dilute, sterile, slightly basic solutions of tyrosine were exposed to sunlight while being bombarded with beta-particles from strontium-90 which had been added to the solutions. After 18 months of observation it was apparent that the D-isomer had been decomposing at a faster rate than the L-isomer. This phenomenon may be responsible for the apparent enrichment of the L-isomer in the amino acids extracted from the Precambrian sediments (Kvenvolden et al. 1969). Certain other natural processes could have also contributed to the accumulation of one optical isomer, e.g. the stereoselective interaction of amino acids with metal coordination complexes already chelated by other amino acids having the same optical configuration (Harada 1965; Gillard et al. 1968). A further example cited by Wald (1957) is the fact that a polymer which assumes the shape of a alpha-helix can only be formed from monomers of the same optical configuration.

At the present time many biologists are enamoured with the idea of a master molecule. This idea is Lamarckian. It is analogous to the architectural engineer being obliged to personally instruct each electrical and communications worker in the arts and tools of their respective trades every time that a building is under construction. All that is necessary is that the job gets done. Some natural scientists are still convinced that life is contrary to the universal increase of entropy with time. Fig. 5 was constructed to answer that question pictorially. Wilson (1968a, b) has answered the question mathematically. From an examination of the experiments discussed in this chapter, it would appear that chemical evolution could not have taken place without microenvironments. Cloud (1968) has succinctly pointed out that biological evolution is also dependent on microenvironments. In point of fact, Darwin would not have conceived of the idea of biological evolution without having had the opportunity to observe microenvironments. If there is any validity to the scientific method, then the data are always there and can be uncovered or brought to light by anyone. The datum is eternal. Knowledge, which is a thing that changes with time, is dependent upon arbitrary limits and the number of sets viewed by the observer.

References

ABELSON, P. H., 1953–1954, Paleobiochemistry, Carnegie Inst. Washington Yearbook 53, 97.

ABELSON, P. H., 1966, Proc. Natl. Acad. Sci. (USA) 55, 1365.

AKABORI, S., 1955, Kagaku (Japan) 25, 54.

AKABORI, S., K. OKAWA and M. SATA, Bull. Chem. Soc. (Japan) 29, 608.

AKERLOF, G. C. and P. W. D. MITCHELL, 1963, Final Report, NASA Contract NASR-88.

ALLEN, W. and C. PONNAMPERUMA, 1967, Currents in Mod. Biol. 1, 24.

ANBAR, M., 1965, Reactions of the hydrated electron; In: Solvated electron, advances in chemistry series 50, E. J. Hart, symposium chairman. (American Chemical Society, Washington, D.C.) 55.

ANBAR, M. and E. J. HART, 1965, J. Phys. Chem. 69, 271.

ANDERS, E., 1968, Accounts of Chem. Res. 1, 289.

AUSLOOS, P., R. GORDEN, JR and S. G. LIAS, 1964, J. Chem. Phys. 40, 1854.

BADAREU, E. and C. POPOVICI, 1960, Acad. Rep. Populare Romaine, Studii Cercetari Fiz., 11, 557; C.A., 58, 6333d (1963).

BADAREU, E., D. STEFANESCU and C. POPOVICI, 1959, Acad. Rep. Populare Romaine, Rev. Phys. 4, 5.

BARTH, C. A., 1961, Chemical reactions in the lower and upper atmosphere. (Wiley-Interscience, New York) 303.

BEADLE, G., 1960, Accad. naz. Lincei, Roma 47, 301.

BECK, A. and L. E. ORGEL, 1965, Proc. Natl. Acad. Sci. (USA) 54, 664.

BERTHELOT, M., 1869, Compt. Rend. 67, 1141.

BLAUSTEIN, B. D., symposium chairman, 1969, Chemical reactions in electrical discharges, Advances in Chemistry Series 80. (American Chemical Society, Washington, D.C.).

BRANDT, J. C. and M. B. MCELROY, eds., 1968, The atmospheres of Venus and Mars. (Gordon and Breach Science Publishers, Inc., New York).

BROWN, H., 1949, Rare gases and the formation of the earth's atmosphere; In: The atmospheres of the earth and planets, G. P. Kuiper, ed. (The University of Chicago Press, Chicago) 260.

BUHLER, F., P. EBERHARDT, J. GEISS, J. MEISTER and P. SIGNER, 1969, Science 166, 1502.

BUTLEROW, A., 1861, Ann. 120, 296.

CADLE, R. D. and E. R. ALLEN, 1970, Science 167, 243.

CHEMICAL BOND APPROACH PROJECT, 1964, Chemical systems. (McGraw-Hill Book Company, Webster Division, St. Louis).

CLOUD, P. E.., JR., 1968, Science 160, 729.

COLE, T. and J. T. HARDING, 1958, J. Chem. Phys. 28, 993.

CRUTCHFIELD, M. M., C. H. DUNGAN, J. H. LETCHER, V. MARK and J. R. VAN WAZER, 1967, Topics in Phosphorus Chemistry, Vol. 5. (Interscience Publishers, New York) Chaps. 3 and 4.

DAVIS, T. and I. ACKERMAN, 1945, J. Am. Chem. Soc. 67, 486.

DOLLE, L., 1958, Comm. energie atom. (France), Rappt. No. 1014.

DORFMAN, L. M. and P. C. NOBLE, 1959, J. Phys. Chem. 63, 980.

DOSE, K. and C. PONNAMPERUMA, 1967, Rad. Res., in press.

DOSE, K. and B. RAJEWSKY, 1957, Biochim. Biophys. Acta 25, 225.

ELIAN, J. and M. LePINGLE, 1953, Belg. Patent 513, 288 (Feb. 2, 1953); C.A., 51, 18573c (1957).

ELLENBOGEN, E., 1958, abstracts, 134th National Meeting of the American Chemical Society, Chicago.

ENRIGHT, R. J., 1959, Oil Gas J. 57, 72.

EREMIN, E. N., 1958, Khim. Prom. 73.

EULER, H. and A. EULER, 1906, Ber. 39, 50.

EVANS, H. G. V., G. R. FREEMAN and C. A. WINKLER, 1956, Can. J. Chem. 34, 1271.

EVANS, H. G. V. and C. A. WINKLER, 1956, Can. J. Chem. 34, 1217.

FERRIS, J. P., J. E. KUDER and A. W. CATALANO, 1969, Science 166, 765.

FERRIS, J. P., R. A. SANCHEZ and L. E. ORGEL, 1968, J. Mol. Biol. 33, 693.

FOX, S. W., 1965, Simulated natural experiments in spontaneous organization of morphological units from proteinoid; In: The origins of prebiological systems and of their molecular matrices, S. W. Fox, ed. (Academic Press, New York) 361.

FOX, S. W. and K. HARADA, 1958, Science 128, 1214.

FOX, S. W. and K. HARADA, 1960, J. Am. Chem. Soc. 82, 3745.

GABEL, N. W., 1965, Life Sci. 4, 2085.

GABEL, N. W., 1966, Unpublished data.

GABEL, N. W. and C. PONNAMPERUMA, 1967, Nature 216, 453.

GABEL, N. W. and V. THOMAS, 1971, J. Neurochem., in press.

GARAGANIS, P. A. and C. A. WINKLER, 1956, Can. J. Chem. 34, 1457.

GARAY, A. S., 1968, Nature 219, 338.

GELPI, E., D. W. NOONER and J. ORÓ, 1969, Geochim. Cosmochim. Acta 33, 959.

GERMAIN, J. E. and C. VANISCOTTE, 1958, Bull. soc. chim. France, 319.

GILLARD, R. D., P. R. MITCHELL and H. L. ROBERTS, 1968, Nature 217, 949.

GLONEK, T., M. LUNDI, M. MUDGETTE and T. C. MYERS, 1970, Submitted manuscript.

GORDON, S., E. J. HART, M. S. MATHESON, J. RABINI and J. K. THOMAS, 1963, Disc. Faraday Soc. 36, 193.

GRIFFIN, J. B., N. M. DAVIDIAN and R. PENNIALL, 1965, J. Biol. Chem. 240, 4427.

GROTH, W. 1957, Photochemistry of the Liquid and Solid State, Symposium, Dedham, Mass., pp. 21–24.

GROTH, W. and H. VON WEYSSENHOFF, 1960, Planet. Space Sci. 2, 79.

HALDANE, J. B. S., 1928, Rationalist Annual, 148.

HALMANN, M., R. A. SANCHEZ and L. E. ORGEL, 1969, J. Org. Chem. 34, 3702.

HANAFUSA, H. and S. AKABORI, 1959, Bull. Chem. Soc. (Japan) 32, 626.

HARADA, K., 1965, Nature 205, 590.

HARADA, K. and S. W. FOX, 1964, Nature 201, 335.

HARE, P. E. and R. M. MITTERER, 1966, Carnegie Inst. Washington Yearbook 65, 362.

HAROLD, F. M., 1966, Bacteriol. Revs. 30, 772.

HART, E. J., S. GORDON and J. K. THOMAS, 1964b, J. Phys. Chem. 68, 1271.

HART, E. J., J. K. THOMAS and S. GORDON, 1964a, Radiation Res. Suppl. 4, 79.

HASSELSTROM, R., M. C. HENRY and B. MURR, 1957, Science 125, 350.

HERZBERG, G. and L. HERZBERG, 1948, Nature 161, 283.

HICKLING, A. and G. R. NEWNS, 1959, Proc. Chem. Soc. 272.

HOCHSTIM, A. R., 1963, Proc. Natl. Acad. Sci. (USA) 50, 200.

HODGSON, G. W. and B. C. BAKER, 1967, Nature 216, 29.

HODGSON, G. W. and C. PONNAMPERUMA, 1968, Proc. Natl. Acad. Sci. (USA) 59, 22.

HOLLAND, H. D., 1961, J. Geophys. Res. 66, 2356.

HOROWITZ, N. H., 1945, Proc. Natl. Acad. Sci. (USA) 31, 153.

HOROWITZ, N. H. and MILLER, 1962, Fortschr. Chem. Organ. Naturstoffe 20, 423.

HUNTRESS, W. T., JR., J. D. BALDESCHWIELER and C. PONNAMPERUMA, 1969, Nature 223, 468.

JOLLY, W. L., 1965, The structure and kinetics of metal-ammonia solutions; In: Solvated electron, Advances in chemistry series 50, E. J. Hart, symposium chairman. (American Chemical Society, Washington, D.C.) 27.

JOSAN, V., K. K. TEWARI and P. S. KRISHNAN, 1965, Indian J. Biochem. 2, 134.

JUCKER, H. and E. K. RIDEAL, 1957, J. Chem. Soc. 1058.

KARAGUNIS, G. and G. DRIKOS, 1934, Z. Physic. Chem. 26, B, 428.

KLISS, R. M. and C. N. MATTHEWS, 1962, Proc. Natl. Acad. Sci. (USA) 48, 1300.

KNOX, J. H. and A. F. TROTMAN-DICKENSON, 1957, Chem. and Ind. (London) 268.

KREWSON, C. and J. COUGH, 1943, J. Am. Chem. Soc. 65, 2256.

KVENVOLDEN, K. A., E. PETERSON and G. E. POLLOCK, 1969, Nature 221, 141.

LAMPE, F. W., 1957, J. Am. Chem. Soc. 79, 1055

LANGENBECK, W., 1942, Naturwiss. 30, 30.

LATIMER, W. M., 1950, Science 112, 101.

LEE, T. D. and C. N. YANG, 1956, Phys. Rev. 104, 254.

LEWIS, F. M. and M. S. MATHESON, 1949, J. Am. Chem. Soc. 71, 747.

LOEW, O., 1886, J. Prakt. Chem. 34, 51.

LOEW, O., 1889, Ber. 22, 475.

LOHRMANN, R. and L. E. ORGEL, 1968, Science 160, 64.

LOWE, C. U., M. W. REES and R. MARKHAM, 1963, Nature 199, 219.

MACKIN, R. J., JR. and M. NEUGEBAUER, eds., 1966, The solar wind. (Pergamon Press, London).

MANTON, J. E. and A. W. TICKNER, 1960, Can. J. Chem. 38, 858.

MASON, B., 1962, Meteorites. (John Wiley and Sons, Inc., New York).

MATHESON, M. S., 1962, Ann. Rev. Phys. Chem. 13, 77.

MATTHEWS, C. N. and R. E. MOSER, 1966, Proc. Natl. Acad. Sci. (USA) 56, 1087.

MAYER, R. and L. JASCHKE, 1960a, Angew. Chem. 72, 635.

MAYER, R. and L. JASCHKE, 1960b, Ann. 635, 145.

MCGOVERN, W. E., 1969, J. Atmos. Sci. 26, 623.

MCKELLAR, A., 1960, Isotopes in stellar atmospheres; In: Stellar atmospheres, J. L. Greenstein, ed. (The University of Chicago Press, Chicago) 569.

MEISELS, G. G., W. H. HAMILL and R. R. WILLIAMS, JR., 1966, J. Chem. Phys. 25, 790.

MERRILL, P. W., 1963, Space chemistry. (The University of Michigan Press, Ann Arbor).

MILLER, S. L., 1953, Science 117, 528.

MILLER, S. L., 1957, Biochim. Biophys. Acta 23, 480.

MILLER, S. L. and M. PARRIS, 1964, Nature 204, 1248.

MILLER, S. L. and H. C. UREY, 1959, Science 130, 245.

MIYACHI, S., 1961, J. Biochem. (Japan) 50, 367.

MOORTHY, P. N. and J. J. WEISS, 1965, Formation and reactions of electrons and holes in γ-irradiated ice and frozen aqueous solutions; In: Solvated electron, Advances in chemistry series 50, E. J. Hart, symposium chairman. (American Chemical Society, Washington, D.C.) 180.

MUNDAY, C., K. PERING and C. PONNAMPERUMA, 1969, Nature 223, 867.

MUNSON, M. S. B., F. H. FIELD and J. L. FRANKLIN, 1965, J. Chem. Phys. 42, 442.

NIMIERKO, W., 1962, Colloq. Intern. Centre Natl. Rech. Sci. (Paris) 615.

OPARIN, A. I., 1924, Proiskhozhdenie Zhizni. (Izd. Moskovskii Rabochii, Moscow).

OPARIN, A. I., 1964, The chemical origin of life, Transl. by A. Synge. (Charles C. Thomas Publishers, Springfield, Ill.).

OPIK, E. J., 1962, Icarus 1, 200.

ORÓ, J., 1960, Biochem. Biophys. Res. Commun. 2, 407.

ORÓ, J., 1961, Federation Proc. 20, 352.

ORÓ, J. and C. S. GUIDRY, 1961, Arch. Biochem. Biophys. 95, 166.

ORTHNER, L. and E. GERISCH, 1933, Biochem. Z. 259, 30.

PALM, C. and M. CALVIN, 1962, J. Am. Chem. Soc. 84, 2115.

PAVLOVSKAYA, T. E. and A. G. PASNYSKII, 1959, I.U.B. Sympos. 'im Series 1, 151.

PETERSEN, W. F., 1943, Arch. Biochem. 1, 269.

PEVERA, E. F. and H. V. HESS, 1958, U.S. Patent 2,858, 261 (Oct. 28, 1958).

PFORDTE, K. and G. LEUSCHNER, 1959, Ann. 622, 1.

POLLOCK, G. E. and V. I. OYAMA, 1966, J. Gas Chrom. 4, 126.

PONNAMPERUMA, C., 1965, Abiological synthesis of some nucleic acid constituents; In: The origins of prebiological systems and of their molecular matrices, S. W. FOX, ed. (Academic Press, New York) 221.

PONNAMPERUMA, C. and J. FLORES, 1966, Abstracts, 152nd National Meeting of the American Chemical Society, New York, C-33.

PONNAMPERUMA, C. and N. W. GABEL, 1968, Space Life Sci. 1, 64.

PONNAMPERUMA, C. and P. KIRK, 1963, Nature 203, 400.

PONNAMPERUMA, C., R. H. LEMMON, R. MARINER and M. CALVIN, 1963, Proc. Natl. Acad. Sci. (USA) 49, 737.

PONNAMPERUMA, C. and R. MACK, 1965, Abstracts, 150th National Meeting of the American Chemical Society, Atlantic City, C-44.

PONNAMPERUMA, C. and K. PERING, 1966, Nature 209, 979.

PONNAMPERUMA, C. and K. PERING, 1967, Geochim. Cosmochim. Acta 31, 1350.

PONNAMPERUMA, C. and E. PETERSON, 1965, Science 147, 1572.

PONNAMPERUMA, C., C. SAGAN and R. MARINER, 1963, Nature 199, 222.

PONNAMPERUMA, C. and F. WOELLER, 1964, Nature 203, 272.

PONNAMPERUMA, C. and F. WOELLER, 1967, Currents in Modern Biol. 1, 156.

RABINOWITZ, J., S. CHANG and C. PONNAMPERUMA, 1968, Nature 218, 442.

RABINOWITZ, J., J. FLORES, R. KREBSBACH and G. ROGERS, 1969, Nature 224, 795.

RABINOWITZ, J. and C. PONNAMPERUMA, 1969, Abstracts, 158th National Meeting of the American Chemical Society, New York, BIOL-212.

RASOOL, S. I. and W. E. McGOVERN, 1966, Nature 212, 1225.

REID, C., L. E. ORGEL and C. PONNAMPERUMA, 1967, Nature 216, 936.

RITTMANN, A., 1962, Volcanoes and their activity. (John Wiley and Sons, Inc., New York).

RUBEY, W. W., 1955, Crust of the earth, Geol. Soc. Amer. Spec. Paper 62, 631.

RUSSELL, H. N., 1929, Astrophys. J. 70, 11.

SANCHEZ, R. A., J. P. FERRIS and L. E. ORGEL, 1966a, Science 153, 72.

SANCHEZ, R. A., J. P. FERRIS and L. E. ORGEL, 1966b, Science 154, 784.

SCHIMPL, A., R. M. LEMMON and M. CALVIN, 1965, Science 147, 149.

SCHNEPP, O. and K. DRESSLER, 1960, J. Chem. Phys. 32, 1682.

SCHONLAND, B., 1953, Atmospheric electricity. (Methuen, London).

SCHOPF, J. W. and E. S. BARGHOORN, 1967, Science 156, 508.

SCHOPF, J. W., K. A. KVENVOLDEN and E. S. BARGHOORN, 1968, Proc. Natl. Acad. Sci. (USA) 59, 639.

SCHWARTZ, A. and C. PONNAMPERUMA, 1968, Nature 218, 443.

SELBY, S. and L. SWEET, 1969, Sets – Relations – Functions. (McGraw-Hill Book Company, New York) Chap. 1.

SHERWOOD, P. W., 1960, Gas J. 302, 58.

SNYDER, L. E., D. BUHL, B. ZUCKERMAN and P. PALMER, 1969, Phys. Rev. Letters 22, 679.

STARODUBTSEV, S. V., S. H. A. ABLYAEV, F. BAKHRAMOV, L. G. KEITLIN and E. N. YUSOVA, 1961, Izvest. Akad. Nauk. Uzbek SSR, Ser. Fiz. Nauk., pp. 3–11; C.A., 55, 25223i (1961).

STEINMAN, G., D. H. KENYON and M. CALVIN, 1965b, UCRL-16806.

STEINMAN, G., R. M. LEMMON and M. CALVIN, 1964, Proc. Natl. Acad. Sci. (USA) 52, 27.

STEINMAN, G., R. M. LEMMON and M. CALVIN, 1965a, Science 147, 1574.

SWALLOW, A. J., 1960, Radiation chemistry of organic compounds. (Pergamon Press, New York) 242.

SYMONS, M. C. R., 1959, Quart. Revs. 13, 99.

SZUTKA, A., 1966, Nature 212, 401.

TAUBE, M., K. SAMOCHOCKA, S. Z. ZDROJEWSKI and K. JEZIERSKA, 1967, Inst. Nucl. Res. (Warsaw), Rep. No. 838/c; C.A., 68, 78582 (1968).

TERENIN, A. N., 1959, Photosynthesis in the shortest ultraviolet; In: Proceedings of the first international symposium on the origin of life on the earth, Moscow, 1957. (Pergamon Press, New York) 136.

TEWARI, K. K. and M. SINGH, 1964, Phytochemistry 3, 341.

THOMPSON, B. A., P. HARTECK and R. R. REEVES, JR., 1963, J. Geophys. Res. 68, 6431.

TILTON, G. R. and R. H. STEIGER, 1965, Science 150, 1807.

TUNITSKII, N. N. and S. E. KUPRIYANOV, 1957, Trudy pervogo vsesoyuz. Soveshchaniya Radiatsion. Khim., Akad. Nauk. SSSR, Otdel. Khim. Nauk., Moscow, pp. 7–12; C.A., 53, 12017h (1959).

ULBRICHT, T. L. V., 1959, Quart. Revs. 13, 48.

UREY, H. C., 1952a, The planets: their origin and development. (Oxford University Press, London).

UREY, H. C., 1956, Bull. Geol. Soc. Am. 67, 1125.

VAGABOV, V. M. and I. S. KULAEV, 1964, Dokl. Akad. Nauk. SSSR 158, 218.

VALLENTYNE, J. R., 1964, Geochim. Cosmochim. Acta 28, 157.

VAN 'T HOFF, I., 1908, Die Lagerung der Atome im Raume. (Vieweg, Braunschweig).

VESTER, F., T. L. V. ULBRICHT and H. KRAUCH, 1959, Naturwissenschaften 46, 68.

WALD, G., 1957, Ann. N.Y. Acad. Sci. 69, 352.

WALKER, D. C. and R. A. BACK, 1963, J. Chem. Phys. 38, 1526.

WALKER, J. F., 1964, Formaldehyde, (Reinhold Publishing Corporation, New York) 53.

WATANABE, K., M. ZELIKOFF and E. C. V. INN, 1953, AFCRC Tech. Rept. 53–23.

WEININGER, J. L., 1960, Nature 186, 546.

WEISSLER, G. L., J. A. R. SAMPSON, M. OGAWA and G. R. COOK, 1959, J. Opt. Soc. Amer. 49, 338.

WHELAND, G. W., 1949, Advanced organic chemistry, 2nd ed. (John Wiley and Sons, Inc., New York) Chap. 1.

WIAME, J. M., 1947a, Biochim. Biophys. Acta 1, 234.

WIAME, J. M., 1947b, J. Am. Chem. Soc. 69, 3146.
WILLIAMS, R. R., JR., 1959, J. Phys. Chem. 63, 776.
WILSON, J. A., 1968a, Nature 219, 534.
WILSON, J. A., 1968b, Nature 219, 535.
WRIGHT, A. N. and C. A. WINKLER, 1968, Active nitrogen. (Academic Press, New York).
YAMAMOTO, T., 1954, J. Soc. Org. Syn. Chem. (Japan) 12, 418.

C. Ponnamperuma (ed.), Exobiology. © North-Holland Publishing Company

CHAPTER 5

Electronic factors in biochemical evolution

B. PULLMAN

1. Introduction

The existence of chemical bonds can be accounted for exclusively by quantum-mechanics. It is the quantum theory, therefore, which contains in itself the possible explanation for the formation of the first, undoubtedly small, molecules in the universe and for their development into the numerous and large compounds known today. If only by virtue of this situation, considerations of quantum biochemistry have a natural place in the study of chemical evolution and of the origin of life.

It is well established today that the majority if not the totality of the molecules which represent the essential building blocks of biological compounds could have been formed under the conditions supposedly prevailing on the primitive earth from a small number of simple precursors such as water, ammonia, methane etc. through the action of a variety of physical agents such as ultra-violet radiation, electrical discharges, heat etc. (For an excellent recent review see e.g. Ponnamperuma and Gabel 1968.) This is substantiated by the large number of corresponding synthesis carried out succesfully in a number of laboratories. A long, prebiological period probably existed during which a great number of molecules and aggregates were formed, altered and destroyed, the very agents which were responsible for the formation of molecules being also able to bring about their destruction. This was the period of struggle for the survival and selection of biomolecules, the period of competition for reaction and function networks.

The aim of this chapter will be to outline the electronic factors which could possibly have played a substantial role in the chemical evolution leading to the selection of biomolecules, as indicated by the application to this problem of some basic ideas of quantum biochemistry (B. Pullman and A. Pullman 1963). We shall center our attention on three aspects of the problem: (1) the electronic factors associated with the presence of conjugated systems, (2) the role of the electronic factors in intra- and intermolecular interactions and (3) the role of the electronic factors in mutagenesis.

2. The significance of conjugated systems in biochemical evolution

The starting point for the considerations developed in this section is the observation that a large number of essential biomolecules which are related to, or perform, the fundamental functions of the living matter (purines, pyrimidines, porphyrins, pteridines, flavins, quinones, carotenes, retinals, 'energy-rich' phosphates, practically all coenzymes etc.) are conjugated systems, generally heterocyclic, rich in π electrons, highly delocalizable (B. Pullman and A. Pullman 1962). Following an abundant series of paleobiochemical findings, it seems highly probable that a large number of these substances have been formed and have been associated with the phenomena of life in very remote epochs, when life was still in its most primitive form (Meinschein et al. 1964; Blumer 1965; Rosenberg 1965; Calvin 1969). It is thus extremely tempting to assume that the basic manifestations of life are therefore intimately connected with the existence of at least some of these highly conjugated compounds which for some decisive reason have been chosen by nature for this task.

When looking for such a reason one can hardly think of anything else than *electronic delocalization*, which represents the most specific and at the same time the most important characteristic of such compounds. Obviously, the possibility of utilizing this class of molecules must have represented for the process of biochemical evolution a number of substantial advantages susceptible to facilitate or/and to accelerate the appearance and the development of life.

In fact, a few such advantages, connected with the very nature of conjugated molecules, can easily be imagined. We are going to examine them successively.

2.1 Thermodynamic stability

One of the major and most apparent results of electronic delocalization is the increment of stability which this delocalization confers on compounds in which it occurs. Quantitatively this increment is defined as delocalization or resonance energy, and is nowadays one of the most familiar concepts of quantum chemistry. Particularly strong in pentagonal and hexagonal rings it may amount in the common previously quoted biomolecules from a few tens to a few hundreds of kilocalories per mole (table 1). E.g., it is of the order of 30–40 kcal/mole in the pyrimidines, of the order of 50–60 kcal/mole in the purines and of the order of 150–200 kcal/mole in the porphyrins. Such a stabilization constitutes a very notable advantage in the phenomenon of natural selection. In particular, it represents an unexpected economy for the energy expenses in anabolic processes. Thus, e.g., the de novo synthesis of purines from simple constituents may be considered as a series of ATP-dependent syntheses of carbon-nitrogen bonds (Buchanan 1960). Eight such bonds are established during the formation of hypoxanthine, the first purine ring to be formed in this synthesis, and in every case but one the formation of such a bond is associated with the hydrolysis of an energy-rich bond of ATP. This represents an energy cost for the synthesis of the purine skeleton of about 50–60 kcal/mole. The gain of resonance energy associated with this synthesis, calculated by the molecular orbital method (B. Pullman and A. Pullman 1963), is about 55 kcal/mole and thus compensates this expenditure.

I. Hypoxanthine.

It seems natural to assume that this thermodynamic stabilization must have played a non-negligible role in the evolutionary selection of this type of molecules for biochemical processes. Although the adaptability to a developing function was, of course, highly important, it seems nevertheless probable that this condition must have been subordinated to the requirements of stability and that the molecules which finally gained m_st have been those which were the most stable and whose biochemical possibilities must have, in fact, oriented or at least influenced the function itself.

A confirmation of this point of view may be found both in the extraordinary unity of biochemistry, the same limited number of compounds being used all over the plant and animal kingdoms, and in the multiplicity of functions linked with some of the basic skeletons. This last statement may be illustrated by two striking examples. The first one concerns the case of prophyrins. These very strongly resonance-stabilized molecules (in which four cyclic pyrroles are united by supplementary double bonds into a conjugated supercycle, II) are utilized in photosynthesis, in the respiratory (electron transfer) enzymes, and in the oxygen-carrying enzymes of higher animals. It seems as if nature, having discovered this particularly stable skeleton, has found means of utilizing it in a number of different functions. The same point of view is substantiated by the observations that porphyrin may be synthesized abiogenetically under simulated geochemical conditions (Szutka 1964; Hodgson and Baker 1967; Hodgson and Ponnamperuma 1968) and that the chlorophyll molecule is today manufactured by a sequence of reactions almost identical with the sequence of reactions used to manufacture the heme. The branching between these two syntheses only occurs at the stage when the metal cation (iron or magnesium) is introduced into the porphyrin skeleton (Granick and Mauzerall 1961). Now it seems rather difficult to believe that porphyrin is really the best suited of all the possible molecules for all its functions simultaneously. In fact, it certainly is not. As remarked by Wald (1963) chlorophyl absorbs badly in the region where the solar radiations are most abundant and, similarly, following Gaffron (1962) a number of other dyes could have been utilized with just as much efficiency as porphyrins in the photosynthetic apparatus. This last author infers, therefore, that the porphyrins owe their predominant position essentially to the fact that they came early and were much more stable than other pigments. On the other hand, Calvin (1962) has indicated strong reasons why, once the porphyrins are formed, more of them will have a tendency to be formed by an autocatalytic self-selection mechanism owing to the evolutionary pressure of peroxides. Complementary reasons for their polyutilization in biochemistry have also been suggested to reside in the flexibility of the porphyrin cycle (Webb and Fleischer 1965) and in their

TABLE 1

Resonance energies (kcal/mole).

Adenine	62	Uracil	31	β-carotene	38
Guanine	61	Pteridine	58	Retinal	18
Xanthine	56	2-amino-4-hydroxypteridine	66	Phenylalanine	32
Uric acid	54	Isoalloxazine	90	Imidazole	27
Cytosine	36	Porphine	142	Tryptophane	54
Thymine	33	1,3-divinylporphine	169		

tendency to participate in molecular associations (Mauzerall 1964; Caughey et al. 1969).

II. Porphyrin.

III. Adenine.

The second example of polyfunction concerns the problem of adenine (III). This molecule plays an apparently unique role in nature. It is one of its most important single compounds and it enters into the constitution of a great number of complex biomolecules. Thus, adenine, besides contributing to the structure of the nucleic acids, is also a part of several fundamental coenzymes NAD, NAT, FAD, CoA and other important transfer agents (S-adenosylmethionine), and it is also the carrier of the pyrophosphate chain in the energy-rich ATP and ADP. The clue to this omnipresence may perhaps reside in the fact that, as shown by calculations (B. Pullman and A. Pullman 1960, 1961), adenine has the greatest resonance energy per π electron among all the biochemical purines. It may not be unreasonable to imagine that, inasmuch as nature was induced to utilize a purine or pyrimidine base as a constituent of these fundamental biomolecules, a phenomenon of natural selection played a role, leading to the choice of the energetically most stable base. Very significant in this respect are also the recent results (Oro and Kimball 1961; Ponnamperuma et al. 1963; Ponnamperuma 1965a, b; Oro, 1965a, b; Ferris and Orgel 1965; Sanchez et al. 1968) showing that of the five bases of the nucleic acid, adenine is the one most readily synthesized under prebiotic conditions, by electron irradiation of methane, ammonia and water, or by the action of heat. It may also be pointed out that adenine, which is 6-aminopurine, has a greater resonance energy than the two other possible isomers, 2-aminopurine (IV) or 8-aminopurine (V), which

IV. 2-aminopurine.

V. 8-aminopurine.

are of negligible biochemical importance (B. Pullman and A. Pullman 1963). On the other hand guanine (VI; next after adenine in resonance stabilization among the nucleic acid bases) has been shown to be formed readily in the thermal polymerization of amino acids (Ponnamperuma et al. 1964), while no reports have shown up till now about the similar formation of the energetically less stable and biochemically unimportant isomeric isoguanine VII.

VI. Guanine.   VII. Isoguanine.

2.2. Radio- and photoresistance

Till now we have considered the resonance stabilization of conjugated biomolecules as a thermodynamic advantage of their ground state. In fact, this stabilization seems to be equally effective in a second field in which its role must have been particularly important for the biochemical evolution. This second field concerns the effect of ionizing and ultraviolet radiations. As demonstrated in details in the particular case of biochemical purines and pyrimidines, there is a very close parallelism between the resonance stabilization of biochemicals and their resistance to damage produced by both ionizing and ultraviolet radiations (B. Pullman and A. Pullman 1960; B. Pullman 1961). The intimate nature of this correlation is not entirely elucidated but it seems highly probable that one of its predominant factors is the possibility for resonating molecules to produce, under the influence of the radiation, relatively long-lived excited states, radicals or ions, having themselves a high resonance energy content and therefore stable enough and tending to recover their initial structure, before any important dissociation processes may occur. Thus, we have been able to show (table 2) the existence of a parallelism between the resonance energy of the ground state and that of the first excited states in the group of the nucleic purines and pyrimidines (B. Pullman 1961, 1964). Naturally a correlation between the stability of the excited state and photo- or radioresistance is more directly significant. Whatever it be, the existence of the correlation and its wide occurrence have been recently substantiated by the experimental work of Duchesne, Van de Vorst and their collaborators on a large series of substances through the measurements in the solid phase of the radiolytic yields by means of electron spin resonance (Duchesne and Van de Vorst 1965; Van de Vorst et al. 1966). Thus, these authors have demonstrated in particular the relatively very high radioresistance of the conjugated polyenes, carotenes and retinals, and of phtalocyanines, molecules which have the same general skeleton as the porphyrins (Checcuci et al. 1964a). They have shown that the radioresistance of proteins parallels their content in the conjugated aromatic amino acids (Checcuci et al. 1964b). They have also shown, by comparing the radio-resistance of nucleosides with that of their constituents, the existence of a highly efficient protection of the pentose by the base i.e. of the unconjugated part of the system by its conjugated component (Van de Vorst and Williams-Dorlet 1963; Phillips et al. 1964; Duchesne and Van de Vorst 1965b; Van de Vorst et al. 1966b).

Now, it seems obvious that such photo- and radioresistance must have been particularly advantageous for the evolution of life under the early conditions of the earth in which the energy derived from the sun in terms of radiation or electronic discharges and the energy provided by cosmic rays

TABLE 2

Resonance energies of purines and pyrimidines.

Base	Resonance energy in the ground state (in β units)	Resonance energy in the first exited state (in β units)
Uracil	1.92	2.36
Thymine	2.05	2.58
Cytosine	2.28	2.89
Guanine	3.84	4.48
Adenine	3.89	4.54

and the radioactivity of the earth crust were the primary sources of energy for the synthesis of molecules and even for the polymerization processes leading to macromolecules, as it is substantiated today by the reproduction of similar syntheses in the laboratory. The conjugated systems, whose loosely bound π and lone-pair electrons may be relatively easily raised to the excited states, were among those which could make the most efficient use of the available light energy. It must also be realized that if the different types of radiations appear as the most probable agents for the synthesis of primitive biomolecules, the same radiations could also exert a destructive effect upon the existing molecules. In particular, the more complex compounds synthesized under the influence of radiations frequently absorb towards longer wavelength than do their simpler precursors. Now, a greater quantity of energy being available in the realm of greater wavelength (Miller and Urey 1959), the organic molecules freshly synthesized were in obvious danger of being destroyed quicker than they were formed, so that the total balance of the protoproduction would be nul. This situation underlines the importance of the increment of radio- and photoresistance produced by the manifestation of electronic delocalisation. In direct relation to it, is the experimental observation that carotenoid-containing bacteria are killed more slowly than non-pigmented organisms by X-rays (Fram et al. 1950) and that extremely radiation-resistant microorganisms are highly pigmented (Davis et al. 1963).

The involvement of resonating molecules, retinals (VIII), as primary

VIII. All-trans retinal.

receptors of light in vision is, of course, self-explanatory. It represents, moreover, two complementary aspects interesting from the general view-point developed in this paper. In the first place, phylogenical studies lead to the conclusion that the vitamins A, of which the retinals are the aldehydes, seem to have been repeatedly evolved during the biochemical evolution in order to perform their specific task (Wald 1963). They are thus an excellent example of the fact that powerful selection pressures connected with an important function may bring very diverse organisms to produce or use the same molecule. Second, it appears that, in this particular case, the retinal isomer that has been selected in vision (the 11-*cis* or neo-*b* isomer; IX) is

IX. 11-*cis* retinal.

sterically hindered and thus is not the most stable one. It may, however, be shown that it is the one which confers the highest intrinsic photosensitivity on the visual pigments, being the one which undergoes most easily the s-*cis*-all-*trans* isomerization (Hubbard 1956; A. Pullman and B. Pullman 1961; Langlet *et al.* 1969).

Light is one of the most effective agents for promoting cis-*trans* isomerizations and this may be the general reason for which its action in visual excitation consists essentially of producing such an isomerization and for which conjugated polyenes rather than rigid, conjugated rings have been chosen as pigments of vision (Wald 1963).

2.3. *Functional advantages*

Life is a delicate, dynamic equilibrium between a series of synthesis and decomposition phenomena. Among its main characteristics are the constant mobility of its mechanisms, the interdependence of its constituent biochemicals, the long-range transmission of its manifestations, and its numerous transducing activities. It appears obvious that the mobility, the fluidity, the great polarizability of the electronic cloud of conjugated molecules, and the possibility that the existence of such a unique, large and deformable cloud offers for the rapid and distant transmission of perturbations (which in biological language may mean an order or a warning) represent important functional advantages that make the conjugated systems particularly suited to perform the functions required by the existence of the living matter. The essential fluidity of life agrees with the fluidity of the electronic structure of these compounds. Thus, while the previously discussed advantages of the conjugated systems for the maintainance of life were to some extent of a static nature, the functional advantages that we wish to stress now are essentially dynamic. We may illustrate what is meant by this proposition on two types of example.

The first one concerns the very important field of bioenergetics, in which the utilization of the 'energy-rich' phosphates as the principle coupling agents for the driving of anabolic reactions springs from some advantageous characteristics of these compounds due essentially to the conjugation of their mobile electrons (Oesper 1950; B. Pullman and A. Pullman 1960). As a matter of fact the first immediate observation in this field is that while the 'energy-rich' phosphates are conjugated molecules, the 'energy-poor' ones are not. Thus in the 'energy-rich' phosphates the mobile electrons of the phosphoryl group always interact either with those of another similar phosphoryl group (the case of the pyrophosphates ATP and ADP, (X) and (XI), or with those of a π electron possesing organic radical (the case of the

X. ATP.

XI. ADP.

guanidino, acyl and enol phosphates (XII), (XIII) and (XIV). No such interaction is present in the energy-poor phosphates, in which the phosphoryl

XII. Guanidino phosphates.   XIII. Acyl phosphates.   XIV. An enol phosphate
(2-phosphoenolpyruvic acid).

group is attached to a saturated carbon (XV). As is well known this un-

XV. An 'energy-poor' phosphate (glucose-6-phosphate).

fortunate denomination of 'energy-wealth' corresponds in fact to a measure of the chemical potential of the compounds and the classical properties of ATP as the principal source of energy for the multitude of synthetic reactions necessary for the maintainance of life spring from its (relatively) great free energy of hydrolysis. Now, this comes from the fact that the products of the hydrolysis form a system more stable than the initial molecule and it may be shown that this situation is due to (1) the resonance stabilization of the products of the hydrolysis, (2) the electrostatic repulsions due to the particular electronic distribution in the phosphates (consisting generally of a main chain of at least three and frequently more atoms carrying net positive π charges surrounded by a cloud of negatively charged atoms, as illustrated in fig. 1, for the case of ATP), which is a direct result of the delocalization of the mobile electrons, (3) the keto-enol tautomerism of the products of the hydrolysis, and (4) the free energies of ionization. The 'energy wealth' is therefore directly dependent on the manifestations of electronic delocalization.

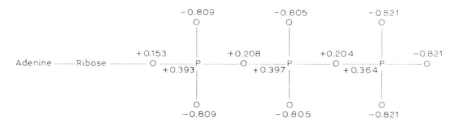

Fig. 1. Net π charges in the pyrophosphate chain of ATP.

However, the most striking example of the functional advantages associated with the utilization of conjugated molecules in the living systems is offered by the coenzymes. Biochemical reactions need enzymes to proceed and enzymes are essentially proteins. However, if one excepts the hydrolytic ones, the great majority of enzymes exert their activity in conjunction with a coenzyme, in which case the essential reaction takes place at the coenzyme, the function of the protein (or apoenzyme) consisting mainly of ensuring the high efficiency and the specificity of the reaction, by suitably orienting and polarizing the reactants. Now, there are hundreds of enzymes but, as already mentioned, there is only a very limited number of coenzymes, and practically all of these coenzymes are conjugated organic molecules. Such is the case, in particular, with the oxidation-reduction coenzymes NAD (XVI), NAT (XVII), FMN (XVIII), FAD (XIX), the heme prosthetic groups of the cytochromes and the quinones. And it is also the case with the

XVI. NAD.

XVII. NADP.

XVIII. FMN.

XIX. FAD.

coenzymes involved in group transfer reactions: tetrahydrofolic acid (XX), pyridoxal phosphate (XXI), thiamine pyrophosphate (XXVI), vitamin B12, etc. This is in no way an accident but a deeply significant situation arising from the simple fact that these conjugated molecules are particularly well adapted for being the sites of biochemical transformations.

XX. Tetrahydrofolic acid.

XXI. Pyridoxal phosphate.

XXII. Thiamine pyrophosphate.

This may be illustrated by examples which we shall take in the two fields of the oxidation-reduction coenzymes and the group-transfer coenzymes.

Thus it has been shown (B. Pullman and A. Pullman 1959, 1960) that the mechanism of functioning of the oxidation-reduction coenzymes in electron transfer may be related to the values of the energies of the lowest empty molecular orbital (lemo) of their oxidized form and of the highest occupied molecular orbital (homo) of their reduced form and to the variation of the energies of these two essential orbitals in the course of the oxidation-reduction. It is observed that, in each case, the oxidized form of the co-enzyme has a very low-lying lemo and the reduced form of the coenzyme a particularly high-lying homo. The oxidation-reduction of these compounds is thus associated with an instantaneous redistribution of these two essential orbitals in such a way that in each case the oxidized form of the coenzyme has a relatively great electron affinity or electron-accepting tendency and the reduced form of the coenzyme a particularly low ionization potential or a particularly strong electron-donating tendency. This oscillation makes these coenzymes particularly suitable for being the vectors of electron transfer. Now, obviously, only conjugated molecules can involve such a specific displacement of molecular orbitals and satisfy the conditions imposed upon the efficient electron-transfer agents. In fact, even among the conjugated molecules themselves, only very specific ones can fulfill these requirements. Thus e.g., if we replace the nicotinamide ring of the pyridine nucleotides by benzene or the isoalloxazine ring of the flavin coenzymes by anthracene, nothing useful will happen. This remark brings us to an essential complement to our general thesis about the importance of conjugated molecules in the processes of life; namely, that what is needed and essential are not simply conjugated molecules but, mostly, conjugated heterocycles disposing of atoms with lone pairs, such as nitrogen or oxygen, which by undergoing suitable changes in their valence states are able to bring about the most economical and spectacular transformations. Thus, the previously discussed redistribution of the molecular orbitals in the respiratory coenzymes, which apparently is so well adapted to the function of these coenzymes as electron transfer agents, is only possible because the oxidation-reduction is associated with changes in the valence state of the nitrogen atoms of these compounds, which alternatively involve their lone pair in the over-all π delocalization or exclude it from such a participation (fig. 5). It may even be observed that, owing to this modification in the valence state of the nitrogen atom, the reduction of NAD^+ to $NADH$ does not involve any loss in the total number of the π electrons of the system (in NAD^+, N_1 contributes one π electron, while in $NADH$ it contributes its lone pair) and that the reduction of FMN to $FMNH_2$ involves even an increase in the total number of π electrons (in FMN, N_1 and N_{10} contribute one π electron each; in $FMNH_2$, they contribute two electrons each, namely their lone pairs). Although the reduction is in each case accompanied by a certain loss of resonance energy, this loss is probably the minimum possible one.

Our second example comes from the field of group transfer coenzymes. We may specifically quote, as an illustration, the case of pyridoxal phosphate, although exactly the same or very similar considerations apply to other coenzymes of this category such as thiamine pyrophosphate or tetrahydrofolic acid.

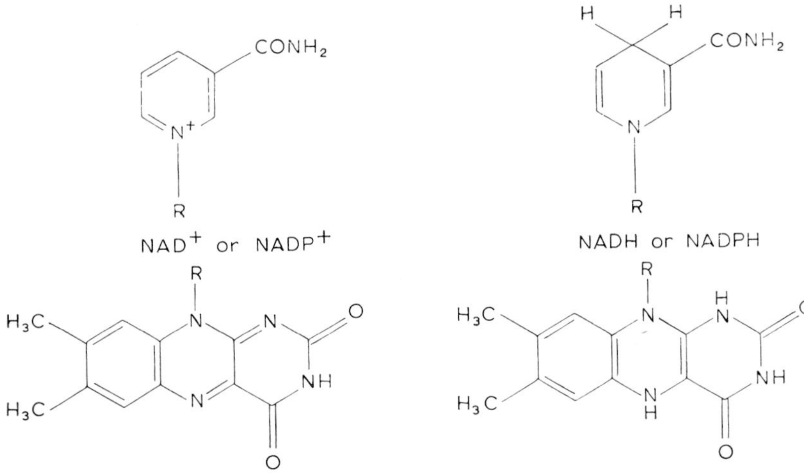

NAD$^+$ or NADP$^+$ NADH or NADPH

Fig. 2. Oxidized and reduced forms of the respiratory coenzymes.

Pyridoxal phosphate, which is a heterocyclic aldehyde, catalyzes a great number of group transfer reactions involving the α-amino acids of the proteins (e.g. Snell 1961). A representative reaction, transamination, is illustrated in fig. 3. The mechanism of functioning involves, in the first place, the formation of a Schiff's base (an imine) between pyridoxal phosphate and the α-amino acid (fig. 4). This formation labilizes the departure of the COOH$^+$, H$^+$ and R$^+$ groups attached to the α carbon, thus permitting

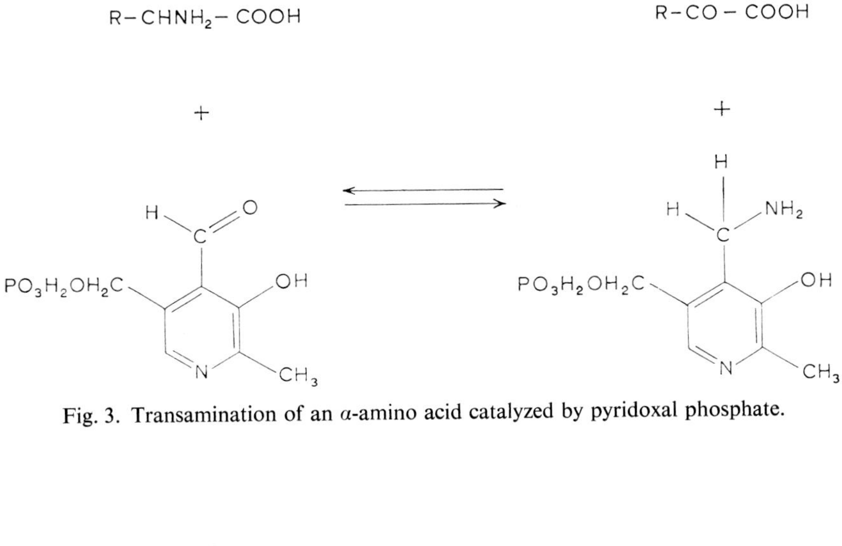

R−CHNH$_2$− COOH   R−CO − COOH

Fig. 3. Transamination of an α-amino acid catalyzed by pyridoxal phosphate.

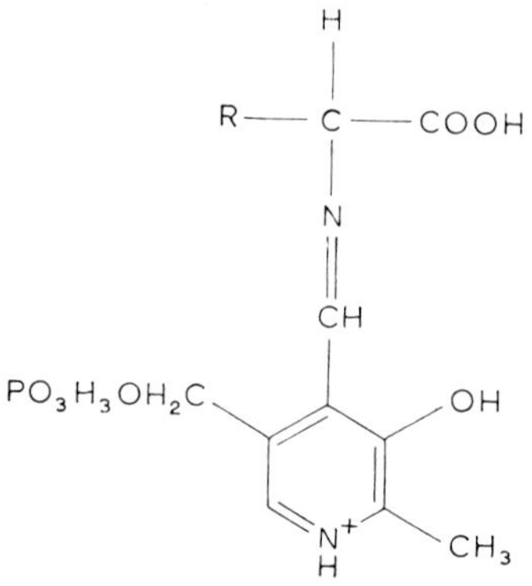

Fig. 4. The Schiff's base between an α-amino acid and pyridoxal phosphate.

the involvement of the α-amino acid in a series of reactions. It may be shown then that the essential electronic factors connected with the catalytic properties of pyridoxal phosphate reside in (Perault et al. 1961; B. Pullman 1963):

(1) The strong resonance stabilization of the transitional forms (XXIII), (XXIV), (XXV), deriving from the initial Schiff's base through the labilization of any of the $COOH^+$, H^+ or R^+ groups, stabilization being due to the increase in dimensions or the reorganization or both of the total conjugated system.

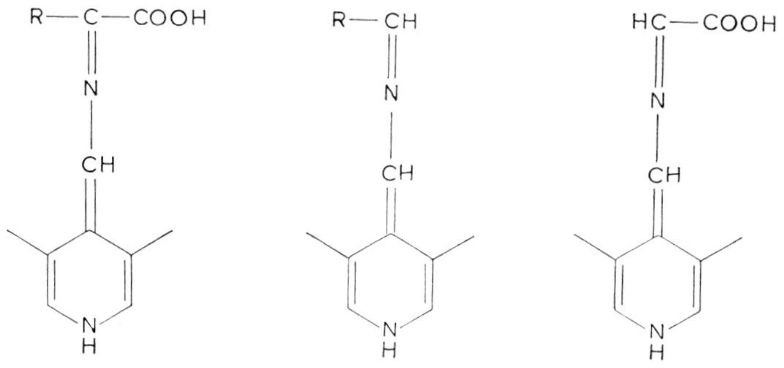

XXIII.   XXIV.   XXV.

(2) The creation in these transitional forms of prominent centers of extremely high reactivity (great local concentrations of electronic charges or of free valences), which ensures the rapid consecutive development of the reaction toward its final products. This is illustrated in fig. 5 for the transitional form (XXIII) whose extracyclic carbons of the C–N–C chain carry very great excesses of π electrons.

Thus, the catalytical activity of these coenzymes, and of a number of others is essentially dependent on the transformation of the initial coenzyme-substrate complex into a reaction intermediate, highly stabilized by resonance and at the same time electronically activated which represents an essential intermediate of the reaction. Only conjugated molecules, endowed with a cloud of mobile and easily polarizable π-electrons and built of atoms capable of oscillating among different valence states may produce such results. It seems highly probable that some of them might have exerted catalytic effects of that kind, to a lesser extent of course, even at the beginning of the biochemical evolution before the elaboration of organized proteins.

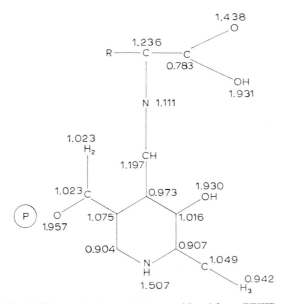

Fig. 5. Electronic charges in the transitional form (XXIII).

3. Intermolecular associations and intramolecular interactions

The role of electronic factors in these phenomena, so essential for the production of the large organized structures without which any elaborated form of life does not seem to be possible, is again particularly evident in conjugated molecules. These molecules have a manifest tendency to form self-associations or molecular or even polymolecular associations, a statement which may be substantiated with numerous examples concerning the formation between such molecules of complexes in solution or of mixed crystals in the solid state. Suffice it to quote in this respect the associations which occur between purines and pyrimidines in solution and which following the nature of the solvent may consist either of specific coplanar associations by hydrogen bonding or of less rigid stacking type associations between parallelly superposed bases (Hoogsteen 1968; Ts'o 1968; Ts'o et al. 1969) and the formation of mixed crystals between purines and pyrimidines where both types of interaction (hydrogen bonding and stacking) coexist (Hoogsteen 1968; Baklagine 1968).

These phenomena may be considered as primings for the formation of miniature nucleic acid. What is particularly striking, however, is the contribution of quantum biochemistry to the understanding, on the one hand, of the nature of the electronic factors and forces operating in such associations and, on the other, of some features of the associations having a direct and fundamental biological significance. The most striking among those is the *complementarity*, or the selectivity of a large number of such associations. The most celebrated among them is, of course, the exclusivity of the hydrogen-bonding pairing between the Watson–Crick complementary purine and pyrimidine bases, particularly as substantiated by a large number of experimental findings even outside the nucleic acids framework. The existence of similar although less well pronounced or less well established exclusivities between other types of biological compounds seems to gain reliability. We shall come back to those later. In the first place we would like to indicate the role of electronic factors in the Watson–Crick complementarity and to point out the contribution of quantum calculations toward their understanding.

3.1. The nature of complementarity

Thus, a large series of recent studies on the co-crystallization of derivatives of the nucleic acid bases, carrying a substituent at their glycosylic nitrogen and on their interaction in nonaqueous solvents, e.g. deutorochloroform, CCl_4, or dimethylsulphoxide point to the striking result that hydrogen-bonded pairs are formed exclusively between bases showing complementarity in the Watson–Crick sense, namely between adenine and thymine (uracil) or guanine and cytosine. In the case of the A–T (or A–U) interactions, the pairing appears to be different from that occurring in the nucleic acids; in the case of the G–C interactions, it is similar to it. On the contrary, no interactions whatsoever seem to occur in these circumstances between guanine and thymine or adenine and cytosine, or between cytosine and thymine or guanine and adenine. This result seems surprising at first sight, because it is easily seen that hydrogen bonds can in principle be established in all these cases (fig. 6).

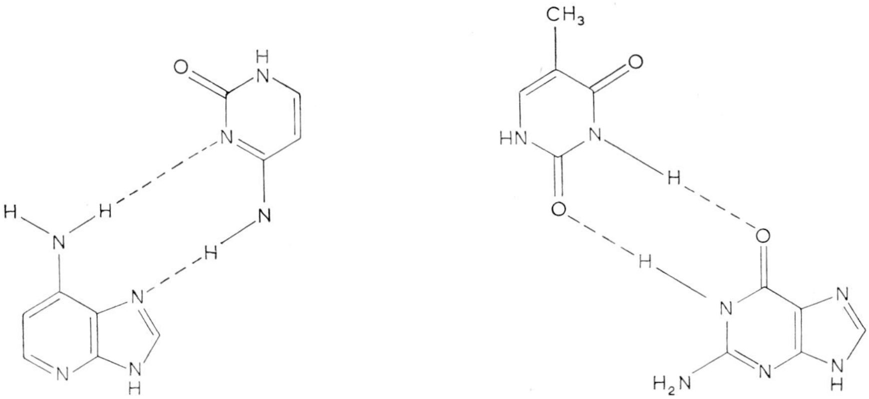

Adenine-cytosine   Guanine-thymine

Fig. 6. Examples of a priori possible hydrogen-bonded base pairs between 'noncomplementary' purines and pyrimidines.

It was the merit of theoretical calculations on *the intermolecular forces* operating in such associations to have proposed a possible explanation for this situation (Pullman et al. 1966a, b) somewhat prior to its substantiation by experimental findings. The forces referred to and which we shall call broadly the Van der Waals–London interaction forces correspond to the summation of the electrostatic (monopole-monopole), polarization, dispersion and short range repulsion components of the general theory of intermolecular forces. (For details see e.g. B. Pullman and A. Pullman 1968; B. Pullman and A. Pullman 1969.) They may be evaluated in different degrees of technical refinements but in so far as our problem is concerned, they all lead, however, to the same *general* result. This result is summarized in table 3, which indicates for each pair of considered bases the total interaction energy corresponding to the strongest conformation of the association.

It can then be observed that, in principle, all these associations, whether complementary or non complementary in the Watson–Crick sense, should involve considerable stabilization energies. A distinct difference appears, however, between the two classes of base pairs when the interaction energies of the cross-associations are compared with those of the self-associations of the constituent molecules. Thus, it may be observed that the stabilization of the A–T association, although relatively moderate, is nevertheless greater than that of the A–A or T–T self-associations. Similarly, although the stabilization of the G–G and C–C self-associations is relatively strong, that of the G–C association is still stronger. On the contrary, all the remaining cross-associations appear always weaker than the self-association of one of their constituents, G or C. It may be easily verified that the same result prevails when the values of the cross-associations are compared with the mean values of the corresponding two self-associations instead of being compared with the values of the strongest self-association.

These results obviously indicate the essentially electronic origin of base complementarity. A more detailed examination shows that in such in-plane interactions it is moreover the electrostatic component of the forces which is by far predominant. Thus it is essentially the coulombic interaction between the electrical charges on the two partners which governs the exclusivity of their association.

Two remarks may perhaps be usefully added to the preceding data:

(1) In the first place it may be useful to indicate that all the essential aforementioned theoretical indications on the *relative* energies of interaction have been confirmed recently by experimental studies on the interaction of bases in non aqueous solvents. The available experimental results are summarized in table 4 from which it may be observed that the only disagreement between theory and experiment concerns the relative order of the U–U versus A–A interaction energies, both of which are relatively small and close to each other, so that this reversal is not significant. It is important to realize that this situation indicates that although the calculations have been carried out for systems in vacuum, they seem nevertheless to be valid also in so far as the relative strength of the interactions is concerned for systems in solution.

(2) Still more recently the values of the enthalpies of association, in chloroform solution, have been established for the principal self- and cross-associations of the nucleic acid bases. The values are listed in table 5. Compared with the theoretical values of table 4 they appear generally appreciably smaller. This situation is in no way astonishing, because one cannot, of course, expect a *numerical* agreement between calculations carried out for systems in vacuum and measurements in solution. One immediately obvious possible reason for the disparity between the two sets of values could simply be the adoption of a dielectric constant of 1 in the calculation of the electrostatic component of the interaction energies, a component which represents the greatest part of this energy in hydrogen-bonded systems.

TABLE 3

In-plane interaction energies for hydrogen-bonded pairs (kcal/mole).

A–A	−5.8	A–T	−7.0
T–T	−5.2		
G–G	−14.5	G–C	−19.2
C–C	−13		
A–A	−5.8	A–C	−7.8
C–C	−13		
G–G	−14.5	G–T	−7.4
T–T	−5.2		
C–C	−13	C–T	−6.5
T–T	−5.2		
A–A	−5.8	A–G	−7.5
G–G	−14.5		

$$A\text{-}T > \frac{A\text{-}A}{T\text{-}T}$$

$$G\text{-}C > \frac{G\text{-}G}{C\text{-}C}$$

$$A\text{-}C < C\text{-}C$$

$$G\text{-}T < G\text{-}G$$

$$C\text{-}T < C\text{-}C$$

$$A\text{-}G < G\text{-}G$$

Following the indication of Poland and Scheraga (1967) a value close to 3 (close to 2 following Ramachandran (1969)) for the dielectric constant would be more appropriate at the interatomic distances under consideration. The adoption of such a value would immediately bring the theoretical values in much closer agreement indeed with the experimental ones.

In no way, as I said previously, are such association limited to the interactions between purines and pyrimidines. One can discuss in similar terms, analogous interactions which occur in solution between purines or pyrimidines and isoalloxazine or nicotinamide derivatives (McCormick 1968; Mantione 1968) possible primers of the complex structure of coenzymes. It may easily be conceived that this type of associations, loose of course and fragile, existed before its constituent parts has become more strongly bound by more solid bridges. Similarly, a number of recent publications have drawn attention and abundantly demonstrated the strong tendency of porphyrins toward forming molecular associations (Mauzerall 1964; Caughey et al. 1969; Hill et al. 1969). There is also no possible doubt that the highly specific interactions which are observed between nucleotides and polypeptides (e.g. Wagner and Arav 1968) or between polyribonucleotides and

TABLE 4

Experimental results on hydrogen bonding between nucleic acid bases.

Result	References	Solvent (and method)	
A–U < A–A or U–U	k	$CDCl_3$	(IR)
	a	CCl_4	(IR)
	b	$CDCl_3$	(IR)
	c	$CHCl_3$	(calorimeter)
G–C > G–G or C–C	d	$CDCl_3$	(IR)
	e		
	f	DMSO	(NMR)
G–C > A–T or A–U	d	$DMSO + CHCl_3$	(NMR)
	g	$CHCl_3$	(IR)
	h	DMSO	(NMR)
	f	DMSO	(NMR)
G–G > C–C	e	$CHCl_3$	(IR)
U–U > A–A	b	$CHCl_3$	(IR)
	i	$CHCl_3$	(calorimeter)
No interaction between	j	No cocrystallization.	
non complementary	h		
nucleic acid bases.	c		

a: E. Köchler and J. Derkosch, Z. Naturf. 21b, 209 (1966).
b: Y. Kyogoku, R. C. Lord and A. Rich, J. Amer. Chem. Soc. 89, 497 (1967).
c: J. S. Binford Jr. and D. M. Holloway, J. Mol. Bio. 31, 91 (1968).
d: K. Katz and S. Penman, J. Mol. Biol. 15, 220 (1966).
e: Y. Kyogoku, R. C. Lord and A. Rich, Science 154, 518 (1966).
f: R. A. Newmark en C. R. Cantor, J. Amer. Chem. Soc. 90, 5010 (1968).
g: J. Pitha, R. N. Jones and P. Pithova, Canad. J. Chem. 4, 1045 (1966).
h: R. R. Shoup, H. T. Miles and E. D. Becker, Biochem. Biophys. Res. Commun. 23, 194 (1966).
i: Y. Kyogoku, R. C. Lord and A. Rich, Biochim. Biophys. Acta 179, 10 (1969).
j: A. E. V. Haschemeyer and H. M. Sobell, Nature 202, 969 (1964).
k: R. M. Hamlin Jr., R. C. Lord and A. Rich, Science 148, 1734 (1965).

TABLE 5

Enthalpies of hydrogen bonding (kcal/mole).*

G–G	8.5–10	G–C	10–11
C–C	6.3		
U–U	4.3	A–U	6.2
A–A	4.0		

* Y. Kyogoku, R. C. Lord and A. Rich, Biochim. Biophys. Acta 179, 10 (1969).

protenoids rich in particular amino acids (Yuki and Fox 1969) may again be attributed, at least to a large extent, to the interplay of favorable electronic factors.

3.2. *The conformational preferences*

I just said 'at least to a large extent'. This apparent limitation is due to the well recognized role of molecular *conformations* in establishing specific associations, in particular when those concern large or even medium sized polymeric systems. It is, however, important to realize that just as it is obvious that the previously discussed tendency of certain fundamental biomolecules towards specific intermolecular agglomerations – a natural outcome as we have seen of their electronic properties and in particular of their charge distribution – was of essential importance in the priming and the mechanism of agglomeration of prebiological molecules into more

$$H_3C \longrightarrow C \longrightarrow N \overset{\overset{\displaystyle H}{|}}{\underset{}{\Big)} C \overset{\overset{\displaystyle H}{|}}{\underset{\underset{\displaystyle CH_3}{|}}{\Big)}} C \longrightarrow N \longrightarrow CH_3$$

XXVI.

complex units, the existence of intramolecular interactions, which have the important effect of generating selective conformations and of stabilizing them, is again attributable to the same electronic properties and to the action of similar factors. Illustrative of this state of affairs are e.g. the results of quantum-mechanical calculations on the *syn* or *anti* conformations of purine and pyrimidine nucleosides and nucleotides which indicate (Tinoco et al. 1968; Jordan and Pullman 1968) in conformity with experimental findings the preference of adenine, thymine and cytosine derivatives for the anti conformation and that of the guanine derivative for the syn conformation (fig. 7), or perhaps still better, the results of calculations, empirical during a time (see e.g. Ramachandran and Sasisekharan 1968; Scheraga 1968; Venkatachalam and Ramachandran 1969) and quantum-mechanical more recently (Hoffman and Imamura 1969; Kier and George 1969; Maigret et al. 1970), on the conformation of aminoacid residues indicating the tendency of these building blocks of proteins to assume naturally, by the sole virtue of intramolecular interactions, some particular conformational forms. The situation is illustrated on the example of the alanyl res due as present in the dipeptide (XXVI) on fig. 8. For this particular case the quantum-mechanical

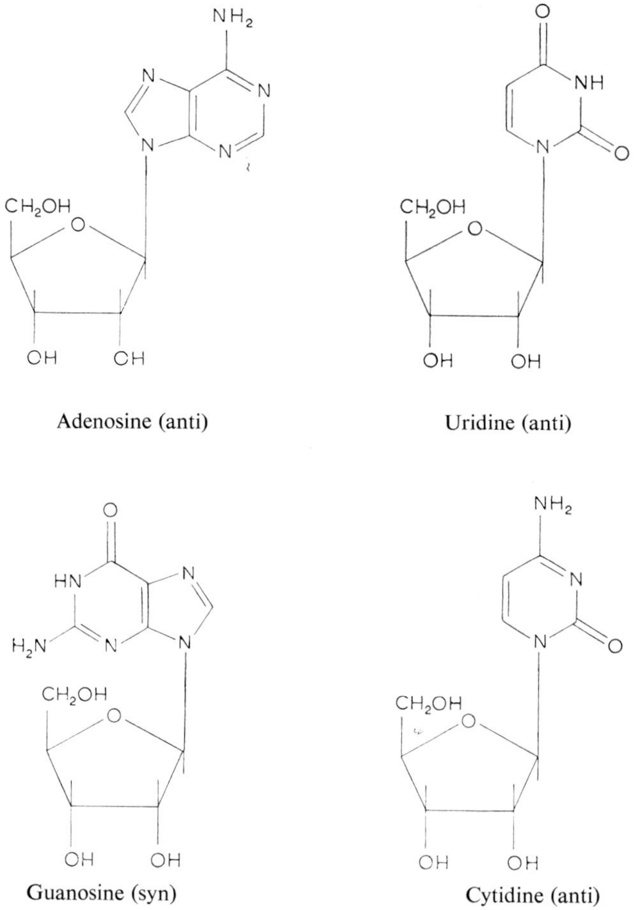

Fig. 7. Conformations found in crystals and assumed for solutions of mononucleosides.

calculations (Maigret et al. 1970) predict, in distinction to earlier more empirical calculations but in conformity with recent experimental findings (Bystrov et al. 1969) a preferential seven-membered hydrogen bonded arrangement, indicated in fig. 9.

One of the general results of such calculations, having possibly a direct significance for some recent experiment studies on the synthesis of biologically pertinent peptides under possible primordial conditions, is the demonstration (Ramachandran and Sasisekharan 1968; Maigret et al. 1970) that the restrictions about the allowed region of the torsional angles (φ and ψ) on a conformational map of the type of fig. 8 increase grosso modo with the

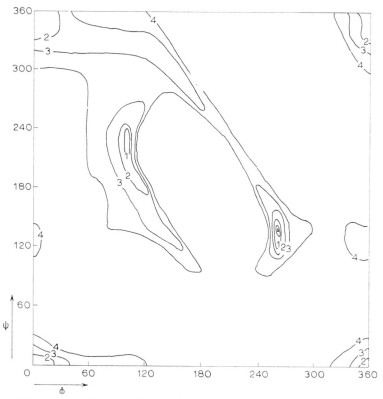

Fig. 8. Conformational map for the alanyl residue (isoenergy curves as a function of the torsional angle φ and Ψ).

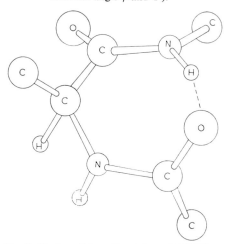

Fig. 9. Preferential conformation of (XXVI).

dimensions of the amino acid residue. The aforementioned synthetic studies (Steinman and Cole 1967) show that with other factors being equal, the probability of union of individual amino acids with one another, under possible prebiological conditions (aqueous solutions of amino acids, HCO and sodium dicyanamide) is determined by the size of the side chains involved, at least in the case of amino acids with nonpolar functions. The possible significance of this result for prebiological evolution being indicated by the apparent parallelism between the yields of the dipeptides in such model synthesis and the frequencies of the presence or formation of the same dipeptides deduced from the known amino acids sequences in a series of proteins. Now, the conformational arrangement and thus the volume of a residue in space being governed by intramolecular electronic factors, it is of course the combination of such factors with the purely steric (dimensional) ones which must be considered as being at the bottom of all tendencies towards specific associations.

4. *Electronic factors in mutagenesis*

Finally, one of the most direct illustrations of the role of electronic factors in the processes of life concerns its role in mutations and, thus, its decisive influence on the biological evolution of species. Thus, as is well known, the purine and pyrimidine bases of the nucleic acids are considered as existing essentially in the lactam and amino tautomeric forms (fig. 10). This situation is in agreement with quantum-mechanical calculations about the relative stabilities of the different possible tautomeric forms of the bases.

The specific pairing of adenine with thymine and of guanine with cytosine, characteristic of the Watson–Crick model of DNA, is dependent on the predominance of these usual tautomers (fig. 11).

The possibility that spontaneous mutations may involve the *rare tautomeric forms* of the bases was advanced already by Watson and Crick in 1953. Such rare tautomeric forms would be *lactim* for uracil and guanine and *imine* for adenone and cytosine (fig. 12). The presence of a rare tautomeric form may give rise to a coupling, through hydrogen bonds, of unusual bases (fig. 13) and may thus lead to a perturbed sequence of base pairs in later generations, that is, to a mutation. Although other mechanisms, such as miscoupling of ionized rather than tautomeric bases or base deletions, may also play a role in spontaneous mutations, the intervention of rare tautomeric forms remains plausible. It may also be considered in connection with mutations induced with base analogues or by the effect of radiation.

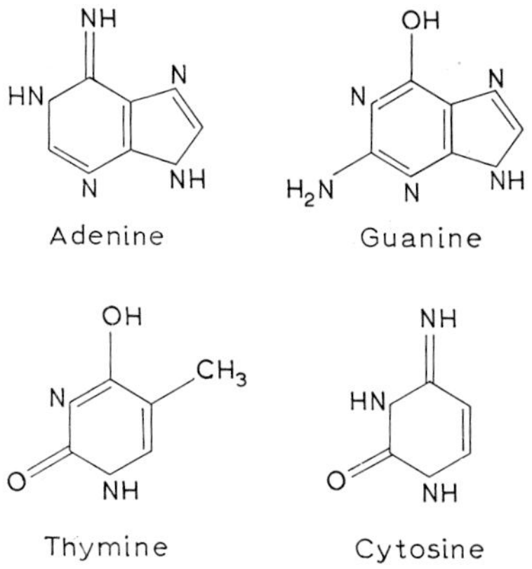
NH
HN
N
N
NH
Adenine
OH
N
N
H2N
N
NH
Guanine
OH
CH3
N
O
NH
Thymine
NH
HN
O
NH
Cytosine

Fig. 10. Usual tautomeric forms of the nucleic acid bases.

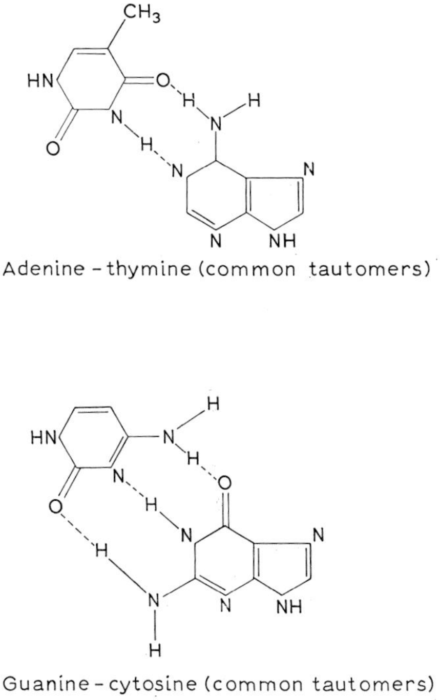
Adenine - thymine (common tautomers)
Guanine - cytosine (common tautomers)

Fig. 11. Usual coupling of the bases of the nucleic acids.

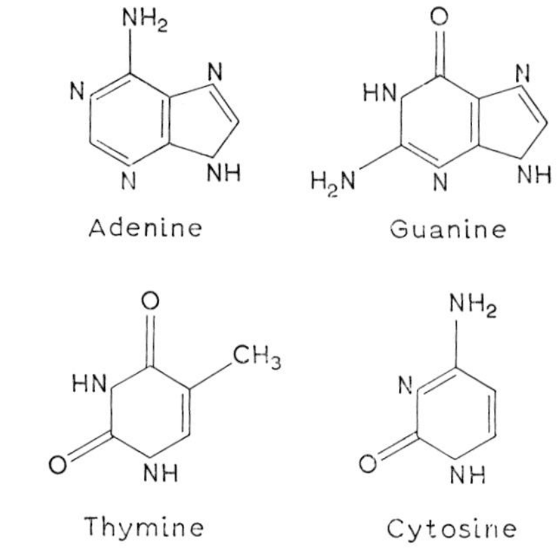

Adenine   Guanine

Thymine   Cytosine

Fig. 12. Rare forms of the nucleic acid bases.

Cytosine (rare imino form) – adenine (normal form)

Guanine (rare enol form) – thymine (usual form)

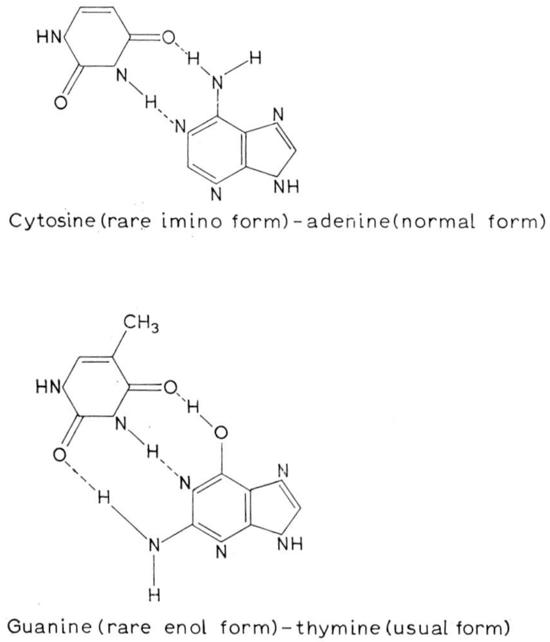

Fig. 13. Examples of miscoupling of the nucleic acid bases.

From the electronic quantum-theoretical point of view, it may be interesting then to investigate a number of questions connected with the possibility of tautomerization of the bases. Among these questions are the problems of which of the purine and pyrimidine bases of the nucleic acids has the greatest probability to exist in a rare tautomeric form and thus to be particularly involved in spontaneous mutations, what are the principle characteristics of the rare tautomeric forms of the bases and finally what are the consequences of their interference, i.e. of a miscoupling on a number of physicochemical properties of the nucleic acid e.g. their stability, relevant to their biological function.

Now, as concerns the essential problem of the *relative tendency* of their bases to exist in their rare forms, for which very limited experimental data are available, but upon which may depend, following the previous consideration the site and the rate of mutations, it can be easily shown that one of the principle factors responsible for it is related to the π-electronic orbitals of the bases (B. Pullman and A. Pullman 1962; A. Pullman 1964). Thus, it can be easily shown (B. Pullman and A. Pullman 1952; Wheland 1955) that, at least at first approximation, when a given type of a tautomeric equilibrium is being studied in a related series of compounds, the essential *varying* factor responsible for the *relative* tendency of the compounds to exist in a rare tautomeric form is the variation of the resonance energy which accompanies the tautomeric transformation. In the case of the purine-pyrimidine bases there are two such transformations to be considered: the *lactam-lactim* and the *amino-imine* transformations. In the lactam-lactim tau omerism the transformation of the lactam form (the most stable) to the lactim form (less stable) is associated with an increase of resonance energy so that the proportion of the lactim form will be the greater, the greater this increase. In the amino-imine tautomerism, the transformation of the amino form (the more stable) to the imino form (less stable) is accompanied by a decrease in resonance energy and will therefore be the greater, the smaller this decrease. Explicit calculations of resonance energies for the different tautomeric forms of the purine and pyrimidine bases of the nucleic acids, lead to the prediction that the bases which, from that point of view, should have the greatest tendency to exist in rare imino and lactam forms are cytosine and guanine, respectively (table 6). These are therefore the bases which have the greatest probability to be involved in spontaneous mutations in so far, of course, as tautomerization may be considered as a cause of such mutations. The transformation G–C → A–T should then be more frequent than the reverse one.

TABLE 6

Resonance energies of the tautomeric forms (in β units $\approx$ 16 kcal/mole).

Compound	Resonance energies of the tautomeric forms		ΔR^*
Guanine	Lactam: 3.84	Lactim: 4.16	0.32
Cytosine	2.28	2.69	0.41
Uracil	1.92	2.14	0.22
Thymine	2.05	2.27	0.22
Guanine	Amine: 3.84	Imine: 3.68	-0.16
Adenine	3.89	3.62	-0.27
Cytosine	2.28	2.15	-0.13

* ΔR = Variation of resonance energies accompanying the transformation from the stable to the less stable form.

It may be interesting to remark that the fact that the G–C pairs constitute the unstable part of the genome and that they mutate spontaneously more frequently than the A–T pairs has been reported in a number of publications (Freeze 1961, 1963; Drake 1966).

When considering the possible role of tautomerization of the bases in mutagenesis one need not restrict oneself to spontaneous mutations. The same factor may be considered as playing a certain role in induced mutations, whether by physical or chemical means. Neither do we need, however, and nobody does, to consider base tautomerization as the only one or even the most important factor generating mutations. In some cases connected in particular with chemical or radiation induced mutagenesis it seems rather obvious that it is not and that other mechanisms are much more likely. The physico-chemical nature of the basic phenomena involved in mutagenesis seems however to become more and more evident every day and the interpretation of these phenomena at the electronic level appears relatively conclusive. The subject is too wide to be described here even in its broad lines, but a recent summary of its present status may be found in B. Pullman and A. Pullman (1969).

5. Conclusion

This short discussion clearly shows to what decisive degree the processes of biochemical evolution and thus finally the nature of life itself depend on factors characteristic of the electronic structure of molecules and more specifically perhaps on the properties of the mobile electronic systems of resonating molecules, in particular of resonating heterocyclics. This is true whether the origin of life itself or its present form are concerned. In fact, the situation is to some extent inherent in the very elementary chemical composition of living matter (Wald 1962); 99% of it is composed of hydrogen, carbon, nitrogen and oxygen. Now, these last three first-row elements are, of all the elements, those which most easily form multiple bonds. The next two most abundant and important elements to enter into the structure of the cell are sulfur and phosphorus and these are practically the only other than first-row elements to form multiple bonds. One can also imagine (Wald 1962) that carbon rather than silicon has been chosen by nature as an essential component of organic matter (although silicon is much more abundant on the surface of the earth than carbon) possibly because carbon may form chains of conjugated double bonds while silicon cannot and possibly because carbon bonds are much stronger than silicon bonds. The choice and utilization of conjugated compounds as structural components of life appears as one of the most important quantum effects in biochemical evolution. Their appearance must have been the turning point, from which the complex necessities of an elaborated form of self-perpetuating system could be satisfied.

References

BAKLAGINA YU, G., 1968, Mol. Biol. (U.R.S.S. Engl. ed.) 2, 507.

BLUMER, M., 1965, Sciences 149, 722.

BUCHANAN, J. M., 1960, Harvey Lectures 54, 104 (1958–1959).

BYSTROV, V. F., S. L. PORTINOVA, V. I. TSETLIN, V. T. IVANOV and Y. A. OVCHINNIKOV, 1969, Tetrahedron 25, 493.

CALVIN, M., 1962, In: Horizons in biochemistry, M. Kasha and B. Pullman, eds. (Academic Press, New York) 23.

CALVIN, M., 1969, Chemical Evolution. (At the Clarendon Press, Oxford).

CAUGHEY, W. S., H. EBERSPACCHER, W. H. FUCHSMAN, S. McCOY and J. O. ALBEN, 1969, Ann. New York Acad. Sci. 153, 722.

CHECCUCCI, A., J. DEPIREUX et J. DUCHESNE, 1964, C.R. Acad. Sc. 259, 1585.

CHECCUCCI, A., J. DEPIREUX et J. DUCHESNE, 1964, C.R. Acad. Sc. 259, 1669.

DAVIS, N. S., G. J. SILVERMAN and E. B. MASUROWSKY, 1963. J. Bacteriol. 86, 294.

DRAKE, J. W., 1966, Proc. Natl. Acad. Sci. U.S. 55, 738.

DUCHESNE, J. and A. VAN DE VORST, 1965. Acad. Roy. Belg. A. Classe Sci. Mem. 51, 778.

DUCHESNE, J. and A. VAN DE VORST, 1965, Acad. Roy. Belg. Bull. Classe Sci. 60, 778.

FERRIS J. P. and L. E. ORGEL, 1965, J. Amer. Chem. Soc. 87, 4976.

FRAM, H., B. E. PROCTOV and C. G. DUNN, 1950, J. Bacteriol. 60, 263.

FREESE E., 1961, Proc. Interm. Congress Biochem. Moscow. (Pergamon Press, London) Vol. I, 204.

FREESE, E., 1963, In: Molecular genetics, J. H. Taylor, ed. (Academic Press, New York, Part I) 207.

GAFFRON, H., 1962, In: Horizon in biochemistry, M. Kasha and B. Pullman, eds. (Academic Press, New York) 59.

GRANICK, S. and D. MAUZERALL, 1961, In: Metabolic pathways, D. M. Greenberg, ed., Vol. 2. (Academic Press, New York) 525.

HILL, H. A. O., A. J. MC FARLANE and R. J. P. WILLIAMS, 1969, J. Chem. Soc. (A), 1704.

HUDGSON, G. W. and B. L. BAKER, 1967, Nature 216, 29.

HUDGSON, G. W. and C. PONNAMPERUMA, 1968, Proc. Natl. Acad. Sci. U.S. 59, 22.

HOFFMAN, R. and A. IMAMURA, 1969, Biopolymers 7, 207.

HOOGSTEEN, K., 1968, In: Molecular associations in biology, B. Pullman, ed. (Academic Press, New York) 21.

HUBBARD, R., 1956, J. Amer. Chem. Soc. 78 4662.

JORDAN, F. and B. PULLMAN, 1968, Theoret. Chim. Acta 9, 242.

KIER, L. B. and J. M. GEORGE, 1969, Theoret. Chim. Acta 14, 258.

LANGLET, J., B. PULLMAN and H. BERTHOD, 1970, J. Chim. Phys. 67, 480.

MAIGRET, B., B. PULLMAN and M. DREYFUS, 1970, J. Theoret. Biol. 26, 321.

MANTIONE, M. J., 1968, In: Molecular associations in biology, B. Pullman, ed. (Academic Press, New York) 411.

MAUZERALL, D., 1964, Biochemistry A.C.S. 4, 1801.

MC CORMICK, D. B., 1968, In: Molecular associations in biology, B. Pullman, ed. (Academic Press, New York) 377.

MEINSCHEIN, W. G., E. S. BARGHOORN and J. W. SCHOPF, 1964, Science 145, 262.

MILLER, S. L. and H. C. UREY, 1959, Science 130, 245.

OESPER, P., 1950, Arch. Biochem. 27, 255.

ORO, J. and A. P. KIMBALL, 1901, Arch. Biochem. Biophys. 94, 217.

ORO, J., 1965, In: The origins of prebiological systems and of their molecular matrices, S. Fox, ed. (Academic Press, New York) 157.

ORO, J., 1965, In: G. Mamikunian and M. H. Briggs; Current aspects of exobiology. (Pergamon Press, Oxford) 13.

PERAULT, A., B. PULLMAN and C. VALDEMORO, 1961, Biochim. Biophys. Acta 46, 555.

PHILLIPS, G. O., F. A. BLOUIN and J. C. ARTHUS Jr., 1964, Nature 202, 1329.

POLARD, D. and H. A. SCHERAGA, 1967, Biochemistry 6, 3791.

PONNAMPERUMA, C., R. M. LEMMON, R. MARINER and M. CALVIN, 1963, Proc. Natl. Acad. Sci. U.S. 49, 737.

PONNAMPERUMA, G., R. S. YOUNG, E. F. MUNOZ and B. K. MC CAW, 1964, Science 143, 1449.

PONNAMPERUMA, C., 1965, Sci. J. 39.

PONNAMPERUMA, C. 1965, In: The origins of prebiological systems and of their molecular matrices, S. W. Fox, ed. (Academic Press, New York) 221.

PONNAMPERUMA, C. and N. W. GABEL, 1968, In: Space Life Sciences, G. Mamikunian, ed. (Reidel Publishing Co., Dordrecht-Holland) 1, 64.

PULLMAN, B. and A. PULLMAN, 1952, In: Les théories electroniques de la chimie organique. (Masson, Paris).

PULLMAN, B. and A. PULLMAN, 1959, Proc. Natl. Acad. Sci. U.S. 45, 136.

PULLMAN, B. and A. PULLMAN, 1960, Rev. Nod. Phys. 32, 428.

PULLMAN, B. and A. PULLMAN, 1960, In: Comparative effects of radiations, M. Burton, J. S. Kirby-Smith and J. L. Magee, eds. (Wiley, New York) 105.

PULLMAN, B. and A. PULLMAN, 1960, Radiation Res. Suppl. 2, 160.

PULLMAN, A. and B. PULLMAN, 1961, Proc. Natl. Acad. Sci. U.S. 47, 7.

PULLMAN, B., 1961, Acad. Roy. Belg. A. Classe Sci. Mem. 33, 175.

PULLMAN, B. and A. PULLMAN, 1961, Nature 189, 725.

PULLMAN, B. and A. PULLMAN, 1962, Biochim. Biophys. Acta 64, 403.

PULLMAN, B. and A. PULLMAN, 1962, Nature 196, 1137.

PULLMAN, B., 1963, In: Chemical and biological aspects of pyridoxal catalysis. (Pergamon Press, Rome) 103.

PULLMAN, B. and A. PULLMAN, 1963, Quantum Biochemistry. (Wiley-Interscience, New York).

PULLMAN, A., 1964, In: Electronic aspects of biochemistry, B. Pullman, ed. (Academic Press, New York) 135.

PULLMAN, B. 1964, Radioresistance of solic conjugated molecules; In: Electronic aspects of biochemistry, B. Pullman, ed. (Academic Press, New York).

PULLMAN, A., 1965, In: Modern quantum chemistry, O. Sinanoglu, ed. (Academic Press, New York) Vol. 3, 283.

PULLMAN, B., P. CLAVERIE and J. CAILLET, 1966, Proc. Natl. Acad. Sci. U.S. 55, 904; 1966, J. Mol. Biol. 22, 373.

PULLMAN, A. and B. PULLMAN, 1968, Advan. Quantum Chemistry 4, 267.

PULLMAN, B. and A. PULLMAN, 1969, Progress Nucleic Acid Res. and Molecular Biol. 9, 327.

RAMACHANDRAN, G. N. and V. SASISEKHARAN, 1968, Adv. Protein Chem. 23, 283.

RAMACHANDRAN, G. N., 1969, Intern. J. Protein Res. 1, 5.

ROSENBERG, E. 1965, Science 146, 1680.

SANCHEZ, R. A., J. P. FERRIS and L. E. ORGEL, 1968, J. Mol. Biol. 38, 121.

SCHERAGA, H. A., 1968, Advan. Phys. Org. Chem. 6, 103.

SNELL, E. E., 1961, Ciba Foundation Study Group No. 11, 18.

STEINMAN, G. and M. N. COLE, 1967, Proc. Natl. Acad. Sci. U.S. 58, 735.

SZUTKA, A., 1965, Nature 202, 1231.

TINOCO, I. Jr., R. S. DAVIS and S. R. JASKUNAS, 1968, In: Molecular associations in biology, B. Pullman, ed. (Academic Press, New York) 77.

TS'O, P. O. P., 1968, In: Molecular association in biology, B. Pullman, ed. (Academic Press, New York) 39.

TS'O, P. O. P., M. P. SCHWEIZER and D. P. HOLLIS, 1969, Ann. N.Y. Acad. of Sci. 158, 256.

VAN DE VORST, A. and C. WILLIAMS-DORLET, 1963, C.R. Acad. Sci. 157, 2183.

VAN DE VORST, A., M. RICHIR and K. V. RAJALAKSHMI, 1966, Acad. Roy. Belg. A. Classe Sci. Mem. 52, 276.

VAN DE VORST, A., M. RICHIR and D. KRSMANOVIC-SIMIC, 1966, C.R. Acad. Sci. 262, 401.

VENKATACHALAM, C. M. and G. N. RAMACHANDRAN, 1969, Ann. Rev. of Biochem. 38, 45.

WAGNER, K. G. and R. ARAV, 1968, Biochemistry 7, 1771.

WALD G., 1963, Proc. 5th Intern. Congr. Biochem. Moscow 3, 12.

WALD, G., 1962, In: Horizons in biochemistry, M. Kasha and B. Pullman, eds. (Academic Press, New York) 127.

WEBB, C. E. and E. B. FLEISCHER, 1965, J. Chem. Phys. 43, 3100.

WHELAND, G. W., 1955, In: Resonance in organic chemistry. (Wiley, New York).

YUKI, A. and J. FOX, 1969, Biochem. Biophys. Res. Comm. 36, 657.

C. Ponnamperuma (ed.), Exobiology. © North-Holland Publishing Company

CHAPTER 6

Origins of molecular chirality

WILLIAM A. BONNER

1. Chirality, optical activity and asymmetric synthesis

Chiral structures are three-dimensional structures having associated with them an intrinsic *handedness* (Whyte 1957; Pirie 1959). The right- and left-handed forms of such structures are called *enantiomers*, and constitute non-superimposable mirror images of each other. Familiar examples include right and left hands, right and left threaded screws, clockwise and counter-clockwise spiral staircases and, of course, 'right- and left-handed' organic molecules containing one or more asymmetric carbon atoms (C*). Chirality and enantiomerism on the molecular level are readily seen on examining molecular models of (R)-lactic acid (*1a*) and (S)-lactic acid (*1b*). Notice in the (R)-enantiomer *1a* that the sequence of groups: OH, COOH and CH_3 is clockwise, while in the (S)-enantiomer *1b* it is counterclockwise. The symbols (R) (Latin, *rectus*, right) and (S) (Latin, *sinister*, left) are currently used in a systematic way to designate the two opposite chiralities which an individual asymmetric center may display.

Chiral molecules of a single chirality are *optically active*, that is, one enantiomer will impart a clockwise rotation (*dextrorotation*) to the plane of polarization of plane polarized light traversing it, while the opposite enantiomer (antipode) will rotate the plane of polarization an equal angle to the left (*levorotation*). A *racemic* mixture (racemate) containing equal numbers each of both enantiomeric molecules is *optically inactive*, since the dextrorotation of one enantiomer is exactly compensated by the levorotation of its antipode.

When chiral molecules are produced synthetically under ordinary laboratory conditions from achiral (symmetrical) molecules, the resulting product

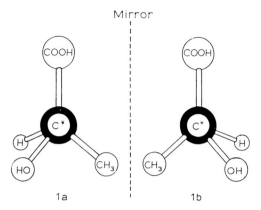

1a   1b

is invariably racemic, since the intrinsic probabilities of forming either enantiomer are equal, and we are dealing with large numbers of molecules. Thus optically inactive products are invariably the result of ordinary chemical syntheses. If, on the other hand, we introduce synthetically a new chiral center B* into an optically active molecule A*X, already possessing an asymmetric center of uniform chirality, e.g. (R)-A*, the two chiralities (R) and (S) of the new asymmetric center B* are not produced with equal probability. Thus if we subsequently remove the original asymmetric center (R)-A* degradatively from the product, the resulting final product B*Y will be optically active, since its two (R)- and (S)-enantiomers, *2a* and *2b*, are present in unequal amounts. Such an overall process, shown schematically below, is called an asymmetric synthesis. Without now examining the

$$(R)-A^*X \xrightarrow[+B^*]{-X} \begin{cases} (R)-A^*-(R)-B^* \xrightarrow[+Y]{-(R)-A^*} (R)-B^*Y \\ \qquad\qquad\qquad\qquad\qquad\qquad\qquad 2a \\ + \qquad \text{unequal amounts} \\ (R)-A^*-(S)-B^* \xrightarrow[+Y]{-(R)-A^*} (S)-B^*Y \\ \qquad\qquad\qquad\qquad\qquad\qquad\qquad 2b \end{cases}$$

mechanistic causes of asymmetric synthesis, we see however that the original chirality of the optically active precursor A*X imposes a preferred chirality on the optically active product B*Y. If the ratio of enantiomer *2a* to enantiomer *2b* were 3:1 starting with (R)-A*X, for example, it would be 1:3 starting with the antipodal precursor (S)-A*X. In such typical asymmetric syntheses performed in the laboratory the overall *stereoselectivity* of the process (percentage of the predominant enantiomer in the total product), however, is usually not impressively high – perhaps 55–65%, though sometimes higher. In any case it is clear that, in the sense of a typical laboratory asymmetric synthesis, optical activity begets optical activity.

2.　*Molecular chirality in nature*

Those organic molecules which make up both living organisms as well as the metabolic products and byproducts of living matter, and which are also capable of chirality – for example, those which contain 4 different atoms or groups on their one or more individual (asymmetric) carbon atoms, (C*ABCD) – are almost invariably found in nature in the optically active and not the racemic state. Furthermore, those polymeric molecules which form the basic structural units of living matter, such as the proteins and polysaccharides, not only consist of optically active monomer subunits, namely amino acids and monosaccharides, but furthermore of subunits which are almost invariably of uniform chirality. Thus the hydrolysis of proteins yields exclusively amino acids of the L-configuration (or (S)-configuration), *3*, except in a few rare instances involving certain bacterial

$$H_2N - \overset{COOH}{\underset{R}{\overset{|}{\underset{|}{C^*}}}} - H$$

3

4

proteins (Stevens et al. 1951; Bricas and Fromageot 1953), proteins from cancerous tissue (Kögl 1939, 1949), or possibly protein from aged organisms (Kuhn 1958). Similarly, hydrolysis of such ubiquitous and biologically important polysaccharides as cellulose, starch and glycogen invariably affords only D-glucose, *4*, as the constituent monomer, and never the enantiomeric L-glucose. Thus the 'primary constituents' of protoplasm are not only optically active, but also are almost invariably of a single chirality. Their 'optical purity' (enantiomeric homogeneity) has been characterized as 'obligatory' (Gause 1941).

　　As one passes from the primary to the secondary chiral constituents of living matter, namely, to the chiral metabolic or other products which are excreted by or stored within a given organism, enantiomeric homogeneity looses its obligatory character (Gause 1941). That is, chiral molecules of such products are usually optically active, but may at times be in the partially or completely racemic state. Furthermore, the same chiral molecule may be found as the dextrorotatory enantiomer – (+)-enantiomer – in one species of organism, and as the levorotatory form – (−)-enantiomer – in a different species. Thus (+)-borneol, *5*, occurs in the *Lavandula spica* plant, while (−)-borneol is found in *Pinus maritima*. Similarly, (+)-limonene, 6, may be obtained from *Pinus serotina*, (−)-limonene from *Eucalyptus staigeriana*, and racemic (±)-limonene from *Pinus silvestris* (Gause 1941).

5   6

We see from the selected examples above that all of the primary and most of the secondary constituents of living matter are in nature uniquely associated with optical activity, that is, with asymmetric molecules of one chirality. It is thus not surprising that since the time of Pasteur scientists have looked upon optical activity as *the* unique criterion of life (Gause 1941). Indeed in 1860 Pasteur himself characterized 'the molecular asymmetry of natural organic products' as 'the great characteristic which establishes perhaps the only well-marked line of demarcation that can at present be drawn between the chemistry of dead matter and the chemistry of living matter' (Japp 1898). Conversely, it is generally accepted (Terent'ev and Klabunovskii 1959; Stryer 1966) that life, at least as we know it, could not exist in the absence of optical activity, that is, on a racemic basis. Thus net molecular chirality, as indicated by measurable optical activity, provides a potential probe, perhaps the only valid probe, for the recognition of life in remote areas of time or space. Marginal optical activity has been reported both in meteorites (Nagy et al. 1964; Hayatsu 1965; Meinschein et al. 1966; Hayes 1967), in petroleum (Terent'ev and Klabunovskii 1959), and in rock samples billions of years old (Kvenvolden et al. 1969), and the validity and implications of these observations have accordingly been subjects of vigorous debate. Similarly, in the present age of space exploration, the detection of optically active compounds on the moon and on other planets has been recently widely proposed (Lederberg 1965; Fox 1966; Halpern et al. 1966; Stryer 1966; Draffan et al. 1969) as perhaps the most valid and sensitive criterion for the recognition of extraterrestrial life.

3.   *The transmission and preservation of molecular chirality*

If the molecular constituents of living matter involve molecules of a single chirality, there is no fundamental mystery as to how these molecules impart a preferred chirality onto those molecules which are the subsequent products of their anabolic (synthetic) and catabolic (degradative) biochemical reactions. In principle, such reactions are qualitatively similar to the simple illustrative example of asymmetric synthesis described above. That is, by ordinary, reasonably-well understood intermolecular interactions involving electrical attractions or repulsions, bulk interference between atoms or groups of atoms, or the transmission of electronic effects within molecules – all operating within an asymmetric molecular environment – the predominant chirality of the product molecule becomes the inevitable consequence of the preexisting chirality of the reactant molecule. Again, molecular asymmetry begets molecular asymmetry. In practice, of course, the complex, enzyme-catalyzed reactions of the polymeric protein, carbohydrate and nucleic acid molecules involved in biochemical processes, which are ordinarily *stereospecific* (100% stereoselective), are vastly more difficult to analyze and understand in mechanistic stereochemical terms than are the far less stereoselective reactions of laboratory asymmetric synthesis. Presumably, however, no qualitatively different phenomena are involved. In other words, although the mechanistic details of the stereospecific enzymatic processes associated with living matter may be most challenging and difficult to understand, there is no reason to believe that the requisite conceptual tools are not substantially at hand.

A question more fundamental, perhaps, than that of the transmission of optical activity is that of the preservation of optical activity. The conversion of one enantiomer into a racemic mixture of both enantiomers, a process called racemization, is thermodynamically a spontaneous process. Therefore, granting that a suitable mechanism is available, any optically active substance will in time racemize spontaneously – slowly if the activation energy of the process is high, but rapidly if the activation energy is low, or can be made low by the action of a catalyst. Thus it has been argued recently that 'Optical purity is bound to deteriorate, since the free energy of racemization is negative' (Stryer 1966). How then does molecular chirality maintain itself in the face of this universal thermodynamic imperative toward racemization?

This important question was answered theoretically and experimentally by Langenbeck and Triem (1936). Suppose that a reaction, $A + B \rightarrow AB$, takes place between two optically active reactants A and B which are 'optically impure', that is, predominantly but not completely of one chirality. Four stereochemically different competing reactions, (1)–(4), may then occur, where R and S designate the chirality of the asymmetric centers involved:

$$(1) \quad A_R + B_R \xrightarrow{k} A_R B_R$$

$$(2) \quad A_S + B_S \xrightarrow{k} A_S B_S$$

$$(3) \quad A_R + B_S \xrightarrow{k'} A_R B_S$$

$$(4) \quad A_S + B_R \xrightarrow{k'} A_S B_R$$

Suppose also that the (R)-enantiomers predominate at the outset, that is $[A_R] > [A_S]$ and $[B_R] > [B_S]$ (where [ ] designates concentration). Suppose finally that the overall reaction is interrupted before all of the reactants have had the opportunity to combine. Since the rate constants k for reactions (1) and (2) are the same, it follows that the product $A_R B_R$ will be produced faster than $A_S B_S$. Finally, because of second order kinetics, it also follows that:

$$\frac{[A_R B_R]}{[A_S B_S]} > \frac{[A_R]}{[A_S]} \quad \text{and} \quad \frac{[A_R B_R]}{[A_S B_S]} > \frac{[B_R]}{[B_S]}$$

In other words, the enantiomeric homogeneity of the product is greater than that of the starting materials. That is, an overall increase in optical purity accompanies one of the products of such an interrupted reaction. The optical purity of the unreacted components, of course, will have deteriorated, while the products of reactions (3) and (4) may or may not be formed with greater enantiomeric homogeneity. Langenbeck and Triem then demonstrated the validity of their arguments by several experiments. In one experiment, for example, a mixture of tyrosine methyl ester, 7, containing 63.7% (−)- and 36.3% (+)-tyrosine, was heated to yield the dimeric product 8, and the reaction was interrupted at 40% completion. The product 8 proved

to be made up of tyrosine residues comprised of 65.4% (−)- and 34.6% (+)-enantiomers, an increase of 1.7% in the optical purity of the (−)-

tyrosine component. Thus the incomplete reaction of an optically impure substance with another optically impure substance leads to a product of greater optical purity than that of the reactants, thereby 'correcting' for any partial racemization which the reactants might previously have suffered. It has been suggested more recently that such processes must have occurred in the forerunners of primitive cells (Böhm and Losse 1967).

4. The fundamental question of origin

We have seen above that simpler optically active substances may react to produce more complex optically active substances. Furthermore, under the right conditions such reactions can, at least at the level of one chiral product, oppose the universal, entropy-increasing tendency toward racemization. We have also seen that life itself is inexorably involved with asymmetric molecules, producing and utilizing optically active substances with complete stereospecificity. We have seen finally that life as we know it would indeed be inconceivable on the racemic basis of optically inactive molecules. A fundamental and philosophically important question thus arises as to the primeval origin of molecular chirality and optical activity, a question intimately connected with that of the origin of life itself. Did optically active inert matter precede optically active living matter on the face of the primitive earth, such that life emerged by utilizing molecules already endowed with a preexisting chirality? Or did primordial protoplasm evolve on a racemic basis, gradually somehow later originating and adapting molecular chirality as a more efficient evolutionary expedient during subsequent stages of development? Is optical activity responsible for life, or is life responsible for optical activity? This is, in all probability, the most fundamental formulation of the classic question: which came first, the chicken or the egg?

It is therefore not at all surprising that the question of the origin of optically active substances in nature has intrigued chemists for over a century, and has been the impetus for both extensive speculation and varied experimentation. It shall be the task of the remainder of this chapter to examine and evaluate briefly the principal arguments and experiments pertaining to the origin of optical activity. Briefly, both sides of the fundamental question have been championed. That is, arguments have been advanced – in some cases with extensive supporting experimentation – upholding one or the other of the following two opposing viewpoints as to the origin of molecular chirality (Ulbricht 1962): (1) optically active molecules arose *abiotically* from inanimate matter as the result of either statistical fluctuations or the action of rational dissymmetric physical and/or chemical processes on racemic matter, and (2) optical activity arose *biotically* only as life itself evolved, by metabolic activity which somehow selected molecules of uniform chirality from racemic mixtures. We shall examine our subject matter within the various aspects of this organizational framework. In the interest of brevity, however, our survey will be representative rather than exhaustive.

5. *Statistical theories of the origin of molecular chirality*

At about the turn of the century Japp (1898) presented a lecture on 'Stereochemistry and Vitalism' which, reviving the obsolete concept of a 'vital force' and applying it to the question of optical activity, prompted a spirited polemic for a period of several months in Nature magazine. Among his arguments, Japp pointed out that the formation of an asymmetric molecule C*HXYZ, from a symmetrical molecule CH_2XY, must always afford a racemic product, since the probabilities of replacing either H-atom of CH_2XY with substituent Z are equal, and 'we are dealing with an infinitely great number of molecules'. Applying similar arguments to other types of reactions producing asymmetric molecules, Japp concluded that 'we cannot produce them singly as long as we have at our disposal only the symmetric forces which we command in the laboratory'. In other words, net chirality and optical activity could never arise from ordinary chemical reactions involving symmetrical molecules.

Just two weeks later Pearson (1898a, b) took issue with Japp's statements, and provided the first explanation for the origin of molecular chirality in purely statistical terms. Pearson emphasized that there are statistical fluctuations about the 50:50 mean of any equal probability (heads–tails) process, and that therefore during countless primeval ages asymmetric molecules having a small degree of net chirality must certainly have been produced by such inevitable deviations from the mean. Furthermore, once produced by such statistical fluctuations, products with slight optical activity could act as 'breeders', which were 'endowed with a power of selecting their own kind of asymmetry from other racemoid substances' (Pearson 1898a). In subsequent months several other authors advanced comparable arguments against Japp's 'vitalism', and supported Pearson's statistical hypothesis in one way or another (Fitzgerald 1898a; Bartrum 1898; Errera 1898; Strong 1898).

The next author to champion the statistical origin of optical activity – over 30 years later – was the well-known American organic chemist Henry Gilman (1929). Gilman used statistical arguments identical to those of Pearson, and further emphasized the fact that on a statistical basis 'the chemist actually does effect unwittingly direct asymmetric syntheses'. He also pointed out an unfortunate human foible: if a chemist were ever by chance to observe the spontaneous appearance of optical activity resulting from a symmetrical chemical process, such a chemist would ('very wisely') display a 'pardonable and understandable reluctance to publish a finding of this type'.

Probably the most detailed theory of the statistical origin of optical activity has been provided by Mills (1932). Mills cited several statistical calculations and concluded, for example, that if 'ten million disymmetric molecules are produced under conditions which favor neither enantiomorph, there is an even chance that the product will contain an excess of more than 0.021 % of one enantiomorph or the other. It is practically impossible for the product to be absolutely optically inactive'. Although ten million molecules is a very small amount of matter indeed, and could be contained within a 'sphere of protoplasm ... 30 μ in diameter', this is 'at least a thousand times as large (in bulk) as the smallest forms of vegetable life', such as green or blue-green algae which 'are only 3 μ in their larger diameter'. Thus Mills concludes that in microscopic forms of life it is 'by no means impossible that the number of molecules of one vitally dominant dissymmetric catalyst ... might be sufficiently low to ensure the probability of ... a small but finite relative excess of one of the enantiomorphous forms, when that quantity of compound was produced under symmetrical conditions'. Mills also amplifies Pearson's 'breeder' concept, arguing that the admittedly minor deviation from numerical equality of such enantiomers produced would 'increase with growth continually, according to a compound interest law until, eventually, the system originally in slight defect was completely swamped by its enantiomorph. From this point of view the optical activity of living matter is an inevitable consequence of its property of growth'.

Gilman's (1929) 'pardonable and understandable reluctance' on the part of chemists to publish their observations of spontaneous asymmetric synthesis was, perhaps unwisely, not shown some years later by Paranjape et al. (1944) in Poona. These investigators reported the 'total asymmetric synthesis' of the optically active natural product santonin, *9*, from optically inactive molecules lacking preexisting molecular asymmetry. They also

concluded that the 'stage at which asymmetric synthesis occurs' was the methylation of the optically inactive β-keto aldehyde, *10*, with sodium and methyl iodide, whereupon the optically active methyl derivative *11* was reportedly produced (optical rotation $[a]_D = -26.22°$). The claim that 'this is the first total asymmetric synthesis, apart from asymmetric syntheses carried out in the presence of polarized light etc.' was somewhat disquieting in chemical circles, and chemists at both Oxford (Cornforth et al. 1944) and at Caltech (O'Gorman 1944) immediately attempted to substantiate the remarkable Indian observations. Both investigators repeated the conversion of *10* into *11*, but in neither case did the product *11* show the slightest optical activity. Puzzled, O'Gorman concluded the report of his negative findings with the admission that 'Considerable speculation on possible sources of their reported results has yielded no reasonable explanation'. Thus the sole report in the literature wherein one might have attributed the origin of optical activity to statistical fluctuations has, perhaps by the nature of statistics, remained unconfirmed.

Due to the large numbers of molecules involved it is, of course, impractical to verify the possible origin of molecular chirality by statistical fluctuations in any ordinary chemical experiment, and criticisms of the theory must rest on theoretical objections. Ritchie (1947) has raised one such objection in pointing out that 'the statistical explanation ... has the weakness that, unless we assume that all living matter can be referred back to one single original microscopic particle with a dextro or a levo asymmetry, this unilateral stereochemistry (of living matter) is difficult to explain'. In other words, if optical activity originated once with molecules produced by a statistical fluctuation favoring (S)-chirality, and these eventually developed into living proto-cells, there is an equal chance that the same process at another time or place must statistically favor molecules of (R)-chirality, which would develop into proto-cells having enantiomeric molecular constituents (e.g. (R)-amino acids). Thus the presence of unique chirality in the primary constituents of living matter constitutes a serious flaw in the theory of the statistical origin of molecular chirality, as indeed it does in some of the theories of physical origin discussed below.

6.   *Physical theories of the origin of molecular chirality*

6.1. *Spontaneous resolution*

Racemic solids are sometimes observed to crystallize from supersaturated solutions in distinctively different fashions, depending upon external conditions (Scheibler 1925; Theilaker 1955). Thus crystallization at a temperature above a characteristic 'transition temperature' may lead to crystals of a *racemic compound*, wherein molecules of each enantiomer occur in equal number in the crystal lattice. Crystallization at a temperature below the transition temperature, on the other hand, may produce a *racemic mixture*, that is, a mixture containing equal numbers of individual crystals of each enantiomer. This phenomenon, which is illustrated graphically in fig. 1, was the basis of Pasteur's (1848) classic discovery of enantiomerism in sodium ammonium tartrate, $NaOOCC*H(OH)C*H(OH)COONH_4$. Below

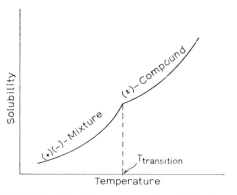

Fig. 1. Transition between racemic modifications.

a transition temperature of 27.2 °C racemic sodium ammonium tartrate crystallized as large enantiomeric crystals whose well-developed hemihedral facets permitted the two crystal forms to be easily distinguished and separated mechanically with tweezers. On doing this, Pasteur noted that one form of crystal proved dextrorotatory and the other levorotatory.

The separation of a racemic compound or mixture into its component enantiomers is a process known as resolution. Resolution, which is the reverse of racemization, may be accomplished by a variety of methods, the most important of which were originally developed by Pasteur. The crystallization of a racemic substance into a mixture of enantiomeric crystals has sometimes been referred to as 'spontaneous resolution' (Scheibler 1925; Ebel 1933; Greenstein 1954; Theilaker 1955), even though subsequent hand sorting is required for actual physical separation of the enantiomers. The overall process, which constitutes Pasteur's classical 'First Method of Resolution', has proved applicable to over a dozen racemic organic sub-stances and inorganic complexes since Pasteur's time. While such phe-nomena have been discussed in detail analytically in terms of appropriate phase diagrams (Jaeger 1930; Ebel 1933), it is sufficient for our purposes to stress that the possibility of such spontaneous resolution of racemates by crystallization depends upon whether the transition temperature (fig. 1) lies within the range of usual laboratory temperatures, and whether one is working at temperatures such that the solubility of the optically active antipodes is less than that of the racemic compound (Werner 1914; Jaeger 1930).

An important practical variation of the spontaneous resolution technique is to bring about by 'seeding' the crystallization of only one enantiomer from a supersaturated solution of a racemate, leaving the other enantiomer behind in the mother liquors. The procedure was first developed by Gernez (1886), who found that seeding a supersaturated solution of racemic sodium am-monium tartrate with either enantiomer of this salt caused selective crystal-lization of that enantiomer which was used as seed. This technique has been employed over a dozen times in the intervening years to 'direct' the course of spontaneous resolution. In some cases, interestingly, a merely iso-morphous substance will act as the requisite seed. Thus Ostromisslensky (1908) found that inoculation of a solution of racemic sodium ammonium tartrate with a seed of $(-)$-asparagine, $HOOCCH_2C^*H(NH_2)CONH_2$, caused preferential crystallization of the $(+)$-tartrate antipode. In addition, he noted that the seeding of solutions of racemic asparagine with glycine, H_2NCH_2COOH, (which is optically inactive) initiated spontaneous crystal-lization of only one of the asparagine enantiomers.

Of greatest interest from our viewpoint are those spontaneous resolutions which involve direct crystallization of one enantiomer from solution without any conscious seeding or inoculation on the part of the experimenter. This mysterious phenomenon was first noted by Pasteur (1852), who found that concentrated solutions of racemic ammonium hydrogen malate, $HOOCCH_2$-$C^*H(OH)COONH_4$, first deposited crystals which resembled the optically active form. Van't Hoff and Dawson (1898) subsequently showed that the product consisted of 3 parts by weight of ammonium hydrogen (−)-malate to 1 part of the (+)-salt, indicating that a partial resolution of the racemic malate had in fact occurred quite spontaneously. Similar truly spontaneous resolutions were subsequently recorded by Jungfleisch (1884), Kipping and Pope (1898a, b, 1909), Soret (1900), Anderson and Hill (1928) and others. Kipping and Pope were the first to suggest that in such spontaneous resolutions dust particles in the laboratory atmosphere might be acting as an asymmetric seed causing preferential crystallization of one enantiomer. They thereupon demonstrated the probable correctness of this hypothesis by conducting crystallizations of sodium ammonium tartrate in the open laboratory and in dust-free desiccators. Thus in the open atmosphere 16 experiments (84%) afforded (+)-tartrate crystals and only 3 (16%) yielded (−)-tartrate, while in the 'dust-free' desiccators 11 out of 19 experiments (58%) produced (+)-tartrate and 8 (42%) gave (−)-tartrate.

The fundamentally statistical nature of such spontaneous resolutions have been convincingly demonstrated by two investigators using sodium chlorate, an inorganic salt which crystallizes in enantiomorphic hemihedral crystal forms whose optical activity depends upon chiral crystal structures alone, and not on molecular chirality. Soret (1900) studied statistically the crystallization of sodium chlorate, and found that crystallization in open containers afforded levorotatory crystals primarily. In sealed tubes, however, 938 such crystallizations gave (+)-crystals in 433 cases (46%), (−)-crystals in 411 cases (44%), and a mixture of both crystal types in 94 cases (10%). Kipping and Pope (1898a) also studied the crystallization of sodium chlorate in dust-protected dishes, counting the individual (+)- and (−)-crystals after crystal growth had proceeded to a size of about 5 mm. They reported that in 46 experiments the percentage of (+)-crystals ranged from a low of 24.14% to a high of 73.36%, but the average number of (+)-crystals deposited overall was almost exactly half (50.82%).

The selected examples above illustrate very briefly the various techniques and results involved with the 'spontaneous resolution' of racemic substances by crystallization. The numbers and types of chemical compounds, both organic and inorganic, which have proved capable of spontaneous resolution by crystallization are imposing indeed (over 63 reports in the literature since 1852), and such techniques have more recently become commercially important. Secor (1963) has reviewed the extensive patent literature on this subject, and he and others have provided comprehensive general reviews as well (Scheibler 1925; Ebel 1933; Greenstein 1954; Theilaker 1955; Terent'ev and Klabunovskii 1959; Klabunovskii 1963).

The spontaneous separation of enantiomers by crystallization in the absence of conscious seeding constitutes, of course, a plausible abiotic mechanism for the origin of molecular chirality. In view of the broad prevalence of the phenomenon, as well as its frequently high degree of effectiveness in producing enantiomeric enrichment, it is not surprising that many authors in this field have specifically championed spontaneous resolution as the most probable mechanism for the primordial origin of optical activity. In answer to the previously discussed 'vitalistic' arguments of Japp (1898), for example, Bartrum (1898), Errera (1898), Pearson (1898b), Kipping and Pope (1898c), and Fitzgerald (1898b) all independently suggested, in one guise or another, that spontaneous crystallization (and not a 'vital force' in living matter) must have been the source of the original optical activity found in nature. Ten years later Ostromisslensky (1908) interpreted his spontaneous resolution of racemic asparagin on seeding with glycine as providing insight into the formation of primeval optically active substances without intervention of 'organized beings'. Jaeger (1930) suggested that such spontaneous resolution by chance seeding might have taken place 'within a vegetable cell whose contained solutions were supersaturated with some organic compound resolvable by spontaneous crystallization' such that 'an optically active antipode might come to be incorporated in the cell, thus determining its one-sided synthesis for all future generations'. Ferreira (1953) argued that his spontaneous partial resolutions of the racemic alkaloids narcotine and laudanosine were 'probably more likely to have occurred in the past than the activation of reactive racemic mixtures by naturally polarized light, even if they lead to a spontaneous decrease in the entropy of the system'. In commenting on Ferreira's paper in the same year, Read (1953) of St. Andrews University pointed out that as early as 1926 he had stressed the possible relationship between such phenomena and the origin of optically active substances in nature as follows: 'A solution of a substance susceptible to spontaneous separation dries up by natural evaporation and deposits a conglomerate of the two kinds of crystals. One of these crystals is introduced fortuitously into an adjacent supersaturated solution of the substance. It is known that under such conditions a crystalline separation may occur of the optical isomer similar to the inoculating crystal. Imagine at this point that the mother-liquor is drained from the crystalline deposit by a further natural process. Two optically active nuclei are now available, namely, the crystalline separation and the substance remaining in the mother-liquor.'

Essentially the same suggestion was advanced by Northrop (1957), who proposed that separate conglomerate crystals from evaporation of a primordial racemic solution 'might easily be scattered by the wind or some other disturbance', and thus provide with 'each individual crystal, separated from the others ... an optically active solution ... available for the first vital reaction'. As late as the mid-1960's Harada (1965) and Harada and Fox (1962), reporting their optical resolution of amino acids and amino acid–copper complexes by the inoculation method, suggested anew that 'In a lagoon or in a shallow pond, some organic compounds might have crystallized out from the primitive sea ('primitive organic soup') by evaporation of water. On crystallization of 'primitive soup', some racemic organic compounds might have been resolved to their optically active isomers'. Harada also argued that dissymmetric inorganic crystals such as quartz might have acted as seed for such crystallizations of organic compounds in the primitive oceans. Unfortunately, actual experiments of this sort have apparently not been reported, wherein chiral inorganic crystals have been used for seeds to induce the spontaneous resolution of organic substances by crystallization.

In 1941 Havinga (1941, 1954) described experiments extending the spontaneous resolution principle which, it was argued, typified 'systems which will have a much higher chance of becoming optically active as a whole than of staying inactive in the course of time'. Such a system might consist of a supersaturated solution of a racemic substance, $(\pm)$-A, having the following properties: (a) crystallization results only in enantiomeric crystals $(+)$-A and $(-)$-A, and no racemic compound $(\pm)$-A; (b) the substance shows rapid catalytic racemization in solution, that is $(+)$-A and $(-)$-A, are in rapid, mobile equilibrium, and (c) formation of crystal nuclei is infrequent; crystal growth, once started, is fast.

When such conditions prevail, Havinga maintained, 'spontaneous asymmetric synthesis' can occur according to a simple process such as the following:

$$
(+)\text{-A} \left\{ \begin{array}{l} \text{supersaturated} \\ \text{in solution} \end{array} \right\} \quad \begin{array}{c} \text{catalyst} \\ \leftrightarrows \\ \text{(fast)} \end{array} \quad (-)\text{-A} \left\{ \begin{array}{l} \text{supersaturated} \\ \text{in solution} \end{array} \right.
$$

$$
\left| \begin{array}{l} \text{chance nucleation;} \\ \text{rapid crystal growth} \end{array} \right.
$$

$$
(-)\text{-A (solid)} \leftarrow\!\!\!\longrightarrow\!\!\rfloor
$$

If the $(-)$-isomer, for example, undergoes chance nucleation in solution, its rate of crystallization will be rapid and the solution will become depleted with respect to the $(-)$-isomer. Due to rapid catalytic racemization, however, $(+)$-isomer will become converted into $(-)$-isomer to compensate for its depletion, thus providing additional $(-)$-isomer for crystallization. Thus, with gradual evaporation of the solvent, all of the original racemate might in principle crystallize as the $(-)$-isomer, provided that the $(+)$-isomer has not itself undergone chance nucleation in the meantime.

As an experimental demonstration of such a process, Havinga studied the crystallization of methylethylallylanilinium iodide, *12*, from chloroform solution. The asymmetric quaternary nitrogen atom in *12* can undergo inversion (leading to racemization) by means of the following reversible dissociation into optically inactive allyl iodide, *13*, and N-methyl-N-ethylaniline, *14*.

When *12* was allowed to crystallize from supersaturated chloroform solutions, the crystals proved to be highly active in many cases, and the amounts of material obtained indicated that one enantiomer had converted into the other during the crystallization process. Havinga argued that these results 'demonstrate the possibility of the spontaneous generation of optically active material starting from an inactive, closed system without interference of any directing, dissymmetric agency'. More recently Calvin (1969) has referred to such processes as 'stereospecific autocatalysis'. Unfortunately, Havinga's experiments have never been repeated, nor have his original novel suggestions for the occurrence of 'spontaneous asymmetric synthesis' been extended to other experimental systems.

Thus we see that it has been abundantly demonstrated experimentally that primeval optically active substances could have originated by the spontaneous resolution of racemates produced in the course of chemical evolution. The process is a highly probable one and, being quite efficient, could readily make available substantial quantities of separated enantiomers for subsequent asymmetric synthesis leading to living proto-cells. Such a theory of the origin of molecular chirality in nature, however, suffers the same flaw inherent in the theory of statistical fluctuations described above.

That is, since there is an equal chance for either of two enantiomers to crystallize spontaneously, one would again expect to find both enantiomers involved in the primary constituents of living matter.

6.2. Asymmetric inclusion compounds

Inclusion compounds are molecular aggregates wherein one or more 'guest' molecules are contained within the open spaces in a crystalline or molecular framework made up of 'host' molecules. The open spaces in the host molecule framework may be in the form of central channels, in the form of cages, or in the form of sandwich layers, and the forces of attraction between the host and guest molecules are at best the weak ones of hydrogen bonding or induced dipole interactions, and not the stronger forces of ordinary chemical bonds. Indeed, the host and guest molecules of inclusion compounds are ordinarily quite incapable of uniting chemically. The existence and stability of inclusion compounds depends instead upon critical relationships in the size and shape of the guest molecules with respect to the geometry of the cavity in the host molecule framework. That is, the guest molecule must be small enough and of the right shape to be accommodated in the host framework, yet not so small as to be able to escape from it. Most inclusion compounds are stable only as crystalline entities, and processes such as dissolving or melting which destroy the crystal structure will liberate the included component. The known types of inclusion compounds and their chemical utility have been reviewed extensively (Powell 1950; Schlenk 1951a; Cramer 1954, 1955, 1956; Swern 1957). In this section we shall examine briefly only those aspects of inclusion compounds which have bearing on the origin of molecular asymmetry.

Urea inclusion compounds. Bengen (1940) discovered that urea yields crystalline inclusion complexes with a variety of straight chain aliphatic compounds, provided their molecules contain 6 or more carbon atoms. Subsequent investigations a few years later by Schlenk (1949), Bengen and Schlenk (1949), and others extended the range of urea adducts (Redlich et al. 1950a), included thiourea as host substance (Redlich et al. 1950b; Schlenk 1951b), and established the crystal structure of such inclusion compounds (Smith 1950, 1952). Urea inclusion compounds crystallize as hexagonal lattices of urea molecules surrounding parallel channels about 5–6 Å in diameter, within and along which the included guest molecules reside (Cramer 1955; fig. 2). Note that the urea molecules in the hexagonal lattice, as shown by the dark circles, may be arranged in the chiral sequences of either a right-handed or left-handed helix.

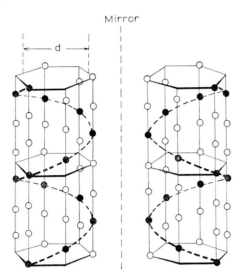

Fig. 2. Crystal lattice in urea and thiourea inclusion compounds. Only host molecules ○● are schown. Diameter (d) of urea is 5–6 Å; of thiourea 7–8 Å.

The right- or left-handed chirality of the hexagonal crystal lattice of urea inclusion compounds permits a novel use of such complexes, which in turn constitutes a possibility for the origin of molecular chirality in nature. Formation of the urea inclusion compound of any suitable guest substance ordinarily affords crystals whose lattices are of either right-handed, (R) lattice, or left-handed, (S) lattice, helicity exclusively. Which lattice forms initially is determined by chance, but the original crystal nuclei act as seed for the remainder of the product. Now if a racemic substance (±)A, is included as guest molecules in a urea lattice, 4 stereochemically different crystalline products, (1) through (4), may arise from the following equilibria:

Crystalline product   *Components in solution*

(1) (+)A-(R) lattice ⇌
(2) (+)A-(S) lattice ⇌   urea + (+)A
(3) (−)A-(R) lattice ⇌
(4) (−)A-(S) lattice ⇌   urea + (−)A

These products have the properties of enantiomers or diastereomers. That is, pairs ((1)–(4)) as well as (2)–(3) represent enantiomers which have identical solubility, while (1)–(3) and (2)–(4) represent diastereomers which have different solubilities. If the inclusion compound by chance crystallizes in the (R) lattice, the products are limited to (1) and (3). If (1) happens to be the least soluble diastereomer it will crystallize preferentially, leaving (3) in solution. Since the solubilities of (1) and (3) are different, and particularly if the amount of urea present is insufficient to react with all of the (±)A, (1) and (3) will crystallize in unequal amounts, with (1) predominating. If finally the product is decomposed, a spontaneous partial resolution of (±)A into (+)A will have been achieved.

The principles of this novel resolution technique were first described and put in o practice by Schlenk (1952a, b), who resolved racemic 2-chlorooctane, $C_6H_{13}C^*H(Cl)CH_3$, into 95.6% of the (+)-isomer by repeated formation of its urea complex, using the initially formed nuclei as seed crystals on each repetition. Schlenk (1957, 1960a, b) later obtained several patents for improvements in this basic technique, and Klabunovskii et al. (1960) in Moscow have also reported the partial resolution of racemic hydrocarbons through such urea complexes. Lastly, Schlenk (1955, 1960c) has claimed to be able to employ optically active compounds of known chirality to establish the lattice chirality of urea inclusion complexes, and in turn to use the latter to correlate the chiralities of optically active compounds of unknown configuration.

The spontaneous resolution of racemic substances by the ready formation of urea inclusion compounds, of course, clearly represents a possible mechanism for the prebiotic appearance of optical activity on the surface of the earth, a suggestion originally made by Schlenk (1952a). This mechanism is made even more plausible by the fact that urea has been found abundantly among the products resulting on subjection of aqueous solutions or gaseous mixtures of such 'primordial' compounds as ammonia, hydrogen, methane, water and hydrogen cyanide to energy sources such as heat (Lowe et al. 1963), electrical discharge (Miller 1957a, b; Miller and Urey 1959), and electron irradiation (Palm and Calvin 1962).

Tri-o-thymotide inclusion compounds. Baker et al. (1952) observed that dehydrating agents such as phosphoric anhydride P_2O_5 act upon 2-hydroxy-6-methyl-3-isopropylbenzoic acid (o-thymotic acid), *15*, with formation of the dimeric cyclic ester *cis*-di-o-thymotide, *16*, and the trimeric cyclic ester tri-o-thymotide, *17*. Dipole moment studies on *17* as well as on simpler analogs revealed that these 12-membered cyclic esters have the planes of

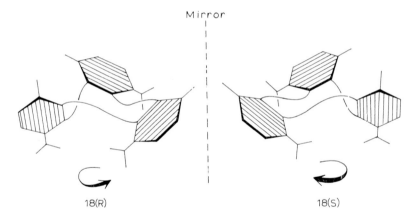

15   16   17

their three benzene rings intersecting at the apex of a trigonal pyramid, causing the molecule to have a propeller-like stereochemical structure, (Baker et al. 1952). Such a propeller structure may have either right-handed, *18*(R), or left-handed, *18*(S), chirality, as shown in the simplified perspective drawings of fig. 3, which emphasize only the planes of the benzene rings.

Mirror

18(R)   18(S)

Fig. 3. Chiralities of tri-*o*-thymotide 'propeller structures'.

Molecular models indicate that the enantiomeric molecular structures *18*(R) and *18*(S) may be readily interconverted by slight rotations about the single bonds making up their 12-membered central rings. Baker et al. (1952) also found that in several inert solvents *17* crystallized as stable solvated inclusion complexes containing 0.5 to 1 mole of solvent per mole of *17*.

A few months after the structure elucidation of *17*, Newman and Powell (1952) made the significant finding that the crystalline inclusion compounds of *17* were optically active, although *17* itself was not. That is, *17* alone crystallized as a racemate (equal amounts of *18*(R) and *18*(S)), but in the presence of an included guest molecule the crystal lattice of the host had a uniform chirality (*18*(R) or *18*(S)). Again the particular chirality which the crystal lattice develops is determined by chance, but once it is established in a seed nucleus it prevails throughout the remainder of the crystallization. Thus by slow cooling of its solution, single crystals of the benzene inclusion compound of *17* weighing up to 1 g could be obtained, which were sometimes dextrorotatory and sometimes levorotatory in chloroform solution ($[a]_D \pm 83°$). For such uniform chirality of the tri-*o*-thymotide lattice in its inclusion compounds, it is obvious that the enantiomeric molecular conformations *18*(R) and *18*(S) must be readily interconvertible in solution at room temperature, a conclusion which also follows from the rapid racemization of such *17* adducts in solution.

Powell (1952) was also able to achieve the partial resolution of racemic 2-bromobutane, $CH_3C^*H(Br)CH_2CH_3$, by means of its inclusion compound with *17*. When a single crystal was grown from a seed crystal of the 2-bromobutane inclusion compound, the final adduct showed a large levorotation due to the tri-*o*-thymotide lattice. This rotation decayed rapidly in solution through subsequent racemization of the *17* lattice, leaving a much smaller permanent levorotation due to the partially resolved 2-bromobutane. The principles involved in this resolution are precisely similar to those previously described for urea inclusion compounds, that is, one enantiomer of the racemic guest molecule forms a more stable inclusion compound with one enantiomer of the host lattice, such that when the latter crystallizes spontaneously the former is automatically selected. Powell's (1956) analogy for the phenomenon is apt: 'Such behavior is to be expected when the empty space (in the lattice cavity) is, like the space in a boot, not superimposable on its mirror image and the enclosed molecule, like the normal foot, fills most of the space'. Powell (1952) also suggested that 'It is possible that somewhat similar processes may be responsible for the original dissymmetry in the chemistry of living matter and for its perpetuation'.

It is interesting that Schlenk's initial disclosure of spontaneous resolution through urea inclusion compounds is dated May 12, 1952, whereas Powell's disclosure of similar resolutions with tri-*o*-thymotide is dated May 20, 1952. This novel resolution method, the first new crystallization procedure since Pasteur's three classical methods, thus appears to have been discovered simultaneously by the two investigators.

Cyclodextrin inclusion compounds.   a-, β- and γ-cyclodextrins (Schardinger dextrins) are formed by the degradation of starch amylose with the amylase enzyme of *Bacillus macerans* (French 1957). These cyclodextrins are small cyclic oligosaccharides having cyclic chain lengths of 6, 7 and 8 a-D-glucopyranose units joined head to tail through glycoside linkages, *19*, as in starch. The schematic structures and the interior hole diameters of the three cyclodextrins (Cramer 1955, 1956) are shown in fig. 4.

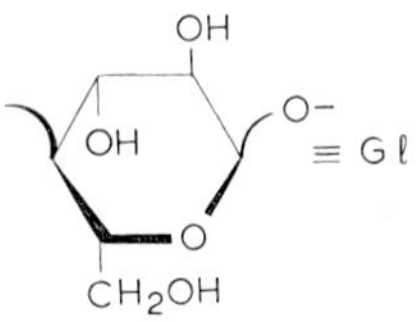

19

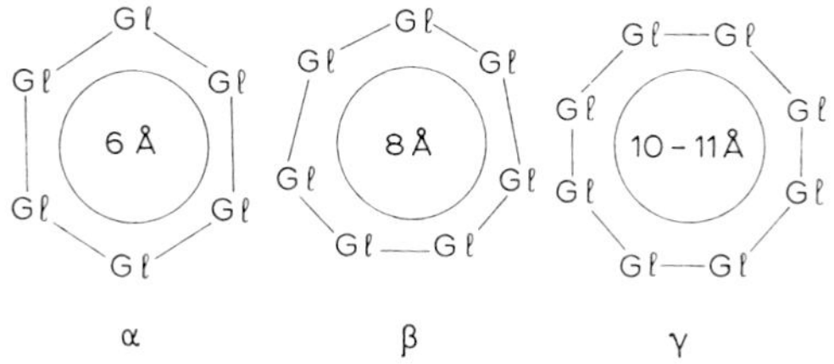

α   β   γ

Fig. 4. Structures of the cyclodextrins (Gl shown in 19).

Owing to their doughnut structures, the cyclodextrins can form inclusion compounds with a wide variety of organic and inorganic molecules. Small guest molecules become included within the central doughnut hole of a-cyclodextrin, while increasingly larger molecules require first the β- and finally the γ-cyclodextrin as the host. Thus in contrast to urea and tri-*o*-thymotide complexes, the inclusion cavity of a cyclodextrin lies within each molecule itself, and not within a particular crystal lattice which the molecules form. For this reason cyclodextrin inclusion compounds may exist not only in the solid state but also in solution.

In view of the spontaneous resolutions with urea and tri-*o*-thymotide inclusion compounds, it is not surprising that cyclodextrins are found to distinguish not only between the structural features of guest molecules, but also between their stereochemical features. Thus partial resolutions of such racemic substrates as mandelic and phenylbromoacetic esters ($C_6H_5C^*H$-$(OH)COOC_2H_5$; $C_6H_5C^*H(Br)COOC_2H_5$) have been achieved by formation of their inclusion compounds with cyclodextrins (Cramer and Dietsche 1959a). In contrast to the previous urea and tri-*o*-thymotide spontaneous resolutions, however, such cyclodextrin resolutions cannot be looked upon as 'abiotic' processes, since the host molecules of the inclusion complexes are themselves made up of preexisting asymmetric molecular components, namely D-glucose units.

Cramer and Dietsche (1958, 1959b) have also observed that asymmetric syntheses and asymmetric degradations may be carried out if certain symmetrical or racemic reactant molecules are held captive within cyclodextrin complexes while undergoing reaction. As before, such processes cannot be regarded as 'abiotic'. If, on the other hand, analogous asymmetric syntheses or degradations could be achieved using symmetrical guest molecules trapped within urea or tri-*o*-thymotide inclusion complexes, a new potential mechanism for the abiotic origin of molecular chirality might be demonstrated.

While not identical, an abiotic origin closely related to that of the above suggestion, however, has in fact been demonstrated recently. Penzien and Schmidt (1969) found that when homogeneous crystals of 4,4'-dimethyl-chalcone, *20*, which crystallizes in enantiomeric crystal forms, were allowed to react in the solid state with bromine vapor, the dibromo product *21* was optically active to an extent ($[a]_D = 9.8°$) indicating an 'average optical

yield of 6%'. In this novel absolute asymmetric synthesis the chirality of the crystal lattice occupied by the symmetrical reactant molecules of *20* has imposed a preferred chirality on the asymmetric molecules of the product, *21*.

6.3. *Asymmetric adsorption on chiral adsorbents*

Adsorption is a heterogeneous equilibrium process in which molecules distribute themselves between a gaseous or liquid phase and a suitable adsorbing surface. The process involves ionic, dipolar or Van der Waals' forces of attraction between the otherwise chemically inert surface of the adsorbent and the molecules of the adsorbate (adsorbed substrate), which may be in the gaseous, liquid, or dissolved state. Variations in the strengths of these attraction forces from one adsorbate to another with respect to a given adsorbent are, of course, the fundamental basis for the widely applied and powerful analytical techniques of adsorption chromatography. If the surface itself of an adsorbent has an intrinsic chirality, or is made up of asymmetric molecules of a single chirality, moreover, one expects that the strength of the attraction forces between this chiral adsorbent and the two enantiomers of a racemic adsorbate would be slightly different. Since geometrical and steric factors interact in the fitting together of adsorbent and adsorbate during the adsorption process, an optically active adsorbent should therefore in principle adsorb one enantiomer of a racemic adsorbate more tenaciously than it does the other, and thereby permit optical resolution of the adsorbate. Thus it is not surprising that the literature records numerous examples of the resolution of racemic compounds by the application of adsorption techniques.

Willstätter (1904) was apparently the first to explore the possibility that optical antipodes might be adsorbed (or absorbed) to different extents by proteins. He allowed wool to interact with aqueous solutions of several racemic alkaloids, but in no case did he observe selective adsorption of one enantiomer. Fifteen years later, however, Porter and Hirst (1919) synthesized several racemic dyestuffs, and reported their partial resolution, with concomitant dyeing, by wool protein. Other workers subsequently reported additional examples of the selective adsorption of one enantiomer of various racemic dyestuffs by wool (Ingersoll and Adams 1922; Porter and Ihrig 1923; Morgan and Skinner 1925), but in other similar cases such selective dyestuff adsorption could not be duplicated (Brode and Adams 1926; Brode and Brooks 1940). Selective adsorption of a number of non-dyestuff racemic substrates by wool and other proteins, however, appears to be well documented (Martin and Kuhn 1941; Bradley et al. 1951, 1953, 1954a, b). Other optically active organic adsorbents besides proteins have also been employed, often with striking effectiveness, in the resolution of both organic and inorganic racemates. Such adsorbents include lactose (Henderson and Rule 1938; Lecoq 1943; Prelog and Wieland 1944; Dimodica and Angeletii 1952; Leonard and Middleton 1952), cellulose (Kotake et al. 1951; Klingmüller and Maier-Sihle 1957; Lederer and Lederer 1957; Contractor and Wragg 1965), starch (Krebs et al. 1954a, b, 1955, 1956a, b, 1957a, b; Ohara et al. 1962, 1964; Gillard et al. 1967), and even optically active synthetic polymers (Pino et al. 1962, 1965; Manecke and Lamer 1968).

An interesting variation in the preparation of chiral adsorbents for chromatographic resolution involves the use of achiral adsorbents such as alumina or silica, which have been appropriately altered by pretreatment with an optically active agent. Such experiments stem from the work of Dickey (1949, 1955), who prepared silica gel adsorbents in the presence of several specific dyestuffs, and then demonstrated that these adsorbents showed enhanced adsorption capability for the dye molecule in whose presence they had been prepared. Dickey suggested at the outset the possible applicability of such 'tailored' adsorbents to the problem of optical resolution. This suggestion received experimental attention three years later by Curti and Colombo (1952), who succeeded in the partial resolution of several racemic organic acids by chromatography on silica gel columns prepared in the presence of optically active antipodes of these acids. Numerous additional optical resolutions with analogous 'stereospecifically tailored' chromatographic adsorption columns have been described in the more recent literature (Beckett and Anderson 1957, 1959, 1960; Karagounis et al. 1959; Klemm and Reed 1960; Klabunovskii et al. 1961).

Important though the above resolutions are, however, they are of only secondary interest from our viewpoint, since the chirality of the natural or synthetic adsorbents employed was in all cases determined by the preexisting asymmetric carbon atoms present in the adsorbent molecules or in the agents used to 'tailor' them. If, on the other hand, analogous resolutions could be demonstrated using inorganic chiral adsorbents, a truly abiotic process would be at hand for the potential spontaneous generation of optically active substances by resolution through differential adsorption.

Several dozen inorganic substances are known which are optically active in the crystalline state (but optically inactive when molten or dissolved) as a result of the unique and homogeneous chirality of their crystal structures (Lowry 1935). Any of these, in finely powdered form, might in principle be a candidate for an abiotic chiral adsorbent, which could act as a resolving agent precisely as the optically active organic adsorbents described above. With one exception, however, only *dextro*- and *levo*-quartz have actually been investigated experimentally in this context.

The first use of quartz as an asymmetric adsorbent dates to 1935, when Tsuchida et al. (1935) of Osaka Imperial University observed the greater effectiveness of *d*-quartz over *l*-quartz in adsorption ability toward one enantiomer of a number of racemic complex cobalt salts, such as $[Co(en)_3]Br$ and $[Co(en)_2Cl_2]Cl$, where (en) is ethylenediamine. A year later these same authors (Tsuchida et al. 1936) used partial chromatographic resolution on quartz to establish the stereochemical configuration of chlorobisdimethyl-glyoximoamine-cobalt, $[Co(dg)_2NH_3Cl]$, a non-electrolyte complex which could not be resolved by conventional methods. They also studied samples of right- and left-handed quartz crystals from several localities, finding that the right-handed quartz invariably adsorbed the (+)-enantiomer of this complex preferentially, and the left-handed quartz the (−)-enantiomer. In one experiment with a left-handed quartz sample from Brazil, however, no rotation could be observed in the solution containing the product. The authors therefore concluded that the quartz sample itself was racemic, in spite of its asymmetric external appearance, and suggested that the utmost caution must accordingly be used in the choice of quartz crystals for such resolution purposes.

Such observations were confirmed two years later by Karagounis and Coumoulos (1938) in Athens, who percolated a solution of triethylenedia-mine chromichloride, $Cr(en)_3Cl_3 \cdot 3\frac{1}{2} H_2O$, through a chromatographic column containing quartz powder made from *d*- or *l*-crystals 'activated' by heating, then collected the eluted solution in successive fractions. In a typical experiment with *l*-quartz the 5 eluted fractions had rotations, respectively, of $-0.058°$, $-0.121°$, $-0.018°$, $+0.046°$, $+0.034°$, indicating that the (+)-enantiomer was more strongly adsorbed. With *d*-quartz the opposite was observed, the initially eluted fractions being dextrorotatory. In the intervening years optically active quartz has been employed as an asymmetric adsorbent on a number of additional occasions to bring about the partial resolution of racemic inorganic complexes of such metals as beryllium, chromium, cobalt and platinum (Bailar and Peppard 1940; Schweitzer and Talbott 1950; Kuebler and Bailar 1952; Busch and Bailar 1953, 1954; Nakahara and Tsuchida 1954; Das Sarma and Bailar 1955).

The chromatographic resolution of racemates using optically active quartz has not been limited exclusively to inorganic metal complexes. Klabunovskii and Patrikeev (1951), for example, passed racemic 2-butanol through a column of *l*-quartz, percolating the adsorbate downward against a rising stream of its vapor. They reported obtaining, after 30 passes, a product having a rotation of $0.05°$, corresponding to 0.22% resolution of the racemate. They also showed that active amyl alcohol, as a dilute solution in decalin, was 2.3% more adsorbed by *d*-quartz than by the same surface area of *l*-quartz.

The only optically active inorganic substance other than quartz which appears to have been tested as a chromatographic adsorbent for optical resolution purposes has been sodium chlorate. Following the earlier reports of successful resolutions using quartz, Ferroni and Cini (1960) employed carefully ground and dried sodium chlorate crystals to resolve racemic bis-(benzoylacetonato)-beryllium-(II), which had been previously resolved chromatographically by Busch and Bailar (1954) using quartz. Observed rotations as high as $0.071 \pm 0.003°$ were noted using a photoelectric polari-meter.

As mentioned above, the resolution of racemic substances by selective adsorption of one antipode on the surface of a chiral inorganic adsorbent would constitute a plausible mechanism for the abiotic genesis of optically active compounds in nature. This viewpoint has, in fact, been championed on several occasions. As early as 1938 Karagounis and Coumoulos, in reporting their partial resolution of a chromium complex by adsorption on quartz, suggested that such processes might have been responsible for the formation of the first optically active substances in nature, by 'many successive adsorptions and elutions of a racemic compound on optically asymmetric surfaces of minerals'. Bernal (1951) has more recently emphasized the probable importance of clay deposits as concentrating agents and catalytic surfaces for mediation of the primitive processes of chemical evolution, and has pointed out that quartz can occur along with clay. Regarding molecular chirality he argued: 'it seems more plausible that the particular twist was given at one time by the preferential adsorption of a pair of asymmetric molecules on quartz, and ..., once one asymmetric isomer was produced, even locally, it would produce a situation in which ultimately only one kind could be formed'.

Since *d*- and *l*-quartz both occur in nature, however, and presumably to approximately equal extents (Lemmlein 1939; see also Goldschmidt 1952), we would again expect that a particular molecular chirality engendered in one locality by selective adsorption on *d*-quartz would be counterbalanced by the opposite chirality resulting from selective adsorption on *l*-quartz in another locality. Asymmetric adsorption as a hypothesis for the origin of molecular chirality thus suffers the same flaw we have encountered in other theories, namely, failure to account for the uniqueness of chirality observed among the primary constituents of living matter.

Furthermore, serious experimental doubt has recently been cast by Amariglio et al. (1968a) on the dozen or so earlier reports, summarized briefly above, claiming resolution of racemic inorganic or organic substances by selective adsorption on quartz. These investigators attempted to duplicate 4 of the quartz column resolutions previously reported in the literature, employing very sensitive equipment for measuring optical rotation and taking extreme precautions to eliminate optically active artifacts (residual quartz dust in suspension). It is therefore impressive that no optical activity beyond experimental error was observed in any of the eluate fractions from any of the 4 resolutions attempted, one of which was even conducted at dry ice temperature.

Their failure to duplicate earlier experimental claims thereupon prompted

Amariglio et al. (1968a) to undertake a critical analysis of these previously reported resolutions. They point out, for example, that most of the optical rotations previously observed were so small as to be within experimental error of zero, and that a number of the claims as to which enantiomer was preferentially adsorbed were actually contradictory. They emphasize that most of the inorganic complexes studied were highly colored, making polarimetric readings on their solutions very difficult and open to error, and that the dilutions necessary to permit such readings caused spuriously high values for observed rotations corrected for dilution. They calculate (for studies where sufficient experimental data were given) that the estimated surface areas of the quartz samples employed would never permit the rotations observed, even assuming adsorption over the total available surface and assuming 100% stereoselectivity for the adsorption of one enantiomer. Finally, they suggest that the small rotations observed in earlier studies may have been due to previously overlooked causes, such as dichroism or double refraction caused by minute quartz particles which were washed from the columns and suspended in the liquid samples being observed.

It is hard to find minor flaws in the paper of Amariglio et al. (1968a), since their work is unquestionably the most careful to date. Yet one feels uneasy that the positive findings of a dozen earlier investigations over a period of several dozen years should be so hastily and summarily invalidated. Perhaps the most serious inadequacy in the French study is the lack of variety in the quartz samples employed. These were trimmings from synthetic samples furnished by the National Center for the Study of Telecommunications, and were not characterized further than *droit* and *gauche*. It will be recalled that as early as 1936 Tsuchida et al., noting that a particular Brazilian sample of left-handed quartz showed no preferential adsorption towards either enantiomer of a cobalt complex under conditions where other samples of right- or left-handed quartz did, cautioned future workers as to the care necessary in the selection of the quartz adsorbents to be employed. One wishes that this care were more evident in the work of Amariglio et al. Thus it would seem safe to conclude that the possibility of abiotic resolution by means of preferential adsorption of one antipode on the surface of optically active quartz may still remain an open question.

6.4. *Asymmetric catalytic processes on chiral catalysts*

Heterogeneous catalytic reactions involve the formation and/or rupture of chemical bonds within one or more reacting substrates which are adsorbed on the surface of a solid catalyst. The above question of asymmetric adsorption is thus intimately involved with the phenomenon of *asymmetric heterogeneous catalysis*, wherein a suitable chiral catalytic surface mediates the degradative or synthetic reaction of an optically inactive reactant to form an optically active product. Under proper conditions asymmetric heterogeneous catalysis might provide yet another plausible mechanism for the origin of molecular chirality.

The first asymmetric synthesis by way of a heterogeneous catalytic reaction was reported by Bredig and Gerstner (1932), who prepared a solid catalyst of 'diethylaminocellulose' fibers by the suitable chemical treatment of purified cotton. This catalyst, like natural emulsion, brought about the predominant synthesis of (−)-mandelonitrile, PhC*H(OH)CN, on the reaction of HCN with benzaldehyde, PhCH = 0. The mandelic acid, PhC*H(OH)COOH, obtained on subsequent hydrolysis of the nitrile product had an optical rotation (+33°) indicating a 22% excess of one enantiomer. In a later study (Bredig et al. 1935) other amines, similarly coupled to cellulose fiber, were also found capable of catalyzing the synthesis of (−)-mandelonitrile as well as other optically active hydroxynitriles.

In 1956 Akabori et al. initiated an extensive study of asymmetric catalytic hydrogenations using metal catalysts supported on protein fibers. Their initial catalyst, prepared by the reduction of palladium chloride in the presence of silk fibroin fibers, was employed to convert ethyl acetoximinoglutarate, *22*, into glutamic acid, *23*; ethyl *a*-acetoximinophenylpropionate, *24*, into phenylalanine, *15*; 4-benzylidene-2-methyloxazol-5-one, *26*, into phenylalanine; and *a*-benzil dioxime, *27*, into 1,2-diphenylethylenediamine, *28*. In each case the product was optically active, with the specific rotation indicated in the equations below. Palladium supported on acetylated silk fibroin was subsequently found to be a slightly more effective asymmetric catalyst for such reactions (Akabori et al. 1957). It is interesting that such palladium–silk catalysts failed to show their characteristic asymmetric hydrogenation capability if the silk fibroin had been previously dissolved in cupra-ammonium solution and then was reprecipitated (Akabori 1959). This suggests that the secondary structure of the protein must be intact to allow for its stereochemical effectiveness in such protein–metal catalysts. A somewhat similar type of catalyst for asymmetric hydrogenations was prepared by Balandin et al. (1959) using platinum or palladium in a colloidal state in gum Arabic.

Fukawa et al. (1962) in Akabori's laboratories later developed still another type of asymmetric hydrogenation catalyst. Following a previous lead of

$$\underset{22}{\text{EtOOCCH}_2\text{CH}_2\overset{\overset{\displaystyle \text{NOAc}}{\|}}{\text{CC}}\text{OOEt}} \xrightarrow[\text{2) Hydrol.}]{\text{1) H}_2/\text{Pd}-\text{silk}} \underset{23,\ [\alpha]_D^{15}\ +\ 2.25°}{\text{HOOCCH}_2\text{CH}_2\overset{\overset{\displaystyle \text{NH}_2}{|}}{\text{C}}\text{*HCOOH}}$$

$$\underset{24}{\text{PhCH}_2\overset{\overset{\displaystyle \text{NOAc}}{\|}}{\text{C}}\text{COOEt}} \xrightarrow{\quad ''\quad} \underset{25,\ [\alpha]_D^{15}\ +\ 9.25°}{\text{PhCH}_2\overset{\overset{\displaystyle \text{NH}_2}{|}}{\text{C}}\text{*HCOOH}}$$

$$\underset{26}{\text{PhCH}=\overset{\overset{\displaystyle }{|}}{\underset{\underset{\displaystyle \text{CO-O}}{|}}{\text{C}}}\text{—N}\diagdown_{\text{CCH}_3}} \xrightarrow{\quad ''\quad} \underset{25,\ [\alpha]_D^{20}\ +12.5°}{\ }$$

$$\underset{27}{\overset{\displaystyle \text{HO—N\ \ N—OH}}{\underset{\displaystyle }{\text{PhC—CPh}}}} \xrightarrow{\quad ''\quad} \underset{28,\ [\alpha]_D^{15}\ +431°}{\text{PhC*HC*HPh}}$$

Isoda (1958), Fukawa and coworkers modified the ordinary hydrogenation catalyst, Raney nickel, by treating it with solutions of various amino acids, and found that such modified catalysts had the ability to catalyze the asymmetric reduction of ketones to optically active alcohols (Izumi et al. 1963a). For example, treatment of Raney nickel with 2% L-glutamic acid solution at 0° gave a catalyst which catalyzed the hydrogenation of ethyl acetoacetate, $CH_3COCH_2COOC_2H_5$, to ethyl β-hydroxybutyrate, CH_3C*$H(OH)CH_2$-$COOC_2H_5$, having a specific rotation of $[a]_Ds\ -2.00°$. Such asymmetric hydrogenations were later extended to include catalysts modified with optically active samples of other amino acids (Izumi et al. 1963b) and hydroxy acids (Tatsumi 1964). These findings have been recently confirmed by Petrov et al. (1966, 1967) in Moscow, who make the assumption that the modification of the catalyst consists in the formation of a surface complex between the nickel and the optically active modifying agent. The bound agent thereupon alters in an asymmetric manner the active sites of the catalyst responsible for carbonyl reduction, thus bringing about the asymmetric reduction of such compounds.

Again, while intrinsically interesting, the above examples of asymmetric heterogeneous catalysis concern us only peripherally, since their success depends again upon the ability of preexisting asymmetric organic molecules to provide a chiral catalyst surface capable of mediating asymmetric synthesis. For truly abiotic asymmetric catalysis, we must turn to analogous processes involving the asymmetric surfaces of inorganic catalysts. Here, as was the case with asymmetric adsorption, optically active quartz has been the mineral receiving essentially exclusive experimental attention.

In the same year that Bredig and Gerstner (1932) conducted their first asymmetric synthesis using an organic heterogeneous catalyst, Schwab and Rudolph (1932) in Munich reported the asymmetric decomposition of racemic 2-butanol, $CH_3CH_2C^*H(OH)CH_3$, using d- or l-quartz covered with a very thin coating of copper. The vapors of the racemic alcohol were passed over the heated catalyst, whereupon gas evolution (H_2 and C_4H_8) commenced and the unreacted alcohol, collected by condensation of the vapors, showed optical activity up to $0.10-0.13° \pm 0.03°$. Thus one enantiomer of the racemic alcohol had been catalytically decomposed (by dehydration and dehydrogenation) more rapidly than the other. Copper coated l-quartz produced a levorotatory product, and d-quartz/Cu a dextrorotatory product. Optically inactive quartz, also copper coated, gave an optically inactive product, as did also optically active quartz/Cu when used with isopropyl alcohol, $(CH_3)_2CHOH$, which is incapable of optical activity. Shortly thereafter Schwab et al. (1934) extended their observations to include catalysts consisting of quartz covered with 'uni-atomic' layers of both platinum and nickel, and noted further that thicker layers of metal on the quartz surface gave less effective asymmetric decompositions.

Tsuchida et al. (1936) were unable to duplicate these catalytic experiments of Schwab and coworkers, even while reporting positive results with asymmetric adsorption on quartz. Stankiewicz (1938), however, was able to repeat the Schwab experiments, and extended them to other racemic alcohols as well. The catalysts employed, prepared with 100–200 atomic layer thicknesses of Cu, Ni or Pt on quartz, showed greater stereoselectivity than those with uni-atomic layers, in contradiction to the report of Schwab. Thus Stankiewicz reports that the decomposition of 2-butanol on d-quartz/ Cu at $530°C$ yielded a product having a rotation of $+0.25°$.

An extensive series of studies of asymmetric heterogeneous catalysis using metal-coated quartz was inaugurated by Terent'ev et al. (1950) at Moscow State University. At the outset, these workers confirmed and extended the earlier reports of the asymmetric decomposition of 2-butanol by quartz catalysts covered with copper, palladium, nickel or silver. Optical rotations as high as $+0.18°$ were observed for the undecomposed 2-butanol product, and the catalysts generally yielded products having a rotation identical in sign to that of the quartz employed.

Terent'ev et al. (1950) have also extended their study of metal–quartz asymmetric catalysis to a number of other degradative as well as synthetic reactions, reporting positive findings in each case. These reactions, illustrated along with maximum product rotations in the following equations, include: (1) the asymmetric isomerization of propylene oxide, *29*, where optical activity is developed in the unreacted reactant; (2) the dismutation of 2-methylcyclohexanone, *30*, into 2-methylcyclohexanol, *31*, and *o*-cresol, *32*, wherein a racemic substance is destroyed with the formation of new centers of asymmetry; (3) the catalytic hydrogenations of *a*-pinene, *33*, and; (4) of ethyl *a*-phenylcinnamate, *34*, which are straightforward syntheses producing a new asymmetric carbon atom.

1) $(\pm)-CH_3C^*H-CH_2 \longrightarrow CH_3COCH_3 + CH_3CH_2CHO$

    29 (−0.055°)

2) 3 (±)

    30 → 2 *31* (−0.042°) + *32*

3) (±)

    33 + H₂ → (+0.046°)

4) $PhCH=CCOOEt + H_2 \longrightarrow PhCH_2C^*HCOOEt$

    34   (−0.084°)

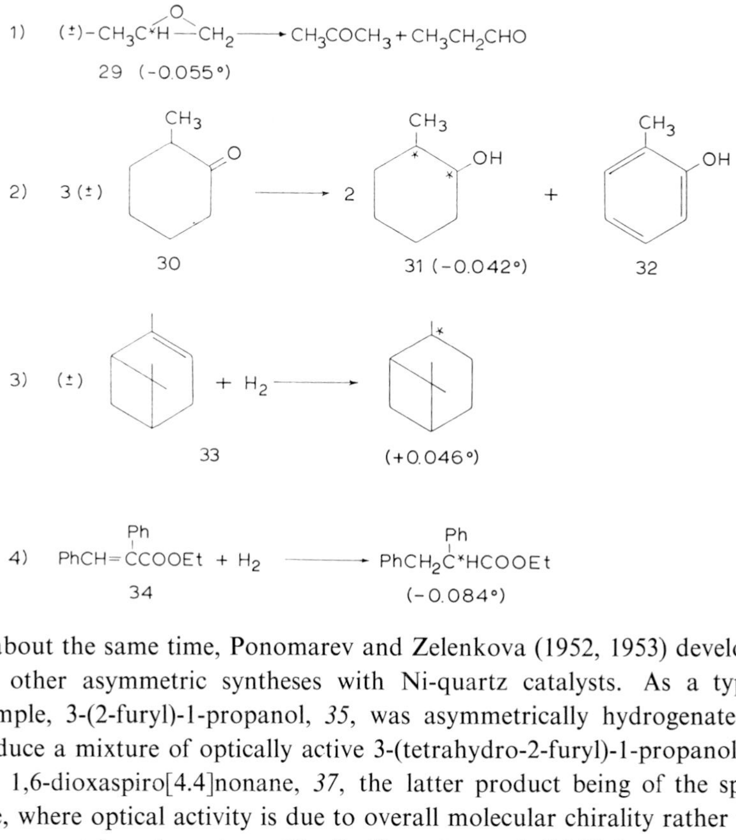

At about the same time, Ponomarev and Zelenkova (1952, 1953) developed still other asymmetric syntheses with Ni-quartz catalysts. As a typical example, 3-(2-furyl)-1-propanol, *35*, was asymmetrically hydrogenated to produce a mixture of optically active 3-(tetrahydro-2-furyl)-1-propanol, *36*, and 1,6-dioxaspiro[4.4]nonane, *37*, the latter product being of the spiran type, where optical activity is due to overall molecular chirality rather than an asymmetric carbon atom. Finally Terent'ev et al. (1953) have reported a

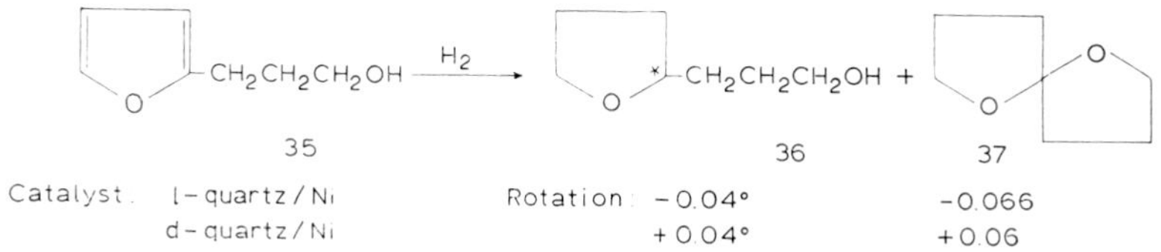

$-CH_2CH_2CH_2OH \xrightarrow{H_2}$ $*-CH_2CH_2CH_2OH$ +

35   *36*   *37*

Catalyst: l−quartz / Ni   Rotation: −0.04°   −0.066
           d−quartz / Ni   +0.04°   +0.06

completely new application of optically active quartz, namely, its use in conjunction with soluble metal alkoxide catalysts (EtOLi, EtONa, EtOK) to promote cyanoethylation reactions between acrylonitrile, *38*, and the cyclohexanone compounds *39* and *40*. The products from these reactions, as seen in the equations below, were again optically active to a slight degree.

The most recent hypothesis regarding a possible mechanism for asymmetric degradations and syntheses mediated by heterogeneous quartz catalysts has been advanced by Amariglio et al. (1968a, b). Amplifying an earlier mechanistic suggestion of Ingersol (1944), these investigators suggest a simple and plausible mechanism for all asymmetric processes brought about by heterogeneous catalysts. For asymmetric decompositions, for example, they postulate a 'diastereomeric surface association' between the (+)- and (−)-enantiomers of the substrate (Sub) and the surface of the quartz–metal catalyst (*d*-quartz, dQ, for example), forming two diastereomeric catalyst-substrate intermediates, (*a*) and (*b*). As with other diastereomers, (*a*) and (*b*) possess different energies,

$$dQ + (+)Sub \rightarrow dQ\text{–}(+)Sub \rightarrow (+)Products$$
$$(a)$$
$$dQ + (-)Sub \rightarrow dQ\text{–}(-)Sub \rightarrow (-)Products$$
$$(b)$$

different rates of formation, and different rates of conversion to products, and the overall interrelation of these rates and energies leads to the ultimate proponderance of one enantiomer in the final product. Similar catalyst-substrate intermediates are thought to prevail in catalytic asymmetric syntheses, such that rate differences in the formation and decomposition of such 'surface diastereomers' again lead to an optically active product.

In view of the above well-documented experimental demonstrations of absolute catalytic asymmetric syntheses, it follows that various authors have speculated theoretically on the possible role of quartz, both as asymmetric catalyst and asymmetric adsorbent, in the primordial origin of optical activity in the organic world. Schwab et al. (1934) first suggested that asymmetric crystal lattices in catalytic processes might provide 'a contribution to the ancient problem of the origin of the first asymmetry in living nature'. Fifteen years later Bernal (1949) emphasized the possible importance of quartz as a natural chiral adsorbent and catalyst, and in 1952 Goldschmidt (1952) argued that 'asymmetric syntheses in adsorbates on asymmetric crystal faces or on crystals which, like quartz, are asymmetric in tridimensional space is the most reasonable conception of a-biotic asymmetric synthesis in nature ...'. More recently, Vol'kenshtein (1959) has agreed with Terent'ev and Klabunovskii (1959), in essence that 'catalysis on the surface of asymmetric crystals might lead to the appearance of pure antipodes of simple organic molecules, in particular, amino acids'. Such hypotheses involving the asymmetric catalytic action of quartz, of course, suffer the same weakness as do hypotheses involving quartz as an asymmetric adsorbent, namely, the previously discussed occurrence of two opposite chiralities in natural quartz crystals, but only one chirality in living organic matter.

Again, as with asymmetric adsorption on quartz, the experimental and theoretical validity of asymmetric catalysis by quartz has recently been brought into serious question by Amariglio et al. (1968a, b). These authors have reinvestigated 3 catalytic reactions involving quartz: (1) the asymmetric dehydration and (2) dehydrogenation of 2-butanol, and (3) the asymmetric reduction of methyl ethyl ketone,

$$\underset{\displaystyle CH_3CCH_2CH_3,}{\overset{\displaystyle O \atop \displaystyle \|}{}}$$

using optically active quartz, either pure or covered with nickel or platinum. They stress the most appropriate experimental conditions, 'ordinarily neglected previously', for which stereoselectivity might be expected in such reactions. Their numerous catalytic experiments, conducted under carefully controlled conditions, however, led consistently to negative results, and the apparent optical rotations which they sometimes observed were shown to be due to extraneous effects caused by minute quartz particles carried into the reaction products.

As with their similar negative studies on asymmetric adsorption, one cannot fault the care and precision with which these catalytic studies were apparently conducted. Again, however, the major experimental fault would appear to involve the limited synthetic quartz samples which the French workers utilized. Certainly additional experimental work will be required before reconciliation can be made between the conflicting negative reports of Amariglio et al. and the positive findings of almost all earlier investigators of absolute asymmetric heterogeneous catalysis with optically active quartz.

6.5. Absolute asymmetric processes with circularly polarized light

Pasteur, the discoverer of enantiomerism, was the first to seek experimentally the origin of optical activity in absolute asymmetric processes. By this term one means an asymmetric chemical synthesis or degradation mediated by an asymmetric 'physical force' of some sort. (Whether or not 'absolute asymmetric processes' extend to include the previously discussed asymmetric syntheses, degradations and adsorptions on chiral mineral surfaces, as well as resolutions by spontaneous crystallization is, of course, only a semantic question.)

Pasteur sought to demonstrate the effect of such asymmetric physical forces by means of various ingenious experiments, including the selective crystallization of racemic compounds in magnetic fields, the synthesis of potentially asymmetric molecules in rapidly rotating tubes, and even the growth of plants under 'reverse' diurnal conditions, using a clockwork heliostat and reflectors. All of these experiments were failures, as were many similar attempts by subsequent investigators. It later became apparent to Pierre Curie that these unsuccessful attempts were all based on the false assumptions that such phenomena as magnetic fields, mechanical motions and the like were, in fact, asymmetrizing forces. Thereupon Curie (1894) formulated precisely the fundamental conditions under which a physical force could suffice to influence a chemical process in an asymmetric fashion. Among the few truly asymmetric (chiral) physical forces in nature, according to Curie, was circularly (and elliptically) polarized light (for a definitive review of this early literature, see Klabunovskii 1963).

Circularly polarized light has its electric vector rotating in a clockwise (right) or counterclockwise (left) direction as the light advances along its axis of propagation. Right-circularly polarized light can thus be represented as a clockwise spiral, and left- a counterclockwise spiral, analogous to right- and left-handed screws. It is important to realize that right- and left-circularly polarized light represent 'enantiomeric physical forces', whose chirality is potentially capable of interacting with the chirality of asymmetric molecules.

It has long been recognized that sunlight, reflected from the surfaces of the sea, is always elliptically or circularly polarized to a slight extent as a result of the earth's magnetic field (Jamin 1850; Becquerel 1899). Before the turn of the century Van't Hoff (1894) suggested that optically active substances in nature might be formed 'by transformations, for example, which occur through the action of right- or left-circularly polarized light'. These two concepts were subsequently connected by Byk (1904a, b), who argued that 'since *dextro*-circularly polarized light predominates at the earth's surface in reflected sunlight, an unsymmetrical form of solar photochemical energy has been available, during countless ages of evolution, for the multitude of vital syntheses which are carried out in living cells under the influence of sunlight' (Ritchie 1933). Thus shortly after the turn of the century the stage was set for a host of experiments, spanning several decades, designed to test the hypothesis that circularly polarized light might act as an asymmetric physical agent for the mediation of absolute asymmetric degradations and/or asymmetric syntheses. This section briefly summarizes these experiments.

The first investigator to show that right- and left-circularly polarized light did in fact interact differently with enantiomers was Cotton (1896), who discovered that right- and left-circularly polarized light (from the red end of the spectrum) was absorbed unequally by alkaline solutions of copper or chromium tartrate. The phenomenon of unequal absorption of right- and left-circularly polarized light of a wavelength near that characteristic of an absorption band of an optically active molecule is called circular dichroism (or the 'Cotton effect'). Since copper tartrate solutions are decomposed by sunlight, Cotton (1909) therefore attempted to effect such decomposition asymmetrically by exposing racemic copper tartrate solutions to right- and left-circularly polarized sunlight. No optical activity could be detected, however, even after prolonged irradiation. This failure was subsequently shown to be due to the fact that the photochemical decomposition of alkaline copper tartrate solutions is brought about only by ultraviolet light (Byk 1904a), while the circular dichroism of such solutions is found only at the red end of the spectrum (Mitchell 1930). Subsequently Byk (1904a, b), Henle and Haakh (1908), Freundler (1909), Padoa (1909), Jaeger and Berger (1921), Pirak (1922) and Bredig et al. (1923) each attempted to conduct analogous absolute asymmetric degradations as well as asymmetric syntheses

under the influence of circularly polarized light, but in no case was an optically active product obtained.

In 1930 Jaeger pointed out that the force being used to promote such
asymmetric reactions must be an 'essential condition for initiating the
chemical process'. That is, the reaction being attempted under the influence
of circularly polarized light must be photochemically mediated, otherwise
asymmetry cannot be developed. Failure to realize this requirement, as
well as the requirement for circularly polarized light of a wavelength permitting circular dichroism, was presumably the reason for the above
experimental failures. Bearing these factors in mind, subsequent investigators have been almost uniformly successful in a variety of attempts to en-
gender optical activity abiotically by absolute asymmetric degradations or
syntheses performed with circularly polarized light. The following is a brief
chronological summary of such successful degradative and synthetic experiments.

The first positive attempt at absolute asymmetric degradation was recorded by Kuhn and Braun (1929) at Heidelberg, who irradiated alcoholic
solutions of racemic ethyl α-bromopropionate, $CH_3C^*H(Br)COOEt$, with
circularly polarized light of 2800 Å wavelength. After about 50% decomposition the solution became dextro- or levorotatory, depending upon
whether right- or left-circularly polarized light was employed. The maximum
rotation achieved was 0.05°. Kuhn and Knopf (1930a, b) subsequently
carried out a more significant experiment, the photolysis of the dimethyl-
amide of α-azidopropionic acid, $CH_3C^*H(N_3)CON(CH_3)_2$. After about
40% decomposition, the residual undecomposed product had developed
rotations of $-1.04°$ or $+0.78°$; depending upon the chirality of the circularly
polarized light used.

In 1930, Mitchell subjected the blue sesquiterpene derivative humulene
nitrosite, $R_1R_2C^*(NO)C(ONO)(CH_3)_2$, to photochemical decomposition
with circularly polarized light of 7000 Å. After 32 hours, maximum rotations
of $+0.21°$ or $-0.20°$ were developed, depending again upon the chirality of
the light. Mitchell and Dawson (1944) later extended such asymmetric
photolyses of racemic nitroso compounds to include 2-chloro-2-nitroso-1,4-
diphenylbutane, $PhCH_2C^*Cl(NO)CH_2CH_2Ph$, where rotations of about
$\pm0.11°$ were developed after 90% decomposition. A novel type of photochemical asymmetric decomposition was more recently demonstrated by
Berson and Brown (1955), wherein a racemic center of carbon asymmetry
was destroyed with concomitant generation of a new center of biphenyl
type asymmetry. Thus the photolytic dehydration of the dihydropyridine

derivative *41* to the extent of 39.5% under the influence of circularly polarized light of 3660 Å yielded the pyridine derivative *42* (optical rotation
−0.022 ± 0.004°), whose chirality depends upon restricted rotation about
the bond between the benzene and pyridine rings.

41   42

As regards racemic inorganic complexes, Tsuchida et al. (1935) have
brought about the asymmetric photochemical decomposition of the racemic
cobalt complex $K_3[Co(C_2O_4)_3]$ using circularly polarized light of 5890 Å,
while more recently Stevenson and Verdieck (1965) have irradiated several
racemic chromium complexes with right-circularly polarized light and
likewise observed the induction of optical activity.

The first absolute asymmetric *synthesis* under the influence of circularly
polarized light was reported by Karagounis and Drikos (1933a, b, 1934),
who allowed chlorine vapor to interact with free triarylmethyl radicals of
the unsymmetrical type ArAr′Ar″C·, while exposing them to circularly
polarized light of 4300 Å wavelength. In the case of the phenyl-*p*-tolyl-*p*-
ethylphenylmethyl radical, *43*, the chlorinated product developed a maximum
rotation of 0.1°, while with the phenylbiphenyl-*a*-naphthyl radical, *44*, a
rotatory power of 0.2° was observed. It is interesting that illumination of

43   44

44 with circularly polarized light of 5890 Å, the wavelength of the absorption band of the radical in the red region, resulted in the development of the opposite sign of optical activity.

Shortly thereafter Davis and Heggie (1935a) at MIT reported an absolute asymmetric photochemical synthesis during the addition of bromine to the double bond of an olefin. 2-,4-,6-Trinitrostilbene, *45*, in carbon tetrachloride was treated with bromine, then exposed to right-circularly polarized light of 3600–4500Å, resulting in the formation of the dextrorotatory dibromide, *46*, which had the rotation +0.021°. Similar results were obtained in benzene solution, where a rotation of +0.078° was developed.

45   46

A short time later Davis and Heggie (1935b) reported similar results on the chlorination of trinitrostilbene during analogous irradiation with 3600–4500 Å circularly dipolarized light. It was observed, however, that no optical activity was developed using 5890–5896 Å circularly polarized light, since trinitrostilbene does not absorb light of this wavelength. Analogous asymmetric additions of chlorine and hydrogen chloride to other olefins in the gas phase were later reported also to yield optically active products (Betti and Lucchi 1938).

The first and only absolute asymmetric synthesis of a natural organic product with circularly polarized light has been described by Davis and Ackerman (1945), who conducted the photochemical hydroxylation of ethyl fumarate, *47*, with hydrogen peroxide under the influence of right-circularly polarized light of wavelength band 2535-7-9 Å. The product, ethyl (+)-tartrate, *48*, showed a maximum rotation of +0.073° in the reaction solution. In a separate experiment, using non-polarized ultraviolet radiation, a 6.6% isolated yield of ethyl tartrate was realized. Since the rotations

47   48

developed were very low, the authors admit that a similar 'hydroxylation with left-circularly polarized light, with the development of levorotation, remains to be carried out before the asymmetric synthesis can be considered to be proved conclusively'.

On a theoretical basis, absolute asymmetric processes mediated by circularly polarized light have been one of the most vigorously championed mechanisms for the abiotic origin of molecular chirality. After describing one of the first absolute asymmetric photochemical degradations, Kuhn and Knopf (1930a) suggested that 'it appears not excluded that an assemblage of optically active substances played a role, photochemically, in preparation for the origin of biological events'. Shortly thereafter, Davis and Heggie (1935a) proposed, on the basis of their asymmetric photochemical bromination of trinitrostilbene, that such absolute asymmetric syntheses might have been responsible for the origin of optically active substances in nature. Later Spiers (1937) championed the asymmetric photochemical origin of optical activity over the previously discussed statistical origin suggested by Langenbeck and Triem (1936), and in 1959 Klabunovskii and Patrikeev (1959) argued that 'The preponderance of one structural isomer of amino acids in the proteins of protoplasm is, therefore, not a chance occurrence. It is explained by the action of the right-circularly polarized component of scattered sunlight throughout the long years of the evolution of the living world'. Davis and Ackerman's (1945) absolute asymmetric synthesis of the ester of naturally occurring (+)-tartaric acid, *48*, by the photochemical hydroxylation of ethyl fumarate, *47*, under the influence of 'naturally predominant' right-circularly polarized light, may be taken as an experimental confirmation of this type of hypothesis.

The circularly polarized light hypothesis for the origin of molecular chirality, however, has not lacked persuasive opponents. Major criticisms have centered about (a) the trivial extent (never actually demonstrated) to which right-circularly polarized light occurs in nature, and (b) the trivial optical enrichment afforded by such absolute asymmetric photochemical processes even when conducted under optimal laboratory conditions. Thus Mills (1932) argues that: 'When one considers, however, the minuteness of the proportion of the total illumination received by an organism under natural conditions that can be circularly polarized, and the difficulty that has been experienced in demonstrating the optically activating effect of this form of light, even under favorable conditions of the laboratory, it is impossible not to feel a certain skepticism of an explanation based on the action of circularly polarized light thus produced ...'.

Perhaps the major feature in favor of this theory, however, is the fact that it argues for the uniform molecular chirality actually observed in nature. That is, even though the asymmetrizing effect of presumably preponderant right-circularly polarized light is miniscule and only marginally demonstrable experimentally, at least it must have been consistent throughout time – and it has had billions of years to exert itself. In contrast to previous hypotheses, therefore, this is the first theory which is able to predict a unique chirality for the molecular constituents of living matter.

6.6. *Molecular chirality from parity violation in β-decay*

The most recent and ingenious physical explanation for the universal existence of optically active organic compounds in nature involves the consequences of parity violation during the β-decay of certain radioactive isotopes. The *principle of parity* maintains that the laws of nature are invariant under spatial reflection (Rodberg and Weisskopf 1957; Ulbricht 1959; Wigner 1965). Thus, any process which occurs in nature can also occur as seen when reflected in a mirror, and the mirror image of an actual object is also a possible object in nature. In 1956 Lee and Yang of Columbia University and Brookhaven National Laboratory, considering certain anomalies in the decay patterns of θ and τ mesons, were led to the conclusion that the parity principle might be invalid for certain weak interactions such as those involved in β-decay. Lee and Yang also suggested experiments which would detect violations of the parity principle, and one of these critical experiments was performed with positive results a few months later by Wu et al. (1957) of Columbia and the National Bureau of Standards. When radioactive cobalt-60 disintegrates into nickel-60 a β-particle (e^-) and a neutrino (ν) are emitted. Now the ^{60}Co nucleus has a spin associated with it,

$$^{60}\text{Co} \rightarrow {}^{60}\text{Ni} + e^- + \nu$$

and in its normal state this spin is random due to thermal motions. Electrons are thus emitted symmetrically in all directions during β-decay. In their now famous experiment, Wu and coworkers undertook to modify this normal picture and to orient the spins of the ^{60}Co nuclei into parallel axial and uniform rotational orientations by placing the ^{60}Co sample in a powerful magnetic field and chilling it below 0.1 °K. They then measured the directional distribution of emerging β-decay electrons with respect to the direction of the external magnetic field. Parity conservation would require that this distribution be symmetrical backwards and forwards along the direction of the magnetic field, since this is the only possibility permitting mirror symmetry (fig. 5). On the other hand, any excess of electrons emerging parallel (or antiparallel) to the direction of the magnetic field would clearly constitute a parity violation, since such a situation is not mirror symmetric (fig. 5).

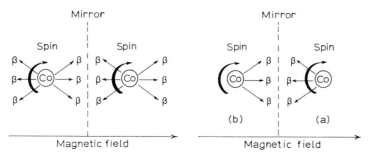

Fig. 5. Parity consequences of ^{60}Co β-decay. Left: Parity conserved . Experiment pictured on either side of mirror could be realized in our world. Right: Parity violated. *a*: actual condition existing in our world. *b*: condition excluded in our world.

The remarkable finding in Wu's experiment was that the number of electrons emerging in the direction antiparallel to the magnetic field was some 40% larger than the number emerging in the parallel direction, thus clearly confirming Lee and Yang's prediction of parity violation in β-decay. Since that time additional violations of the parity principle have been documented experimentally by other investigators (Rodberg and Weisskopf 1957; Ulbricht 1959), and in 1957 Lee and Yang received the Nobel prize for their important theoretical contribution to nuclear physics.

The spin of the ^{60}Co nucleus has associated with it only a sense of rotation. Yet its β-decay electron is emitted in a preferred direction with respect to the spin. There is thus a definite chirality (direction connected with a sense of rotation) exhibited during the β-decay of ^{60}Co. Moreover, this chirality is further reflected in the fact that the electron and the neutrino ejected in β-decay themselves take along some of the spin of the emitting nucleus and – traveling in a preferred direction – are thus themselves chiral entities. Thus the 'longitudinally polarized' electrons emitted in β-decay have been characterized as left-handed, analogous to a left-handed screw (Rodberg and Weisskopf 1957; Ulbricht 1959). The chirality of these left-handed β-decay electrons in turn manifests itself in an additional important phenomenon. As electrons slow down they lose some of their energy by emitting γ-ray photons, referred to as Bremsstrahlung (brake-radiation). Lee and Yang (1956) also predicted that the γ-rays associated with β-decay electrons should be circularly polarized, thereby reflecting the intrinsic chirality of such left-handed electrons. In 1957 Goldhaber et al. at Brookhaven National Laboratory investigated the degree and sense of the circular polarization of external Bremsstrahlung (produced after the electron has left the atom) using a β-ray source of $^{90}Sr + {}^{90}Y$ encased in Monel metal alloy. They found that at the high energy end of the spectrum the photons emitted were almost completely circularly polarized, with their spin antiparallel to their direction of propagation, i.e. in the same direction as the intrinsic spin of the β-rays which generated them.

Because of parity violation, we thus find an inherent chirality associated with the phenomenon of β-decay. The mirror image of this chirality is not prevalent 'in our part of the universe', and would be found, presumably, only in a universe made up of anti-matter, whose corresponding antiparticles possessed chirality opposite to that of our local particles (Rodberg and Weisskopf 1957). Now the almost total prevalence of unique chirality among enantiomers in living matter (e.g. the L-amino acids in proteins) itself has the characteristics of a parity violation. Why does not the 'unnatural' mirror image enantiomer of opposite chirality appear 'in our part of the universe' – that is, in our world? Such questions led Vester (1957) to suggest an intrinsic causal relationship between the unique chirality recently associated with β-decay and the long known unique chirality prevailing in molecules associated with living matter.

How could molecular chirality be induced by polarized β-radiation? Since interactions are negligibly small when energy levels are widely separated, any interactions between molecules and high energy β-rays would have to be mediated by secondary effects of lower energy, such as the circularly polarized Bremsstrahlung described above. Relying on the demonstrated ability of circularly polarized light to induce absolute asymmetric syntheses and degradations (section 6.5), Ulbricht (1959) has accordingly proposed the following scheme for utilizing the chirality inherent in β-decay for the absolute photochemical production of molecular chirality:

$$\begin{array}{ccc} & & \text{absolute asymmetric} & \\ \text{longitudinally} & \text{circularly} & \text{organic syntheses} & \text{chiral} \\ \text{polarized} \rightarrow & \text{polarized} & \xrightarrow{\hspace{3cm}} & \text{organic} \\ \beta\text{-rays} & \text{photons} & \text{or degradations} & \text{molecules} \end{array}$$

Ulbricht and Vester (1962) calculated, however, that even in favorable circumstances such a scheme might not lead to measurable optical activity in reasonable time spans. Thus a 'source of 1 Curie, complete absorption of the β-radiation by the reactants, a product with a specific rotation of 100°, and a quantum yield of 1, might produce measurable optical activity in a period on the order of one year'. Leverage might be obtained, however, by utilizing reactions with a quantum yield considerably greater than 1 – a radiation-initiated isotactic polymerization, for example – wherein measurable optical activity might result within the duration of an ordinary chemical reaction (Ulbricht and Vester 1962).

Despite these anticipated experimental pitfalls, Vester et al. (1959) and Ulbricht (1962) undertook the examination of a number of synthetic and degradative chemical reactions which would yield optically active products if the above effects were valid and of sufficient magnitude. Without discussing the detailed chemistry, some 10 different reactions were conducted in the presence of a variety of β-emitting sources (e.g. ^{32}P, ^{90}Sr-Y, ^{90}Y, ^{152}Eu, ^{108}Ag-^{110}Ag) of various radioactivity levels (30–2500 mc), for a number of specified time intervals (5–2900 min) and temperatures (-50 to $+50$°C). Controls using unpolarized electrons from a linear accelerator were employed in several of the reactions. Optical activity in the products of some 36 separate experiments were examined using two different precision polarimeters, averaging a minimum of 10 readings for each measurement. The maximum rotations observed were on the order of 0.055 $\pm$ 0.033°, while the average measurable rotation in 31 experiments using the β-decay sources at hand was 0.0154 $\pm$ 0.0161°. Since the observed rotations were within experimental error of zero, the authors concluded that any optical activity produced by the relatively β-ray weak sources employed was less than 0.02%, and that definitive results would at the least require stronger sources and longer exposure times.

A puzzling result bearing on the above type of experiment was reported in 1959 by Starodubtsev et al. who irradiated benzene with γ-rays from a ^{60}Co source and then noted changes in the optical properties of the benzene substrate. They report (in translation by Spialter and Futrell 1960) that 'Interesting is the fact of rotation of the plane of polarization. Although the rotation is not large (0.7–0.9° at a dosage of 10^7 r and a sample thickness of 10 cm) it nevertheless indicates the occurrence of asymmetric molecules not possessing a center or plane of symmetry'. The theoretical significance of this claim led Spialter and Futrell (1960) to undertake a very careful duplication of Starodubtsev's experiment. However, after irradiation of carefully purified benzene in a 20,000 c ^{60}Co source to an integrated dose of 3.3×10^7 r, the sample proved to be totally void of optical activity.

The only remotely positive experiments bearing on the origin of molecular chirality by polarized β-rays are those described by Garay (1968) in Hungary, who irradiated racemic D,L-alanine, D,L-tryptophan and D,L-tyrosine with a 0.5 mc ^{32}P source under sterile conditions. After one month, however, no measurable optical activity was noted and the experiment was modified. D- and L-tyrosine in aqueous ethanol solution containing alkali were irradiated separately with ^{90}Sr-Y as the β-ray source, whereupon the ultra-violet absorption spectrum of each solution was measured on a monthly basis over a period of time. Each enantiomer displayed a similar spectrum (weak absorption band at 300 mμ, strong band at 242 mμ) at the outset of experiment, as well as 12 months later. After 18 months, however, the 242 mμ band for the D-tyrosine had been eradicated to a significantly greater extent than had the same band for the L-tyrosine, suggesting that the latter had undergone less radiolysis than the D-enantiomer. No such spectral differences (over an 18-month period) were noted for the D- and L-tyrosine in control experiments using non-radioactive ^{88}Sr, or on irradiation of each enantiomer in acidic solution. The author thus suggested that decomposition of the tyrosine in alkaline solution was actually an oxidative degradation which was enhanced by the β-rays or γ-rays employed. Although his control experiments gave the anticipated negative results, Garay's critical experiment itself was unfortunately not run in duplicate, and the possibility of an artifact effect thus cannot be discounted. Similarly, differences in the shapes of two ultraviolet absorption curves hardly themselves constitute a demonstration of optical activity! Nevertheless, though it can be criticized on minor grounds, Garay's efforts constitute the only positive experimental evidence to date possibly bearing on the induction of optical activity by polarized β-rays, and certainly merit critical duplication and expansion.

Yamagata (1966) has recently used the observation of parity violation to develop a general 'hypothesis for the asymmetric appearance of bio-molecules on earth'. He postulates a sequence of n reactions leading to the real, polymeric DNA molecule (A $\rightarrow$ B $\rightarrow$ C $\rightarrow$ --- $\rightarrow$ DNA), and a similar sequence of mirror symmetric reactions leading to 'imaginary DNA', the mirror image of real DNA, (A$'$ $\rightarrow$ B$'$ $\rightarrow$ C$'$ $\rightarrow$ --- $\rightarrow$ imaginary DNA). He then defines p_k as the ratio of reaction rates for the kth reaction in the real sequence relative to that in the imaginary sequence, whereupon the ratio of the quantity of final DNA in the real series (N_r) to that of the imaginary series (N_i) becomes:

$$N_r/N_i = p_1 p_2 p_3 \ldots p_n = p^n \text{ (assuming for simplicity that all } p_k\text{'s} = p).$$

Now, because of breakdown of the parity principle in weak interactions, p

will not be unity. It will instead exceed unity by a small increment ε (where $\varepsilon = f^2/he = 10^{-6} \sim 10^{-7}$, the coupling constant of such weak interactions). That is, $p = (1 + \varepsilon)$, whereupon:

$$N_r/N_i = p^n = (1 + \varepsilon)^n \cong e^{\varepsilon n}$$

Now, the number of nucleotides in a cell is thought to be on the order of $10^8 \sim 10^9$. Thus for the polymerization of n nucleotides to form DNA, a typical ratio of real to imaginary DNA might be:

$$N_r/N_i \cong e^{10^{-6} \cdot 10^9} = 10^3$$

In other words, there exists an 'accumulation principle' – whereby parity violation leads to a consistent rate bias in favor of the real reactions over the corresponding imaginary ones. This eventually results in the effective swamping out of imaginary DNA and the prevalence of the real chiral polymer we know.

It has been suggested (Swallow 1960) that β-rays and γ-rays, mainly from the radioactive decay of potassium-40, constituted a significant fraction of the energy available for the primordial synthesis of organic compounds. Such an energy source would be particularly effective in local regions of high ^{40}K concentration, and such abiotic syntheses might be mediated by minerals acting as catalysts. Experimental verification of the efficacy of such energy sources has been achieved by Hasselstrom et al. (1957) who obtained glycine, aspartic acid and diaminosuccinic acid on exposing aqueous solutions of ammonium carbonate to β-radiation from a Van De Graaf accelerator, Palm and Calvin (1962), who produced urea and other substances by exposing 'primordial' gas mixtures (CH_4, H_2, NH_3, H_2O) to 5 meV electrons in a linear accelerator, and by Paschke et al. (1957) who obtained glycine, ammonium formate and alanine after subjecting solid ammonium carbonate to γ-radiation. In connection with the above parity violation mechanism for the origin of molecular chirality, it would be of considerable interest to duplicate such 'primordial syntheses' using longitudinally polarized β-rays and circularly polarized γ-rays as energy sources, to see if the amino acid products were formed in an optically active state.

7. Biotic theories of the origin of molecular chirality

The above described searches for the origin of net molecular chirality in abiotic physical or statistical events which might have occurred upon the primitive earth have as their underlying (and frequently explicit) assumption, of course, that the prior availability of optically active molecules in nature was a prerequisite for the subsequent origin of life. That is, optically active molecules of unique chirality were first formed abiotically, after which the stage was set for their further chemical evolutionary development into the more complex chiral molecules which we now find associated with life.

Not all authors have subscribed to these views, however, and more recently the argument has been advanced that 'configurational one-sidedness may have arisen during the biological era rather than before it' (Fox 1957). Such 'biotic' theories of the origin of molecular chirality assume that the very simplest life may have originated on a racemic basis, and that optical activity gradually and subsequently emerged as life forms developed. That is, 'the optical activity of living matter is an inevitable consequence of its property of growth' (Mills 1932). Arguments in favor of this hypothesis include the facts that (1) contemporary 'microorganisms that are relatively low on the phylogenetic scale are rich in (unnatural) D-amino acids, whereas the D-form is all but unknown in higher organisms', and that (2) 'the puzzling occurrence of D-amino acid oxidase in mammals can be explained as chemical evolutionary vestige, inasmuch as the lower forms do contain D-amino acid residues' (Fox et al. 1956). More recently Kuhn (1958) has emphasized the stereobiological necessity that 'the higher organisms possess a mechanism which actively eliminates the 'wrong' amino acids. This is done by the D-amino acid oxidase which is found especially in the liver, and which obviously serves to eliminate by oxidation the D-amino acids from the body. The presence of D-amino acid oxidase in these organisms is almost equivalent to a direct proof for the occurrence of D-amino acids ...'. and that therefore primordial organisms must in fact have had to develop corrective enzymes to cope with such 'unnatural' enantiomers. Rush (1957) has argued that 'On present evidence, a preponderance of right- or left-handed molecular symmetry before the advent of life seems unlikely', and Horowitz and Miller (1962) have more recently pointed out a number of theoretical difficulties attending earlier hypotheses of the abiotical physical origin of optical activity. Finally Wald (1957), in making the most detailed critique to date of abiotic origins, has concluded that 'Indeed, all the inorganic sources of optical activity share the same disabilities: very restricted conditions, a very limited field of operation, poor yields, and the overwhelming tendency to result in only local and temporary asymmetry in what is otherwise a racemic continuum. For the origin of optical activity in living organisms, I think one must look elsewhere'.

After these criticisms, some at least of which are unquestionably debatable, Wald (1957) proceeded to develop the most detailed as well as the only original explanation to date for the origin of molecular chirality under – comparatively speaking, at least – 'post-biotic' conditions. Wald attempted to elaborate the thesis that 'optical activity appeared as a consequence of intrinsic structural demands of key molecules of which organisms were eventually composed, through selection of optical isomers from racemic mixtures', during the process of molecular evolution. He argued that adoption of the coiled *a*-helix secondary structure for proteins plays an integral part in the formation and maintenance of protein molecules, and that this structure in turn is most effectively promoted by the utilization of amino acids of a single configuration. Thus a few D-amino acids might be tolerated in an *a*-helix composed of a sequence of L-amino acids, but the more prevalent this was the less stable the helical structure would be, and anything approaching a random assortment of enantiomeric amino acids might make *a*-helix formation impossible. According to Wald, this 'in itself should provide sufficient basis for the selection of one configuration out of mixtures of enantiomorphs'. Experimental support for Wald's views arose almost immediately in the course of investigations on amino acid and polypeptide derivatives by Wald's colleagues at Harvard, Blout, Doty and their coworkers.

Earlier investigators, however, had already noted marked differences in the physical properties of polypeptides and other polymers depending on the stereochemical homogeneity of their monomer units. Thus Astbury et al. (1948) reported that poly-D,L-alanine, *49*, was readily soluble in water, whereas poly-L-alanine was insoluble. Brewster (1951) has prepared

$$-(NHC^*HCO)_{\overline{n}} \quad -(NH(CH_2)_6NHCOC^*H(CH_2)_2C^*HCO)_{\overline{n}} \quad -(OC^*HCH_2)_{\overline{n}}$$

with CH_3 groups attached at the starred carbons.

| 49 | 50 | 51 |

'dimethyl nylon' polyamides, *50*, from hexamethylenediamine and *meso*-, (+)-, and (±)-*a,a'*-dimethyladipic acids. All of the polymers formed oriented fibers which were weaker than nylon, but the polymer from the racemic acid was the weakest of all. Price et al. (1956a, b) found that the polypropylene oxide, *51*, prepared from (+)-propylene oxide was a crystalline solid, mp 55.5–56.5°, while that prepared under similar conditions from racemic propylene oxide was a liquid. Such studies all tended to indicate that polymers produced from configurationally homogeneous monomer units are more stable as regards their crystal structures than are those prepared from racemic monomer units.

Subsequent studies showed that similar considerations also applied to the *a*-helix secondary structure of high molecular weight polypeptides. Blout and Idelson (1956a) showed by infrared spectral data that the reversible transformation from a helical structure to a random-coil non-helix and back could be readily accomplished for poly-*a*-L-glutamic acid, *52*, depending upon whether its -COOH groups were ionized (-COO$^\ominus$) or not, thus demonstrating that the *a*-helix configuration may be assumed spontaneously. This

$$
\begin{array}{cc}
\overset{\displaystyle (CH_2)_2\,COOH}{\underset{|}{}} & \overset{\displaystyle (CH_2)_2 COOCH_2C_6H_5}{\underset{|}{}} \\[4pt]
-(NHC^*HCO)_{\overline{n}} & -(NHC^*HCO)_{\overline{n}} \\[6pt]
52 & 53
\end{array}
$$

conclusion also followed from investigations by Doty et al. (1957) and by Doty and Yang (1956), who also studied poly-*γ*-benzyl-L-glutamate, *53*, from the viewpoint of its analogous helix $\rightleftharpoons$ random coil transformation. Interestingly, by optical rotatory dispersion measurements Yang and Doty (1957) deduced that the secondary *a*-helical coil structure of *53* 'exists with only one screw sense, probably right-handed'· Blout et al. (1957) have lastly shown that the *a*-helix structure of the L-polypeptide *53* is progressively weakened as D-enantiomer units replace L-units in the polypeptide chain, and have suggested that the *a*-helix of *53* could not be maintained in aqueous media 'if significant amounts of D-residues were present'. Thus it was established experimentally that the introduction of configurational randomness into a polypeptide chain clearly decreases the stability of its secondary *a*-helix structure.

Even more convincing evidence for the advantages of configurational homogeneity became apparent in several investigations concerned with the rate of synthesis of the polypeptide *53* by means of base catalyzed polymerizations of *γ*-benzyl-*N*-carboxy-L-glutamate anhydride, L-*54*, containing varying quantities of the D-glutamate enantiomer, D-*54*. Thus Blout and

$$
\begin{array}{c}
(CH_2)_2\,COOCH_2C_6H_5 \\
|
\end{array}
$$

$$
n\ NHC^*HCO \underset{\displaystyle \underset{|}{}}{\overset{\diagdown}{}} O \xrightarrow{\ \text{base}\ } 53\ +\ nCO_2
$$

$$
\underset{\displaystyle \quad\ \ CO}{} \diagup
$$

54

Idelson (1956b) found that the rate of polymerization of D,L-*54* was only 5% as great as the polymerization rate for enantiomerically homogeneous

L-*54*, and concluded that 'the presence of the opposite isomer has (a rate retarding) effect far beyond that which would be predicted on a simple infinite preference of a growing chain of one isomer for its own isomer, since this would only diminish the rate by 1/2'. Additional rate studies by Doty and Lundberg (1956) showed that the polymerization of L-*54* occurred at a relatively slow rate up to about 8 units, where the *a*-helix secondary structure first becomes viable, after which the polymerization rate abruptly increased 5- to 6-fold. The authors concluded, 'This coincidence strongly suggests that the higher propagation rate is associated with the helical configuration'. Again, Lundberg and Doty (1956) found that the rate of polymerization of D,L-*54* showed a pronounced lowering relative to that of L-*54*, particularly so after the point where the growing chain should adopt its helical configuration. Lastly, Idelson and Blout (1958) discovered that when even as little as 5% of the L-*54* was added to 95% of D-*54* the polymerization rate was reduced to $\frac{1}{3}$ the value for pure D-*54*, and that the degree of polymerization achieved by D,L-*53* was only 20% that achieved by the optically pure polypeptides. Similar observations with regard to both rate and mechanism were made by Weingarten (1958) during polymerizations of analogous *N*-carboxy-*a*-amino acid anhydrides derived from L- and D,L-lysine. In later polypeptide experiments Kulkarni and Morawetz (1961) found that optically pure methacrylylglutamic acid polymerized more rapidly in 0.1 *N* HCl than did the racemic monomer. It was thus abundantly clear that complete optical homogeneity of the monomer units leads not only to larger polypeptide chains, but to vastly more rapid polymerization. Steinman (1967) has undertaken a series of interesting radiochemical experiments to see if such stereoselectivity in polypeptide synthesis might also depend upon primary (neighbor) interactions at the lower oligopeptide level, as well as in the secondary (*a*-helix) interactions which first appear around the octapeptide level. Steinman concluded that there was negligable stereoselectivity at the oligopeptide level and that 'on the primitive Earth ... the synthesis of stereohomogeneous polypeptides would have to depend on chance associations at the simple peptide level and then on stabilization of homopolymers by the *a*-helix at higher degrees of polymerization'. Wald (1957) has also argued that analogous specific configurational requirements for the components of helical macromolecules can explain the unique occurrence of *β*-D-ribofuranose in ribonucleic acid (RNA) and of *β*-D-deoxyribofuranose in deoxyribonucleic acid (DNA).

An interesting non-peptide polymerization experiment bearing on Wald's (1957) suggested 'selection of optical isomers from racemic mixtures' was performed recently by Ciardelli et al. (1969). These investigators found that the copolymerization of an optically active alkene of one configuration, *55*, with a racemic alkene, *56*, led to a copolymer, *57*, made up of the two alkenes of the same absolute configuration, but a homopolymer only, *58*, with the enantiomer in *56*. That is, the optically active monomer *55* specifically selected its own configuration from the racemic monomer *56* to produce the copolymer *57*. In keeping with this is the observation of Farina (1969) that

$$-(CH_2CH\!-\!\!-\!CH_2CH)_{\bar{n}}$$
$$(S)A \qquad (S) B$$
$$57$$

$$CH_2 = CH + CH_2 = CH -$$
$$(S) A \qquad (RS) B$$
$$55 \qquad 56$$

$$-(CH_2CH\!-\!\!-\!CH_2CH)_{\bar{n}}$$
$$(R) B \qquad (R)B$$
$$58$$

asymmetric polymerizations result in products of greater optical purity than do ordinary asymmetric syntheses. Along these same lines Pasynskii and Pavlovskaya (1964) have made the suggestion that 'It is possible that optical asymmetry developed not by the initial formation of optically active monomers, but by the reverse process; high-molecular-weight asymmetric catalysts might have been formed by stereospecific polymerization and then the secondary optical (a)symmetry of low-molecular compounds might have developed on the basis of the polymers. The appearance of optically asymmetric and catalytically active polypeptides was important for accelerating the reaction and was therefore secured by evolution'.

More recently Klabunovskii (1968) has summarized additional experimental evidence in support of this view.

Mechanisms other than direct polymerization with Wald's enantiomeric selection have also been suggested for obtaining optically active hetero-polypeptides. Thus Akabori et al. (1956) have showed that a number of the glycyl residues of polyglycine, *59*, dispersed on kaolinite could be readily converted into seryl, *60*, and threonyl, *61*, residues by condensation of the CH_2 groups of *59* with formaldehyde or acetaldehyde. The polyglycine

$$\begin{array}{cccc} & CH_2OH & CH(OH)CH_3 & NH \\ & | & | & \| \\ -(NHCH_2CO)_{\bar{n}} & -NH\overset{*}{C}HCO- & -NH\overset{*}{C}HCO- & -(NHCH_2\overset{}{C})_{\bar{n}} \\ & & & \\ 59 & 60 & 61 & 62 \end{array}$$

precursor for this process would presumably be readily available on the primitive earth by the polycondensation of glycine in aqueous ammonia (Oro and Guidry 1961). Kliss and Matthews (1962) have suggested that polymers such as *59*, as well as the related imide type, *62*, derived from hydrogen cyanide dimer, should adopt a right- or left-hand helix. The repeating units in the helix would then have one of their -CH$_2$- hydrogen atoms directed toward the center of the coil and one away from the center. The latter would be sterically more accessible for replacement by other groups, such that substitutions by the above mechanism established by Akabori, or by other mechanisms, should convert the polyglycine chain into a random heteropolypeptide whose amino acid residues were all of the same configuration. The absolute configuration of these would be determined by whether the *a*-helix was right- or left-handed. To date these interesting speculations have received no experimental confirmation.

In the above section we have seen that there are a number of recent experimental observations which, while not proof for, are at least consistent with Wald's (1957) viewpoint that 'single optical configurations are chosen out of racemic mixtures because they work best in forming structures of higher order ...'. No subsequent author has advanced this hypothesis of 'natural selection at the molecular level' more convincingly.

8. Uniqueness and frequency; optical activity and the origin of life

Whether molecular chirality arose abiotically or concomitantly with the emergence of life, we are left with the puzzling question of why unique enantiomeric configurations prevail today. That is, why do we find only L-amino acids, D-ribofuranosides and the like among living organisms, and not their enantiomers? Since the occurrence of enantiomeric structures throughout all of nature would provide mirror image stereochemistry but equally efficient biochemistry for life, one can only conclude that the choice on our world was made by chance. Wald (1957) even postulated two separate populations of primeval organisms, one based on L- and one on D-amino acids, for example. After an indefinite period of coexistence and after the initiation of food-chain patterns, however, 'it would become highly advantageous not only for each individual organism, but for all of them collectively, to utilize single configuration series of molecules'. As Pirie (1959) put it, 'In a competitive world, a workshop that standardizes itself on one type of screw is likely to prevail; and organisms live in a competitive world'. The selection of the organism – or the choice of the screw – is a matter of chance, as suggested by Wald as well as Blum (1957) '... selection of the 'left-handed' form depended upon some unrelated aspect of the replicating system – some independent handle for natural selection having no relationship to optical isomerism'. Rush (1957), viewing the equal efficiency of either enantiomeric structure, concluded that 'the existence of only one symmetry today is the result of an accident ... a statistical or chance situation'. Thus with either biotic or abiotic origins, we are left with a chance explanation for the particular chirality we find associated with life's molecules – except in those experimentally unsubstantiated explanations involving right-circularly polarized light (section 6.5) or parity violation in β-decay (section 6.6).

To the best of our knowledge, no author has speculated as to exactly when optical activity first appeared on the primordial earth, either biotically or abiotically. There are simply no observational data (see section 2). A number of authors have commented, however, on the frequency with which life – and its concomitant optical activity – might have originated. The generally agreed upon frequency is once! Thus Mills (1932) argued that 'The development of the organic kingdom from a single germ would provide a simple explanation of the configurational relationship ... between the optically active components of the most diverse forms of life'. Kuhn (1958), discussing the question of primeval molecular chirality, concluded that whenever 'a particular optically active substance was chosen, the selection then was made decisive for the whole system of optically active substances used by today's living organisms'. Speaking more mechanistically, Frank (1953) pointed out that 'if the production of living molecules is an infrequent process, compared with the rate of multiplication of living molecules, the whole earth is likely to be extensively populated with the progeny of the first before another appears'. Schlovskii and Sagan (1966) have concluded that 'The inner workings of terrestrial organisms – from microbes to men – are so similar in their biochemical details as to make it highly likely that all organisms on Earth have evolved from a single instance of the origin of life'. Finally, regarding the uniqueness of molecular chirality on earth and of the random event which produced it, Wald (1957) has made an interesting prediction: 'If the choice of optical isomers is arbitrary as proposed, one should expect that a survey of life throughout the universe would reveal approximately equal numbers of planetary populations in which ... life is based upon L- and D-amino acids and, similarly, for the other molecules'.

After systematically reviewing the literature on the origin of optical activity and the origin of life, one can only affirm the statement made by Errera (1898) over 70 years ago: 'asymmetry begets asymmetry, as life begets life'. Which came first? – is probably a meaningless question. In the context of our world, molecular chirality and life become indistinguishable, if we pragmatically define life as the totality of those completely stereospecific biochemical processes by which living organisms sustain and reproduce themselves. Without net molecular chirality such life could not exist, and it seems probable that without life widespread molecular chirality would likewise be an impossibility. One must conclude by agreeing with Briggs (1960) that 'the origins of optical activity present problems to the hypothesis of chemical evolution that are at present insoluble'.

References

AKABORI, S., 1959, A comment to the paper of Prof. Klabunovskii; In: The origin of life on earth, F. Clark and R. L. Synge, eds. (Pergamon Press, N.Y.) 183.

AKABORI, S., Y. IZUMI and Y. FUJII, 1957, Nippon Kagaku Zasshi 78, 886; 1960, Chem. Abstr. 54, 9889d.

AKABORI, S., K. OKAWA and M. SATO, 1956, Introduction of side chains into polyglycine dispersed on solid surface. Bull. Chem. Soc. Japan 29, 608.

AKABORI, S., S. SAKURI, Y. IZUMI and Y. FUJII, 1956, An asymmetric catalyst. Nature 178, 323.

AMARIGLIO, A., H. AMARIGLIO and X. DUVAL, 1968a, Asymmetric reactions on optically active quartz. Helv. Chim. Acta 51 (8), 2110.

AMARIGLIO, A., H. AMARIGLIO and X. DUVAL, 1968b, Asymmetric synthesis. Ann. Chim 3, 5.

ANDERSON, L. and D. W. HILL, 1928, A further case of the spontaneous resolution of externally compensated mixtures. J. Chem. Soc., 993.

ASTBURY, W. T., C. E. DALGLIESH, S. E. DARMON and G. B. B. M. SUTHERLAND, 1948, Studies of the structure of synthetic polypeptides. Nature 162, 596.

BAILAR, Jr., J. C. and D. F. PEPPARD, 1940, Stereochemistry of complex inorganic compounds. VI. Study of the stereoisomers of dichloro-diamino-ethylenediamine cobaltic ion. J. Am. Chem. Soc. 62, 105.

BAKER, W., B. Gilbert and W. D. OLLIS, 1952, Eight- and higher-membered ring compounds. VI, cis-Di- and tri-o-thymotides. J. Chem. Soc., 1443.

BALANDIN, A. A., E. I. KLABUNOVSKII and Y. I. PETROV, 1959, Configurational relationships in stereospecific catalysis. Doklady Akad. Nauk. S.S.S.R. 127, 557; 1960, Chem. Abstr. 54, 241a.

BARTRUM, C. O., 1898, Chance or vitalism. Nature 58, 545.

BECKETT, A. H. and P. ANDERSON, 1957, A method for the determination of the configuration of organic molecules using stereo-selective adsorbents. Nature 179, 1074.

BECKETT, A. H. and P. ANDERSON, 1959, Footprints in adsorbents. J. Pharm. and Pharmacol. 11, 258T–260T; 1960, Chem. Abstr. 54, 12484c.

BECKETT, A. H. and P. ANDERSON, 1960, The determination of the relative configuration of morphine, levorphanol and 1-phenazocine by stereoselective adsorbents. J. Pharm. and Pharmacol. 12, 228T–236T; 1961, Chem. Abstr. 55, 9784i.

BECQUEREL, H., 1899, The influence of terrestrial magnetism on atmospheric polarization. Compt. Rend. 108, 997.

BENGEN, F., 1940, German patent appl. OZ12438 (March 18, 1940).

BENGEN, F. and W. SCHLENK, Jr., 1949, New type addition compounds of urea. Experientia 5, 200.

BERNAL, J. D., 1949, The physical basis of life. Proc. Phys. Soc. 62A, 537.

BERNAL, J. D., 1951, The physical basis of life. (Routledge and Paul, London) pp. 32-39.

BERSON, J. A. and E. BROWN, 1955, Studies on dihydropyridines, III. An absolute asymmetric synthesis and an attempted conversion of carbon atom asymmetry to biphenyl asymmetry. J. Am. Chem. Soc. 77, 450.

BETTI, M. and E. LUCCHI, 1938, Asymmetric catalysis: absolute asymmetric syntheses. Atti X° Congr. Intern. Chim. 2, 112; 1939, Chem. Abstr. 33, 7273.

BLOUT, E. R., P. DOTY and J. T. YANG, 1957, Polypeptides. XII. The optical rotation and configurational stability of a-helices. J. Am. Chem. Soc. 79, 749.

BLOUT, E. R. and M. IDELSON, 1956a, Polypeptides. VI. Poly-a-L-glutamic acid: preparation and helix-coil conversions. J. Am. Chem. Soc. 78, 497.

BLOUT, E. R. and M. IDELSON, 1956b, Polypeptides. IX. The kinetics of strong-base initiated polymerizations of amino acid-N-carboxyanhydrides. J. Am. Chem. Soc. 78, 3857.

BLUM, H. F., 1957, On the origin of self-replicating systems; In: Rythmic and synthetic processes in growth, D. Rudnick, ed. (Princeton Univ. Press, Princeton, N.J.) pp. 155-170.

BÖHM, R. and G. LOSSE, 1967, Chemical bases for the origin of life on the earth. Zeitschr. für Chem. 7 (11), 409.

BRADLEY, W. and R. A. BRINDLEY, 1954a, Selective absorption of optical antipodes by proteins. Chem. and Ind., 579.

BRADLEY, W., R. A. BRINDLEY and G. C. EASTY, 1954b, The selective absorption of optical antipodes by wool. Disc. Faraday Soc. 16, 152.

BRADLEY, W. and G. C. EASTY, 1951, The selective absorption of optical antipodes by proteins. J. Chem. Soc., 499.

BRADLEY, W. and G. C. EASTY, 1953, The selective absorption of optical antipodes by proteins. Part II. J. Chem. Soc., 1519.

BREDIG, G. and F. GERSTNER, 1932, Asymmetric catalysis with organic fibers. A new enzyme model. Biochem. Z. 250, 414.

BREDIG, G., F. GERSTNER and H. LANG, 1935, Catalysis with organic fibers. II. Biochem. Z. 282, 88.

BREDIG, G., P. MANGOLD and T. G. WILLIAMS, 1923, Absolute asymmetric syntheses. Z. Angew. Chem. 36, 456.

BREWSTER, J. H., 1951, The effect of structure on the fiber properties of linear polymers. I. The orientation of side chains. J. Am. Chem. Soc. 73, 366.

BRICAS, E. and C. L. FROMAGEOT, 1953, Naturally occurring peptides; In: Advances in protein chemistry, vol. 8, M. L. Anson, K. Baily and J. T. Edsall, eds. (Academic Press, N.Y.) pp. 6-7.

BRIGGS, M. H., 1960, The origins of life on earth: a review of the experimental evidence, Sci. and Cult. (Calcutta) 26, 160.

BRODE, W. R. and R. ADAMS, 1926, Optically active dyes. III. Physical properties, dyeing reactions and mechanism of dyeing. J. Am. Chem. Soc. 48, 2193.

BRODE, W. R. and R. E. BROOKS, 1940, Optically active dyes. V. Molecular asymmetry in dyes and their dyeing properties. J. Am. Chem. Soc. 63, 923.

BUSCH, D. H. and J. C. BAILAR, Jr., 1953, The stereochemistry of complex inorganic compounds. XVII. The stereochemistry of hexadentate ethylenediaminetetraacetic acid complexes. J. Am. Chem. Soc. 75, 4574.

BUSCH, D.H. and J.C.BAILAR, Jr., 1954, The optical stability of beryllium complexes. J. Am. Chem. Soc. 76, 5352.

BYK, A., 1904a, The question of the resolution of racemic compounds by circularly polarized light, a contribution to the primary origin of optically active substances. Z. Physik. Chem. 49, 641; 1904b, Ber. 37, 4696.

CALVIN, M., 1969, Chemical evolution. (Oxford Univ. Press, N.Y. and Oxford) p. 149–152.

CIARDELLI, F., C. CARLINI and G. MONTAGNOLI, 1969, Stereochemical aspects of the copolymerization of asymmetric *a*-olefins by stereospecific catalysts. Macromolecules 2 (3), 296.

CONTRACTOR, S. F. and J. WRAGG, 1965, Resolution of optical isomers of D,L-tryptophan, 5-hydroxy-D,L-tryptophan and 6-hydroxy-D,L-tryptophan by paper and thin layer chromatography. Nature 208, 71.

CORNFORTH, J. W., R. H. CORNFORTH and M. J. S. DEWAR, 1944, Reported asymmetric synthesis of santonin. Nature 153, 317.

COTTON, A., 1896, Researches on the absorption and dispersion of light by media endowed with rotatory power. Ann. Chim. Phys. 8 (7), 347.

COTTON, A., 1909, Resolution of compounds inactive by compensation by use of circularly polarized light. J. Chim. Phys. 7, 81.

CRAMER, F., 1954, Einschlussverbindungen. (Springer-Verlag, Berlin).

CRAMER, F., 1955, Inclusion compounds. Rev. Pure Appl. Chem. 5 (3), 143.

CRAMER, F., 1956, Inclusion compounds. Angew. Chem. 68, 115.

CRAMER, F. and W. DIETSCHE, 1958, Asymmetric catalysis by inclusion compounds. Chem. and Ind. (London), 892.

CRAMER, F. and W. DIETSCHE, 1959a, Occlusion compounds. XV. Resolution of racemates with cyclodextrins. Chem. Ber. 92, 378.

CRAMER, F. and W. DIETSCHE, 1959b, Occlusion compounds. XVI. Stereospecific reactions with inclusion compounds. Chem. Ber. 92, 1739.

CURIE, P., 1894, In: 1963, E. I. Klabunovskii, Asymmetric Synthesis. (Veb Deutscher Verlag der Wissenschaften, Berlin), p. 197, ref. [479].

CURTI, P. and U. COLOMBO, 1952, Chromatography of stereoisomers with 'Tailor made' compounds. J. Am. Chem. Soc. 74, 3961.

DAS SARMA, B. and J. C. BAILAR, Jr., 1955, The stereochemistry of metal chelates with polydentate ligands. Part I. J. Am. Chem. Soc. 77, 5476.

DAVIS, T. L. and J. ACKERMAN, Jr., 1945, Asymmetric synthesis. III. Experiments toward a total asymmetric synthesis of tartaric acid. J. Am. Chem. Soc. 67, 486.

DAVIS, T. L. and R. HEGGIE, 1935a, A total asymmetric synthesis by addition of bromine to an ethylenic linkage. J. Am. Chem. Soc. 57, 377.

DAVIS, T. L. and R. HEGGIE, 1935b, Asymmetric syntheses. II. Addition of chlorine to trinitrostilbene J. Am. Chem. Soc. 57, 1622.

DICKEY, F. H., 1949, The preparation of specific adsorbants. Proc. Nat. Acad. Sci. U.S. 35 (5), 227.

DICKEY, F. H., 1955, Specific adsorption. J. Phys. Chem. 59, 695.

DIMODICA, G. and E. ANGELETTI, 1952, Chromatographic separation of derivatives of bitolyls into the optical antipodes. Ric. Sci. 22, 715; 1953, Chem. Abstr. 47, 6918d.

DOTY, P. and R. D. LUNDBERG, 1956, Polypeptides. X. Configurational and stereochemical effects in the amine-initiated polymerization of N-carboxyanhydrides. J. Am. Chem. Soc. 78, 4810.

DOTY, P., A. WADA, J. T. YANG and E. R. BLOUT, 1957, Polypeptides. VIII. Molecular configuration of poly-L-glutamic acid in water-dioxane solution. J. Polymer Sci. 23, 851.

DOTY, P. and J. T. YANG, 1956, Polypeptides. VII. Poly-γ-benzyl-L-glutamate: the helix-coil transition in solution. J. Am. Chem. Soc. 78, 498.

DRAFFEN, G. H., G. EGLINTON, J. M. HAYES, J. R. MAXWELL and C. T. PILLINGER, 1969, Organic analysis of the returned lunar sample. Chem. in Britain 5 (7), 296.

EBEL, F., 1933, The resolution of racemates; In: Stereochemistry, K. Freudenberg, ed. (F. Deuticke, Leipzig and Vienna) pp. 564–567.

ERRERA, G., 1898, Asymmetry and vitalism. Nature 58, 616.

FARINA, M., 1969, Some remarks on macromolecular asymmetric synthesis. Makromolec. Chem. 122, 237.

FERREIRA, R. C., 1953, Resolution of racemic mixtures by symmetrical agents. Nature 171, 39.

FERRONI, E. and R. CINI, 1960, The resolution of complex antipodes by optically active solids. J. Am. Chem. Soc. 82, 2427.

FITZGERALD, G. S., 1898a, Chance or vitalism. Nature 58, 545.

FITZGERALD, G. S., 1898b, Asymmetry and vitalism. Nature 59, 76.

FOX, S. W., 1957, The chemical problem of spontaneous generation. J. Chem. Educ. 34, 472.

FOX, S. W., 1966, The development of rigorous tests for extraterrestrial life; In: Biology and the exploration of mars, C. S. Pittendrigh, W. Vishniac and J. P. T. Pearman, eds. (Nat. Acad. Sci.-Nat. Res. Council, Washington, D.C.) pp. 213.-228.

FOX, S. W., J. E. JOHNSON and A. VEGOTSKY, 1956, On biochemical origins and optical activity. Science 124, 923.

FRANK, F. C., 1953, On spontaneous asymmetric synthesis. Biochim. et Biophys. Acta 11, 459.

FRENCH, D., 1957, The schardinger dextrins. Advan. Carbohydrate Chem. 12, 189.

FREUNDLER, P., 1909, The question of asymmetric synthesis. Ber. 42, 233.

FUKAWA, H., Y. IZUMI, S. KOMATSU and S. AKABORI, 1962, Studies om modified hydrogenation catalyst. I. Selective hydrogenation activity of modified Raney nickel catalyst for carbonyl group and C=C double bond. Bull. Chem. Soc. Japan 35 (10), 1703.

GARAY, A. S., 1968, Origin and role of optical isomery in life. Nature 219, 338.

GAUSE, G. F., 1941, Optical activity and living matter. (Biodynamica, Normandy, Missouri) pp. 19–34.

GERNEZ, M., 1866, Separation of left and right tartrates with the aid of supersaturated solutions. Compt. Rend. 63, 843.

GILLARD, R. D., P. M. HARRISON and E. D. McKENZIE, 1967, Optically active coordination compounds. Part IX. Complexes of dipeptides with cobalt (III). J. Chem. Soc. (A) 618.

GILMAN, H., 1929, The direct synthesis of optically active compounds and an explanation of the origin of the first optically active compound. Iowa State Coll. J. Sci. 3, 227.

GOLDHABER, M., L. GRODZINS and A. W. SUNYAR, 1957, Evidence for circular polarization of Bremsstrahlung produced by beta rays. Phys. Rev. 106, 826.

GOLDSCHMIDT, V. M., 1952, Geochemical aspects of the origin of complex organic molecules on the earth, as precursors to organic life. New Biol. 12, 97.

GREENSTEIN, J. P., 1954, Resolution of racemic *a*-amino acids. Adv. Protein Chem. 9, 129.

HALPERN, B., J. W. WESTLEY, E. C. LEVINTHAL and J. LEDERBERG, 1966, The Pasteur probe: an assay for molecular asymmetry. Life Sci. and Space Res. 5, 239.

HARADA, K., 1965, Total optical resolution of free *a*-amino acids by the inoculation method. Nature 206, 1354.

HARADA, K. and S. FOX, 1962, A total resolution of aspartic acid copper complex by inoculation. Nature 194, 768.

HASSELSTROM, T., M. C. HENRY and B. MURR, 1957, Synthesis of amino acids by beta radiation. Science 125, 350.

HAVINGA, E., 1941, Possibility of spontaneous asymmetric synthesis. Chemisch Weekblad 38, 642; 1942, Chem. Abstr. 36, 5790.

HAVINGA, E., 1954, Spontaneous formation of optically active substances. Biochim. et Biophys. Acta 13, 171.

HAYATSU, R., 1965, Optical activity in the Orgueil meteorite. Science 149, 443.

HAYES, J. M., 1967, Organic constituents of meteorites – a review. Geochim. et Cosmochim. Acta 31, 1395.

HENDERSON, G. M. and H. G. RULE, 1938, A new method for resolving a racemic compound. Nature 141, 917.

HENLE, F. and H. HAAKH, 1908, The question of total asymmetric synthesis. Ber. 41, 4261.

HOROWITZ, N. H. and S. L. MILLER, 1962, Current theories on the origin of life; In: Progress in the chemistry of organic natural products, L. Zechmeister, ed. (Springer, Vienna) pp. 423–459.

IDELSON, M. and E. R. BLOUT, 1958, Polypeptides. XVIII. A kinetic study of the polymerization of amino acid N-carboxyanhydrides initiated by strong bases. J. Am. Chem. Soc. 80, 2387.

INGERSOLL, A. W., 1944, The resolution of alcohols. Org. Reactions 2, 389.

INGESOLL, A. W. and R. ADAMS, 1922, Optically active dyes. I. J. Am. Chem. Soc. 44, 2930.

ISODA, T., A. ICHIKAWA and T. SHIMAMOTO, 1958, Asymmetric synthesis. I. Fundamental conditions for asymmetric reduction of *a*-oxoglutaric acid derivatives. Rikagaku Kenkyusho Hokoku 34, 134; 1960, Chem. Abstr. 54, 287d.

IZUMI, Y., M. IMAIDA, H. FUKAWA and S. AKABORI, 1963a, Asymmetric hydrogenation with modified Raney nickel. I. Studies on modified hydrogenation catalyst. II. Bull. Chem. Soc. Japan 36 (1), 21.

IZUMI, Y., M. IMAIDA, H. FUKAWA and S. AKABORI, 1963b, Asymmetric hydrogenations with modified Raney nickel. II. Studies on modified hydrogenation catalyst. III. Bull. Chem. Soc. Japan 36 (2), 155.

JAEGER, F. M., 1930, Optical activity and high temperature measurement. (McGraw-Hill, N.Y.) pp. 55–76; 203–214.

JAEGER, F. M. and G. BERGER, 1921, The photochemical decomposition of potassium cobalti-oxalate and its catalysis by salts. Rec. Trav. Chim. 40, 153.

JAMIN, M., 1850, On reflection by liquids. Compt. Rend. 31, 696.

JAPP, F. R., 1898, Stereochemistry and vitalism. Nature 58, 452.

JUNGFLEISCH, 1884, The resolution of substances optically inactive by compensation. Bull. Soc. Chim. Paris 41, 222.

KARAGOUNIS, G., E. CHARBONNIER and E. FLÖSS, 1959, Chromatographic resolution of racemic compounds. J. Chromatog. 2, 84.

KARAGOUNIS, G. and G. COUMOULOS, 1938, A new method for resolving a racemic compound. Nature 142, 162.

KARAGOUNIS, G. and G. DRIKOS, 1933a, The stereochemistry of free triarylmethyl radicals. A total asymmetric synthesis. Naturwiss. 21, 607.

KARAGOUNIS, G. and G. DRIKOS, 1933b, Stereochemistry of the free triarylmethyl radicals: A totally asymmetrical synthesis. Nature 132, 354.

KARAGOUNIS, G. and G. DRIKOS, 1934, The stereochemistry of the free triarylmethyl radical. A total asymmetric synthesis. Z. Physik. Chem. B26, 428.

KIPPING, F. S. and W. J. Pope, 1898a, Enantiomorphism. J. Chem. Soc. 73, 606.

KIPPING, F. S. and W. J. POPE, 1898b, The separation of optical isomerides. Proc. Chem. Soc. 113.

KIPPING, F. S. and W. J. FOPE, 1898c, Stereochemistry and vitalism. Nature 59, 53.

KIPPING, F. S. and W. J. POPE, 1909, The crystallization of externally compensated mixtures. J. Chem. Soc. 95, 103.

KLABUNOVSKII, E. I., 1963, Asymmetric syntheses. (Veb Deutscher Verlag der Wissenschaften, Berlin) pp. 175–180.

KLABUNOVSKII, E. I., 1968, Stereospecific catalysis in the field of optically active polymers. Russian Chem. Rev. 37 (12). 969.

KLABUNOVSKII, E. I. and V. V. PATRIKEEV, 1951, Mechanism of the asymmetrizing effect of metal catalysts deposited on right and left quartz. Doklady Akad. Nauk. S.S.S.R. 78 458; 1951, Chem. Abstr. 45, 7860a.

KLABUNOVSKII, E. I. and V. V. PATRIKEEV, 1959, Some questions on symmetry and asymmetry in the animal and plant worlds; In: The origin of life on the earth, F. Clark and R. L. Synge, eds. (Pergamon Press, N.Y.) pp. 175–179.

KLAVUNOVSKII, E. I., V. V. PATRIKEEV and A. A. BALANDIN, 1960, Separation of racemic hydrocarbons into antipodes. Izvest. Akad. Nauk. S.S.S.R., Otdel. Khim. Nauk., 552; 1960, Chem. Abstr. 54, 22312g.

KLABUNOVSKII, E. I., L. M. VOLKOVA and A. E. AGRONOMOV, 1961, A new method of preparation of stereospecific silica gels. Izvest. Akad. Nauk. S.S.S.R., Otdel. Khim. Nauk., 2101; 1962, Chem. Abstr. 56, 9928.

KLEMM, L. H. and D. REED, 1960, Optical resolution by molecular complexation chromatography. J. Chromatog. 3, 364.

KLINGMÜLLER, V. and L. MAIER-SIHLE, 1957, Paper chromatographic resolution of the racemates of histidine and tryptophan. Z. Physiol. Chem. 308, 49.

KLISS, R. M. and C. N. MATTHEWS, 1962, Hydrogen cyanide dimer and chemical evolution. Proc. Natl. Acad. Sci. U.S. 48, 1300.

KÖGL, F., 1949, Chemical and biochemical investigations of tumor proteins. Experientia, 5, 173.

KÖGL, F. and H. ERXLEBEN, 1939, Chemistry of tumors. Etiology of malignant tumors. Z. Physiol. Chem. 258, 57.

KOTAKE, M., T. SAKAN, N. NAKAMURA and S. SENOH, 1951, Resolution into optical isomers of some amino acids by paper chromatography. J. Am. Chem. Soc. 73, 2973.

KREBS, H. and J. DIEWALD, 1957a, Resolution of amino acid racemates into active components. German Patent 1,013,655 (Aug. 14, 1957); 1960, Chem. Abstr. 54, 329c.

KREBS, H. and J. DIEWALD, 1957b, Cleavage of racemic amino acids. German Patent 1,016,713 (Oct. 3, 1957); 1960, Chem. Abstr. 54, 1344g.

KREBS, H., J. DIEWALD, H. ARLITT and J. A. WAGNER, 1956a, Chromatographic resolution of racemates. II. Attempts to resolve octahedral complexes. Z. Anorg. Allgem. Chem. 287, 98.

KREBS, H., J. DIEWALD and J. A. WAGNER, 1955, Chromatographic resolution of racemates by means of starch columns. Angew. Chem. 67, 705.

KREBS, H. and R. RASCHE, 1954a, Chromatographic process for optical activation of racemates. Naturwiss. 41, 63.

KREBS, H. and R. RASCHE, 1954b, Chromatographic resolution of racemates. I. Optically active cobalt complexes with dithio acids. Z. Anorg. Allgem. Chem. 276, 236.

KREBS, H., J. A. WAGNER and J. DIEWALD, 1956b, The chromatographic resolution of racemates. III. Attempts at the activation of organic hydroxy and amino compounds with an asymmetric carbon atom. Chem. Ber. 89, 1875.

KUEBLER, Jr., J. R. and J. C. BAILAR, Jr., 1952, The stereoisomerism of complex inorganic compounds. XIV. Studies upon the stereochemistry of saturated tetravalent nitrogen compounds. J. Am. Chem. Soc. 74, 3535.

KUHN, W., 1958, Possible relationship between optical activity and aging. Adv. Enzymol. 20, 129.

KUHN, W. and E. BRAUN, 1929, The photochemical origin of optically active substances. Naturwiss. 17, 227.

KUHN, W. and E. KNOPF, 1930a, Preparation of optically active substances with the help of light. Z. Physik. Chem. 7B, 292.

KUHN, W. and E. KNOPF, 1930b, Photochemical preparation of optically active substances. Naturwiss. 18, 183.

KULKARNI, R. K. and H. MORAWETZ, 1961, Effect of asymmetric centers on free radical polymerization and the properties of polymers. J. Polymer Sci. 54, 491.

KVENVOLDEN, K. A., E. PETERSON and G. E. POLLOCK, 1969, Optical configuration of amino-acids in pre-cambrian fig tree chert. Nature 221, 141.

LANGENBECK, W. and G. TRIEM, 1936, Theories on the origin and maintenance of optical activity in nature. Z. Physik. Chem. A177, 401.

LECOQ, H., 1943, Resolution of racemic ephedrine by chromatographic analysis. Bull. Soc. Roy. Sci. Liege 12, 316; 1948, Chem. Abstr. 42, 7490a.

LEDERBERG, J., 1965, Signs of life: criterion-system of exobiology, Nature 207, 9.

LEDERER, E. and M. LEDERER, 1957, Chromatography. 2nd Edit. (Elsevier Publishing Co., N.Y.) pp. 32ff.; 420ff.

LEE, T. D. and C. N. YANG, 1956, The question of parity conservation in weak interactions. Phys. Rev. 104, 254.

LEMMLEIN, G. G., 1939, The number of left and right crystals of quartz within a given deposit. Trav. Lab. Biogeochem. Acad. Sci. U.R.S.S. 5, 225; 1940, Chem. Abstr. 34, 4019.

LEONARD, N. J. and W. J. MIDDLETON, 1952, Reductive cyclization. A method for the synthesis of tricyclic compounds possessing a bridgehead nitrogen. J. Am. Chem. Soc. 74, 5114.

LOWE, C. U., M. W. REES and R. MARKHAM, 1963, Synthesis of complex organic compounds from simple precursors: formation of amino acids, amino acid polymers, fatty acids and purines from ammonium cyanide. Nature 199, 219.

LOWRY, T. M., 1964, Optical rotatory power. (Dover Publications, Inc., N.Y.) p. 337–346.

LUNDBERG, R. D. and P. DOTY, 1957, Polypeptides. XVII. A study of the kinetics of the primary amine-initiated polymerization of N-carboxyanhydrides with special reference to configurational and stereochemical effects. J. Am. Chem. Soc. 79, 3961.

MANECKE, G. and W. LAMER, 1968, Racemate separation on optically active polymers. Naturwiss. 55 (10), 491.

MARTIN, H. and W. KUHN, 1941, A multiplication process for separating racemates. Z. Elektrochem. 47, 216.

MEINSCHEIN, W. G., C. FRONDEL, P. LAUR and K. MISLOW, 1966, Meteorites: optical activity in organic matter. Science 154, 377.

MILLER, S. L., 1957a, The formation of organic compounds on the primitive earth. Ann. N.Y. Acad. Sci. 69 (2), 260.

MILLER, S. L., 1957b, The mechanism of synthesis of amino acids by electric discharge. Biochim. et Biophys. Acta 23, 480.

MILLER, S. L. and H. C. UREY, 1959, Organic compound synthesis on the primitive earth. Science 130, 245.

MILLS, W. H., 1932, Some aspects of stereochemistry. Chem. and Ind. 51, 750.

MITCHELL, S., 1930, The asymmetric photochemical decomposition of humulene nitrosite by circularly polarized light. J. Chem. Soc., 1829.

MITCHELL, S. and I. M. DAWSON, 1944, The asymmetric photolysis of β-chloro-β-nitroso-a,δ-diphenylbutane with circularly polarized light. J. Chem. Soc., 452.

MORGAN, G. T. and D. G. SKINNER, 1925, Stereoisomeric azo dyes J. Chem. Soc. 127, 1731.

NAGY, B., M. T. J. MURPHY, V. E. MODZELESKI, G. ROUSER, G. CLAUS, D. J. HENNESSY, U. COLUMBO and F. GAZZARRINI, 1964, Optical activity in saponified organic matter isolated from the interior of the Orgueil meteorite. Nature 202, 228.

NAKAHARA, A. and R. TSUCHIDA, 1954, Synthesis of the tris-(dimethylglyoximo)-cobaltate III. J. Am. Chem. Soc. 76, 3103.

NEWMAN, A. C. D. and H. M. POWELL, 1952, The spontaneous resolution of solvated tri-*o*-thymotide. J. Chem. Soc., 3747.

NORTHROP, J. H., 1957, Optically active compounds from racemic mixtures by means of random distribution. Proc. Nat. Acad. Sci. U.S. 43, 304.

O'GORMAN, J. M., 1944, Attempted repetition of a reported total asymmetric synthesis. J. Am. Chem. Soc. 66, 1041.

OHARA, M., I. FUJITA and T. KWAN, 1962, The selective adsorption of optical antipodes as revealed by the chromatographic techniques. Bull. Chem. Soc. Japan 35, 2049.

OHARA, M., K. OHTA and T. KWAN, 1964, The anomalous adsorption of enantiomers of mandelic acid and atrolactic acid on the optically active adsorbent. Bull. Chem. Soc. Japan 37, 76.

ORO, J. and C. L. GUIDRY, 1961, Direct synthesis of polypeptides. I. Polycondensation of glycine in aqueous ammonia. Arch. Biochem. Biophys. 93, 166.

OSTROMISSLENSKY, I., 1908, Investigations in the realm of mirror image isomerism. Ber. 41, 3035.

PADOA, M., 1909, Attempted asymmetric synthesis with circularly polarized light. Atti Accad. Nazl. Lincei 18II, 390; 1910, Chem. Abstr. 4, 2452.

PALM, C. and M. CALVIN, 1962, Primordial organic chemistry. I. Compounds resulting from electron irradiation of $C^{14}H_4$. J. Am. Chem. Soc. 84, 2115.

PARANJAPE, K. D., N. L. PHALNIKAR, B. V. BHIDE and K. S. NARGUND, 1944, A case of total asymmetric synthesis. Nature 153, 141.

PASCHKE, R., R. W. H. CHANG and D. YOUNG, 1957, Probable role of gamma irradiation in origin of life. Science 125, 881.

PASTEUR, L., 1848, On the relations which can exist between crystalline form, chemical composition, and the sense of rotatory polarization. Ann. Chim. Phys. 24 [3], 442.

PASTEUR, L., 1852, On aspartic and malic acids. Ann. Chim. Phys. 34 [3], 46.

PASYNSKII, A. G. and T. E. PAVLOVSKAYA, 1964, The formation of biochemically important compounds in the pre-biological stages of the earth's development. Russian Chem. Rev. 33, 514.

PEARSON, K., 1898a, Chance or vitalism. Nature 58, 495.

PEARSON, K., 1898b, Asymmetry and vitalism. Nature 59, 30.

PENZIEN, K. and G. M. J. SCHMIDT, 1969, Reactions in chiral crystals: an absolute asymmetric synthesis. Angew. Chem. Internat. Edit. 8 (8), 608.

PETROV, Y. A., E. I. KLABUNOVSKII, and A. A. BALANDIN, 1966, Catalysts of unsymmetric hydrogenation; 1968, Chem. Abstr. 69, 110244q.

PETROV, Y. I., E. I. KLABUNOVSKII and A. A. BALANDIN, 1967, Asymmetric hydrogenation on Raney nickel modified by optically active species. Kinet. Katal. 8 (4), 814; 1968, Chem. Abstr. 68, 58786f.

PINO, P., 1965, Optically active addition polymers. Adv. Polymer Sci. 4, 393.

PINO, P., F. CIARDELLI, G. P. LORENZ and G. NATTA, 1962, Optically active vinyl polymers. VI. Chromatographic resolution of linear polymers of (R)(S)-4-methyl-1-hexene. J. Am. Chem. Soc. 84, 1487.

PIRAK, J., 1922, The question of asymmetric synthesis. Biochem. Z. 130, 76.

PIRIE, N. W., 1959, The position of stereoisomerism in argument about the origins of life. Trans. Bose Research Inst. (Calcutta) 22, 11.

PONOMAREV, A. A. and V. V. ZELENKOVA, 1952, Asymmetric synthesis of spiranes and tetrahydrofuran alcohols. Doklady Akad. Nauk. S.S.S.R. 87, 423; 1954, Chem. Abstr. 48, 663d.

PONOMAREV, A. A. and V. V. ZELENKOVA, 1953, Furan compounds. IV. Asymmetric synthesis of spirans and tetrahydrofuran alcohols. Zhur. Obshchei Khim. 23, 1543; 1954, Chem. Abstr. 48, 10722i.

PORTER, C. W. and C. T. HIRST, 1919, Asymmetric dyes. J. Am. Chem. Soc. 41, 1264.

PORTER, C. W. and H. K. IHRIG, 1923, Asymmetric dyes. J. Am. Chem. Soc. 45, 1990.

POWELL, H. M., 1950, Molecular compounds. Endeavor 9, 154.

POWELL, H. M., 1952, New procedures for resolution of racemic substances. Nature 170, 155.

POWELL, H. M., 1956, Thoughts on optical activity. Endeavor 15, 20.

PRELOG, V. and P. WIELAND, 1944, The resolution of Tröger's base into its optical antipodes; a note on the stereochemistry of trivalent nitrogen. Helv. Chim. Acta 27, 1127.

PRICE, C. C. and M. OSGAN, 1956a, The polymerization of 1-propylene oxide. J. Am. Chem. Soc. 78, 4787.

PRICE, C. C., M. OSGAN, R. E. HUGHES and C. SHAMELAN, 1956b, The polymerization of 1-propylene oxide. J. Am. Chem. Soc. 78, 690.

READ J., 1953, Natural origin of optically active substances and optical resolution by symmetric agents. Nature 171, 843.

REDLICH, O., C. M. GABLE, A. K. DUNLOP and R. W. MILLAR, 1950a, Addition compounds of urea and organic substances. J. Am. Chem. Soc. 72, 4153.

REDLICH, O., C. M. GABLE, L. R. BEASON and R. W. MILLAR, 1950b, Addition compounds of thiourea. J. Am. Chem. Soc. 72, 4161.

RITCHIE, P. D., 1933, Asymmetric synthesis and asymmetric induction. (Oxford Univ. Press, London) p. 44.

RITCHIE, P. D., 1947, Recent views on asymmetric synthesis and related processes. Adv. Enzymol. 7, 65.

RODBERG, L. S. and V. F. WEISSKOPF, 1957, Fall of parity. Science 125, 627.

RUSH, J. H., 1957, The dawn of life. (New American Library of World Literature, Inc., N.Y.) pp. 167–171.

SCHEIBLER, H., 1925, The resolution of optically inactive substances into their active components; In: Methods of organic chemistry, Vol. 2, 3rd Edition, J. Houben, ed. (G. Thieme, Leipzig) pp. 1063–1066.

SCHKLOVSKII, I. S. and C. SAGAN, 1966, Intelligent life in the universe. (Dell Publishing Co., Inc., N.Y.) p. 183.

SCHLENK, Jr., W., 1949, Urea addition of aliphatic compounds. Ann. 565, 204.

SCHLENK, Jr., W., 1951a, Organic occlusion compounds. Fortschr. Chem. Forsch. 2, 92.

SCHLENK, Jr., W., 1951b, Thiourea addition of organic compounds. Ann. 573, 142.

SCHLENK, Jr., W., 1952a, Separation of optical antipodes by complex formation without utilization of an asymmetric molecule. Experientia, 8, 337.

SCHLENK, Jr., W., 1952b, The fractionation of ozocerites and the separation of optical isomers. Analyst 77, 867.

SCHLENK, Jr., W., 1955, An asymmetric inclusion lattice as an instrument of stereochemistry. Angew. Chem. 67, 762.

SCHLENK, Jr., W., 1957, Separation of racemic mixtures of organic compounds capable of forming occlusion compounds with urea. German Patent 1.009,185 (May 29, 1957); 1960, Chem. Abstr. 54, 1308d.

SCHLENK, Jr., W., 1960a, Resolution of racemic compounds. German Patent 1,074,583 (Feb. 14, 1960); 1961, Chem. Abstr. 55, 13318c.

SCHLENK, Jr., W., 1960b, Simultaneous separation of the optical forms from racemic mixtures of organic compounds forming addition products with urea. German Patent 1,076,686 (March 3, 1960); 1961, Chem. Abstr. 55, 13321c.

SCHLENK, Jr., W., 1960c, Investigation of inclusion diastereomerism and the determination of the absolute configuration of hexagonal urea lattices. Angew. Chem. 72, 845.

SCHWAB, G., F. ROST and L. RUDOLPH, 1934, Optically asymmetric catalysis on quartz crystals. Kolloid Z. 68, 157.

SCHWAB, G. M. and L. RUDOLPH, 1932, Catalytic cleavage of racemates by d- and l-quartz. Naturwiss. 20, 363.

SCHWEITZER, G. K. and C. K. TALBOTT, 1950, The resolution of optically active inorganic stereoisomers by adsorption on quartz. J. Tennessee Acad. Sci. 25, 143.

SECOR, R. M., 1963, Resolution of optical isomers by crystallization procedures. Chem. Rev. 63, 297.

SMITH, A. E., 1950, The crystal structures of urea-hydrocarbon and thiourea-hydrocarbon complexes. J. Chem. Phys. 18, 150.

SMITH, A. E., 1952, The crystal structures of urea-hydrocarbon complexes. Acta Cryst. 5, 224.

SORET, C., 1900, The formation of right and left crystals of optically active substances. Z. Krystallog. 34, 630.

SPIALTER, L. and J. H. FUTRELL, 1960, Examination of γ-irradiated benzene for optical activity. Nature 188, 225.

SPIERS, C. W. F., 1937, The question of the origin of optical activity in nature. Naturwiss. 25, 457.

STANKIEWICZ, A., 1938, In: 1963, Asymmetric synthesis; E. I. Klabunovskii. (Veb Deutscher Verlag der Wissenschaften, Berlin) p. 121.

STARODUBTSEV, S. V., M. N. GURSKIY and A. G. SIZYKH, 1959, Changes in optical properties of benzene after exposure to γ-rays. Doklady Akad. Nauk. S.S.S.R. 129, 307; 1961, Chem. Abst. 55, 26626d.

STEINMAN, G., 1967, Stereoselectivity in peptide synthesis under simple conditions. Experientia 23 (3), 177.

STEVENS, C. M., P. E. HALPERN and R. P. GIGGER, 1951, Occurrence of D-amino acids in some natural materials. J. Biol. Chem. 190, 705.

STEVENSON, K. L. and J. F. VERDIECK, 1968, Partial photoresolution. Preliminary studies on some oxalato complexes of chromium (III). J. Am. Chem. Soc. 90, 2974.

STRONG, W. M., 1898, Stereochemistry and vitalism. Nature 59, 53.

STRYER, L., 1966, Optical asymmetry; In: Biology and the exploration of Mars, C. S. Pittendrigh, W. Vishniac and J. P. T. Pearman, eds. (Nat. Acad. Sci.-Nat. Res. Counc., Washington, D.C.) pp. 141–146.

SWALLOW, A. J., 1960, Radiation chemistry of organic compounds. (Pergamon Press, N.Y.) pp. 243–245.

SWERN, D., 1957, Inclusion compounds; In: Encyclopedia of chemical technology first supplement Vol., R. E. Kirk and D. F. Othmer, eds. (Interscience Encyclopedia, Inc., N.Y.) pp. 429–448.

TATSUMI, S., M. IMAIDA, Y. FUKUDA, Y. IZUMI and S. AKABORI, 1964, Asymmetric hydrogenation with modified Raney nickel. III. Studies of the modified hydrogenation catalyst. IV. Bull Chem. Soc. Japan 37 (6), 846.

TERENT'EV, A. P. and E. I. KLABUNOVSKII, 1959, The role of dissymmetry in the origin of living material; In: The origin of life on the earth, F. Clark and R. L. Synge, eds. (Pergamon Press, N.Y.) pp. 95–105.

TERENT'EV, A. P., E. I. KLABUNOVSKII and V. V. PATRIKEEV, 1950, Asymmetric synthesis with the aid of catalysts deposited on right- and left-quartz. Doklady Akad. Nauk. S.S.S.R. 74, 749; 1951, Chem. Abstr. 45, 3798c.

THEILAKER, W., 1955, Methods for the production of optically active from inactive compounds; In: Houben-Weyl, Methods of organic chemistry, 4th Edition, Vol. 4, Part 2. (G. Thieme, Stuttgart) pp. 509–511.

TSUCHIDA, R., M. KOBAYASHI and A. NAKAMURA, 1935, Asymmetric adsorption of complex salts on quartz. J. Chem. Soc. Japan 56, 1339; 1936, Chem. Abstr. 30, 926.

TSUCHIDA, R., M. KOBAYASHI and A. NAKAMURA, 1936, The configuration of chlorobis-dimethylglyoximoamine cobalt. Bull. Chem. Soc. Japan 11, 38.

TSUCHIDA, R., A. NAKAMURA and M. KOBAYASHI, 1935, Asymmetric photochemical decomposition of complex salts. J. Chem. Soc. Japan 56, 1335; 1936, Chem. Abstr· 30, 963.

ULBRICHT, T. L. V., 1959, Asymmetry: the non-conservation of parity and optical activity. Quart. Rev. 13, 48.

ULBRICHT, T. L. V., 1962, The origin of optical activity and the origin of life; In: Comparative biochemistry, Vol. IV, Part B, M. Florkin and H. S. Mason, eds. (Academic Press, N.Y.) pp. 16–25.

ULBRICHT, T. L. V. and F. VESTER, 1962, Attempts to induce optical activity with polarized β-radiation. Tetrahedron 18, 629.

VAN'T HOFF, J. H., 1894, The arrangement of atoms in space. 2nd Edit., p. 30; 1908, 3rd Edit., p. 8.

VAN 'T HOFF, J. H. and H. M. DAWSON, 1898, The racemic transformation of ammonium bimalate. Ber. 31, 528.

VESTER, F., 1957, Seminar at Yale University, Feb. 7.

VESTER, F., T. L. V. ULBRICHT and H. KRAUCH, 1959, Optical activity and parity violation in β-decay. Naturwiss. 46, 68.

VOL'KENSHTEIN, M., 1959, Discussion of asymmetry problem; In: The origin of life on the earth, F. Clark and R. L. Synge, eds. (Pergamon Press, N.Y.) p. 174.

WALD, G., 1957, The origin of optical activity. Ann. N.Y. Acad. Sci. 69, 352.

WEINGARTEN, H., 1958, Kinetics and mechanisms of the polymerization of N-carboxy-*a*-amino acid anhydrides. J. Am. Chem. Soc. 80, 352.

WERNER, A., 1914, The asymmetric cobalt atom. XI. Oxalodiethylenediaminocobalti salts and a new method for splitting racemic inorganic compounds. Ber. 47, 2171.

WHYTE, L. L., 1957, Chirality. Nature 180, 513.

WIGNER, E. P., 1965, Violations of symmetry in physics. Sci. Amer., Dec., pp. 28–36.

WILLSTÄTTER, 1904, An experiment toward the theory of color. Ber. 37, 3758.

WU, C. S., E. AMBLER, R. W. HAYWARD, D. D. HOPPES and R. P. HUDSON, 1957, Experimental test of parity conservation in β-decay. Phys. Rev. 105, 1413.

YAMAGATA, Y., 1966, A hypothesis for the asymmetric appearance of biomolecules on earth. J. Theoret. Biol. 11, 495.

YANG, J. T. and P. DOTY, 1957, The optical rotatory dispersion of polypeptides and proteins in relation to configuration. J. Am. Chem. Soc. 79, 761.

C. Ponnamperuma (ed.), Exobiology. © North-Holland Publishing Company

CHAPTER 7

The origin of membranes and related surface phenomena

DINESH O. SHAH

1. Introduction

It has been recognized that cells and cell organelles are surrounded by a boundary layer which serves as a barrier with selective permeability, thereby controlling many biochemical reactions within the cell. The elucidation of the structure and function of this boundary layer, generally called the membrane, is one of today's most challenging problems. In view of the wide-ranging biochemical and physiological phenomena associated with biological membranes, speculations about the formation of membranous structures and the role they might have played in bringing about a higher level of organization of organic matter during the primitive earth conditions are both interesting and desirable. It is hoped that careful consideration of such speculations and hypotheses may suggest appropriate experimentation which will lead to a greater understanding of the origin of life.

The Oparin-Haldane hypotheses presuppose a long chemical evolution on earth before the appearance of living systems (Oparin 1924; Haldane 1928). During this period of chemical evolution, various organic compounds were formed from primitive atmospheric conditions. These organic compounds may have accumulated until the primitive oceans reached the consistency of a 'hot dilute soup' (Haldane 1928). According to this concept, reactions between various species of organic molecules must have occurred at the air-sea interface, as well as in the aqueous and gaseous environment. These considerations bring the principles of surface and colloid chemistry into the discussion of chemical evolution. The formation of membranous structures resulting from phenomena such as adsorption, micelle formation, solubilization and coacervation, is relevant to the origin of living systems.

2. Current concepts of the structure and function of biological membranes

2.1. Biological membranes

Permeability studies have shown that the exchange of water and solutes between the cell and its environment is controlled by a thin membrane. Observations of electron micrographs have shown that various cell organelles are also surrounded by such membranes. This increases the importance of the role that these membranes play in biological structures and processes. Various enzymes are also known to be associated with membranes. Membranes from various sources may show differences in enzymic and functional properties, presumably due to evolutionary changes or adaptation. The history and development of the study of membranes have been reviewed by several authors: Benedetti and Emmelot (1968a, 1968b), Chapman (1968), Van Deenen (1965), Finean (1966), Glauert and Lucy (1968), Kavanau (1965), Maddy (1966), Robertson (1967), Rothfield and Finkelstein (1968), Sjostrand (1968) and Stein (1967).

The following are some common structural characteristics of membranes. Membranes are flexible structures separating the cell from its environment, or separating the cell-organelles from the cellular matrix. They have a thickness of only a few molecular layers, and can be isolated as separate entities (Brown and Danielli 1964). Lipid and protein are the major components of the isolated membranes. The predominant lipid fraction is phospholipid. Cholesterol is the major neutral lipid, but carotenoids, quinones and glycolipids also may be present. The physical characteristics of many membranes such as thickness, surface tension, electrical resistance, electrical capacitance and water permeability are similar. In electron micrographs membranes generally appear as two dense, parallel, outer bands.

Membranes serve as barriers for solutes and water. They are generally more permeable to substances soluble in nonpolar solvents than to water soluble compounds. The permeability to ions is generally low, as compared with the permeability to water. Membranes are involved in such phenomena as active transport (i.e., energy-coupled transport), facilitated diffusion and the selective transport of ions. Such properties of cells as excitability, impulse conduction, secretion, and cell-cell interactions also may depend upon the presence of membranes.

In general, the structure of membranes should depend upon the lipid-lipid, protein-protein and lipid-protein interactions which can occur in the medium formed by the aqueous environment of the cell. Two major models have been proposed for the structure of membranes: the Danielli, or bilayer model, and the subunit model (Danielli and Davson 1935; Green et al. 1967). Other models have been proposed by Kavanau (1965), Lucy (1964), Lenard and Singer (1966), Benson (1966), and Vandenheuvel (1965, 1966), but these can be considered variations of the two main models.

The Danielli, or bilayer, model is one of the earliest models proposed for the structure of cell membranes. Originally proposed by Gorter and Grendel (1925) and later independently by Danielli and Davson (1935), the model suggests that membrane lipids are arranged in two monomolecular layers in which the polar ends of the lipid molecules are directed outward and the nonpolar ends inward (fig. 1). Danielli originally proposed, on the basis of surface tension measurements, that protein must be adsorbed at the lipid-water interfaces.

EXTERIOR

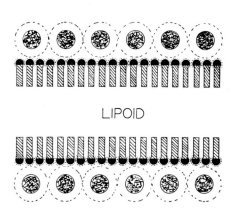

LIPOID

INTERIOR

Fig. 1. A proposed structure of the red cell membrane (Danielli and Davson 1935).

Several investigators have pointed out that the fatty acid chains of the membrane interior are most likely in a disordered state, approaching that of a liquid hydrocarbon (Luzzati 1968; Luzzati and Husson 1962). The surface area per phospholipid molecule in the erythrocyte membrane has been calculated to be 71 ± 4 Å (Engelman 1969). Calculations from X-ray diffraction measurements show an area of 65 Å² per lipid molecule in the lamellar liquid-crystalline phase of mitochondrial lipids at physiological temperatures (Gulic-Krzywicki et al. 1967). Slightly higher areas per molecule have been reported for other phospholipid-water systems (Rand and Luzzati 1968; Small 1967). Birefringence (Rogers and Winsor 1967), infrared, nuclear magnetic resonance (NMR) and calorimetric studies (Steim 1968) have confirmed the liquid-like state of hydrocarbon chains in membranes and model systems.

The role of cholesterol in membranes is still uncertain. However, evidence from monolayer (Shah and Schulman 1967a) and liposome permeability studies (De Gier et al. 1969; Bangham et al. 1967; Demel et al. 1968; Papahadjopoulos and Watkins 1967) suggests that cholesterol can have a dual effect, i.e., either liquefying or solidifying on phospholipids, depending upon the nature of their fatty acid constituents. NMR and differential thermal analysis (DTA) studies on lecithin-cholesterol dispersions are in agreement with this concept (Chapman and Penkett 1966; Ladbrooke et al. 1968). The disorder and fluidity of hydrocarbon chains in biological and model membranes depend mainly upon the temperature and length, the degree of unsaturation and the heterogeneity of the hydrocarbon chains.

The protein in membranes has been considered to be in the form of extended monomolecular layers, layers of globular molecules, or a combination of the two (Maddy 1967). Recent spectroscopic evidence indicates that at least part of the protein in the membrane is in an a-helical configuration (Chapman et al. 1968; Lenard and Singer 1966; Maddy 1967; Steim 1968; Wallach and Gordon 1968). The binding between protein and lipid is generally believed to be predominantly ionic. However, because of the loosely-packed lipid bilayer, hydrophobic interaction between lipid and protein could occur without disruption of the lipid bilayer.

The outer and inner surface of membranes are asymmetric in structure (Schmitt 1959; Robertson 1969; Finean 1957) and function (Adelman and Senft 1968; Hoffman 1962; Singer and Tasaki 1968). Some enzymes and antigens also have been demonstrated to be present only on one side of the membrane (Hasselbach and Elfvin 1967; Benedetti and Emmelot 1968b). The electron microscopy of stained sections often shows differences in the density and width of the two layers of the membrane (Elbers 1964; Sjostrand 1968).

According to the subunit model, the membrane is composed of subunits

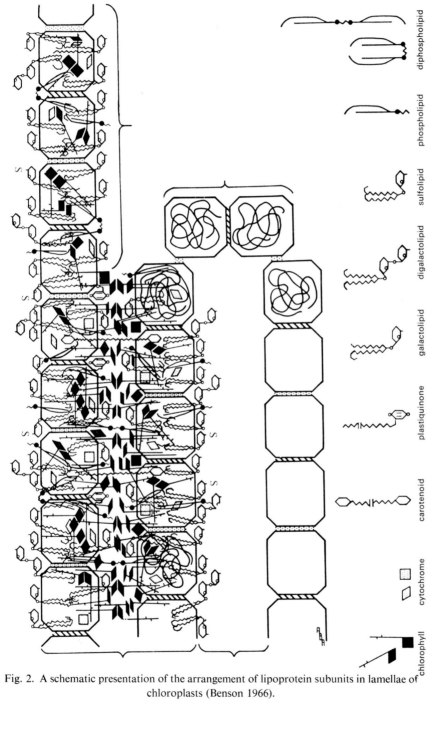

Fig. 2. A schematic presentation of the arrangement of lipoprotein subunits in lamellae of chloroplasts (Benson 1966).

which consist of lipid and protein. Recently, Stoeckenius and Engelman (1969) have pointed out that there is a distinction between structural and functional subunits of membranes. A functional subunit would comprise all the components necessary to carry out a given complex function of a membrane, and different functional subunits may be present in a membrane. The subunit model suggests that membranes may be formed in vivo from lipoprotein particles by a self-assembly process. The membrane of mitochondria (Green and Oda 1961; Lehninger 1959), retinal rods (Nilsson 1965; Omura et al. 1967), and chloroplasts (Benson 1966; Branton 1968) are believed to consist of subunits. The size of the subunit varies between 40–90 Å. A proposed subunit model for chloroplast membranes (Benson 1966; Weier and Benson 1967) is shown in fig. 2. Recent studies on spin-labelled membranes of the sarcoplasmic reticulum have shown that there are regions of low viscosity in the membranes, whereas mitochondrial membranes do not show such regions (Hubbell and McConnell 1968; Keith et al. 1968; Landgraf and Inesi 1969).

In summary, membranes may have an over-all structure represented by one of the above models, but highly differentiated functions of membranes may be attributed to the presence of specific proteins, enzymes, phospholipids, polysaccharides and metal ions in the membranes.

2.2. Model membranes

Considerable work has been done to develop model systems of biological membranes to understand the interactions and physico-chemical reactions which may occur in membranes. Three major membrane models used for this purpose are myelinics or liposomes, bilayers and monolayers. The principal characteristics of these model systems, and the information obtained from them, are summarized as follows:

2.2.1. Myelinics or liposomes

When phospholipids are dispersed in water they form spontaneously multilayered closed structures generally called myelinics, liposomes or liquid crystals. Fig. 3 shows an electronmicrograph of such lecithin dispersion taken by Dr. Anderson of our laboratory. Bangham and his co-workers (Bangham and Horne 1964; Bangham et al. 1965a–c) have investigated the structural and permeability characteristics of phospholipid liposomes or liquid crystals. They have found that the diffusion rate of anions and water across the liposome membranes is several orders of magnitude greater than that of cations.

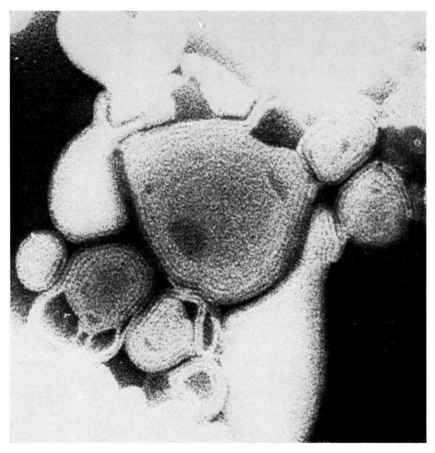

Fig. 3. An electron micrograph of ovolecithin myelinics showing concentric lipid bilayers stained with potassium phosphotungstic aced. The hydrophobic portion of the bilayer appears white, and the aqueous intervening spaces containing the negative stain appear black (by courtesy of Dr. R. O. Anderson).

Furthermore, the diffusion rate of cations is very significantly controlled by the sign and magnitude of the surface charge at the lipid-water interface. In contrast, the diffusion rates of Cl^- and water remains high, irrespective of the surface charge. Although these liposomes do not distinguish between different cations, the diffusion rate of anions is, with certain exceptions, related to their hydrated ionic radius, i.e., $Br^- = Cl^- = HCO_2^- > NO_2^- > F^- > HCO_3^- > H_2PO_4^-$. These model membranes undergo changes in permeability for cations when exposed to lytic or protective steroids (Bangham et al. 1965b).

For liquid crystals of phosphatidylserine, the activation energy for diffusion of Cl^- is 6.4 kcal/mole, whereas, that for Na^+ and K^+ is 14 and 18 kcal/mole, respectively (Papahadjopoulos and Bangham 1966). The intraparticulate water present in the liquid crystals of lecithins is lost by the the addition of an osmotically active compound. The loss of intraparticulate water follows the Boyle-Van't Hoff law (Rendi 1967; Bangham et al. 1967). Liquid crystals of charged phospholipids are freely permeable to methyl and ethyl urea, urea, glycerol, ethylene glycol and ammonium acetate. They are less permeable to malonamide and erythritol, and relatively impermeable to NaCl, KCl, Na acetate, sucrose, glucose and mannitol (Bangham et al. 1967). A discrimination between the diffusion of Na^+ and K^+ is exhibited by the liquid crystals of phosphatidylserine. In this system, the presence of Ca^{2+} increased the diffusion rate of both Na^+ and K^+. However, much greater concentrations of Mg^{2+} were required to produce the effect similar to that of Ca^{2+} on diffusion of monovalent cations through phosphatidylserine liquid crystals (Papahadjopoulos and Bangham 1966). Studies of various phospholipids by optical and electron microscopy, as well as X-ray diffraction, have shown that structural charcteristics of hydrated liquid crystals depend upon the nature of: (1) the polar group, (2) the presence of mono- or divalent cations, (3) the pH of the aqueous solution and (4) the presence of proteins in the solution (Papahadjopoulos and Miller 1967). Recently, Papahadjopoulos and Watkins (1967) have shown that Cl^- diffusion through phosphatidylcholine, phosphatidylserine, and mixtures of phosphatidylcholine, phosphatidylethanolamine and cholesterol is considerably faster than that of K^+ or Na^+, while the cations diffused faster through liquid crystals of phosphatidic acid and phosphatidylinositol. The presence of valinomycin and anaesthetic compounds such as diethyl ether, chloroform and n-butanol increases the permeability of liposomes to K^+ (Johnson and Bangham 1969a, b). Sweet and Zull (1969) have reported that serum albumin also increases the diffusion of glucose from liposomes made of phosphatidylcholine, chloesterol and dicetyl phosphate. In contrast, incorporation of cholesterol into liposomes decreases the rate of diffusion of glucose from the liposomes (Demel et al. 1968). The ability of cholesterol to reduce the glucose permeability of liposomes is markedly influenced by the nature of fatty acyl chains of phospholipids. Permeability of liposomes to glycerol and erythritol also depends upon the temperature, the nature of fatty acyl chains and the presence or absence of cholesterol in the liposomes (De Gier et al. 1968). De Gier et al. (1969) have suggested that cholesterol controls the fluidity of the hydrocarbon chains of phospholipids. In general, cholesterol acts against the effect of temperature on lipid membranes; at low temperatures it has a liquefying or disordering effect, whereas, at high temperatures it has a solidifying (or ordering) effect on phospholipid membranes.

Liposomes have been used extensively to understand the mechanism of action of various steroids, antibiotics and hemolytic agents on membranes (Sessa and Weissmann 1967, 1968a, b; Weissmann and Sessa 1967; Kinsky et al. 1966). Bonting and Bangham (1967) investigated the effect of all-trans retinol and all-trans retinaldehyde on the diffusion of cations from phospholipid liposomes to elucidate the biochemical mechanism of the visual process. Addition of retinaldehyde to phosphatidylethanolamine liposomes increased the leakage rate of Rb^+ and K^+ due to Schiff base formation (Daemen and Bonting 1969). It is evident from the above discussion that liposomes represent a very useful model system for studying the mechanism of permeability changes that may occur in biological membranes.

2.2.2. Bilayers

A method for forming bimolecular lipid membranes between two aqueous compartments was first reported by Mueller and his co-workers (1962, 1963). Fig. 4 schematically shows the process of lipid bilayer formation as suggested

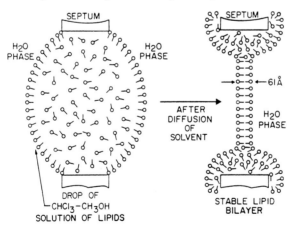

Fig. 4. A proposed mechanism of the formation of lipid bilayers separating two aqueous compartments (Thompson 1964).

by Thompson (1964). Since then, various modifications of this method have been reported to produce: (1) bilayers with large surface areas (Van den Berg 1965), (2) asymmetric bilayer with different phospholipid monolayers on each side of the bilayer (Takagi et al. 1965) or (3) bilayers with proteins adsorbed on both sides (Tsoffna et al. 1966). Danielli (1966) has shown that a bimolecular membrane has the least surface free energy for membranes of any thickness and that, consequently, continuous membranes made of biological phospholipids normally will be bimolecular in thickness. Recently, Good (1969) has reported a thermodynamic analysis of the formation and stability of bilayer films. The energy of formation of the bilayers under a variety of ionic solutions is found to be 3.4 erg per cm² (Coster and Simons 1968). However, Tien (1968a) has reported values ranging from 0.9–5.7 dynes/cm for the bifacial tension of various lipid bilayers. He has further shown (Tien 1968b) that the free energy, entropy and enthalpy of formation are all negative for the formation of bilayers at the water-oil-water interfaces. The order of the specific effect of anions in lowering the resistance of bilayers prepared from cholesterol-dodecane-hexadecyltrimethylammonium bromide was shown to be $I^- > Br^- > SO_4^{-2} > Cl^- > F^-$ (Tien and Diana 1967). Leslie and Chapman (1967) incorporated β-carotene and all-trans retinene in lecithin bilayers and followed the thinning process by measuring the optical absorption of these pigments. Tien (1968c, d) has observed photoelectric effects in bilayers containing small amounts of photopigments such as chlorophyll and xanthophyll. Bilayers of oxidized cholesterol also show photoeffects in the presence of Fe^{3+}, but not Fe^{2+} (Pant and Rosenberg 1970). Iodine, picric acid, 2,4-dinitrophenol and trinitrobenzene have been found to greatly lower the electrical resistance of egg lecithin and oxidized cholesterol bilayers due to formation of donor-acceptor complexes (Jendrasiak 1969; Lauger et al. 1967; Rosenberg and Bhowmik 1969). The presence of phospholipase A or polyene antibiotics in the aqueous phase strikingly reduces the stability of lecithin or lecithin-cholesterol bilayers (Lesslauer et al. 1968; Van Zutphen et al. 1966). The presence of valinomycin produces a marked decrease in the resistance of lipid bilayers when K^+, but not when Na^+, is in the solution (Andreoli et al. 1967; Mueller and Rudin 1967b). The specific direct current resistance of phosphatidylcholine and phosphatidylserine bilayers is dependent upon the pH, salt composition and valency of cations in the solution (Ohki 1969). Electrical conductance of phospholipid bilayers increases upon irradiation with X-rays (Sutton and Rosen 1968). Phospholipid bilayers in the presence of a proteinaceous excitability-inducing material (EIM) show action potentials similar to that of cells (Mueller and Rudin 1967a, 1968). Cass and Finkelstein (1967) have shown in lipid bilayers that the diffusion and hydrodynamic permeabilities for water do not differ. The presence of amphotericin B at low concentrations affects significantly the permeability of cholesterol but not that of phospholipid bilayers (Lippe

1968). The transverse electrical impedence of bilayers decreases markedly and reversibly when antigen-antibody or enzyme-substrate reactions occur at the bilayer interface (Del Castillo et al. 1966). Considerable work on polar group orientation, stability, permeability to water and electrical properties of bilayers has been reported by Haydon and co-workers (Haydon and Taylor 1963; Hanai et al. 1965a–d; Hanai and Haydon 1966; Everitt and Haydon 1968) and by Thompson and co-workers (Huang et al. 1964; Huang and Thompson 1965, 1966; Henn et al. 1967; Pagano and Thompson 1967, 1968; Miyamoto and Thompson 1967; Henn and Thompson 1968). Most of the work on lipid bilayers has been summarized in the recent review articles on this subject (Haydon 1968; Tien and Diana 1968; Thompson and Henn 1970; Henn and Thompson 1969).

2.2.3. Monolayers

Lipid monolayers at the air–water or oil–water interface offer a unique model system with which to study of occurrence of lipid–lipid, lipid–protein and lipid-metal ion interactions that may occur at the cell surface. Monolayers are very well-defined systems in which the molecular area, molecular orientation, intermolecular spacing and surface charge can be precisely controlled. For transport or permeability studies, monolayers may not be as good a model membrane as bilayers, but for studies on electrostatic and van der Waals interactions they offer several advantages over bilayers. Moreover, from the viewpoint of chemical evolution and the origin of membranes, monolayers are relevant to the formation of membraneous structures by adsorption at air-water and solid-liquid interfaces. The most striking contribution of monolayer studies has been in the development of the concept of the bilayer structure of biological membranes. Gorter and Grendel (1925) were the first to extract lipids from erythrocytes and spread them as a monomolecular film. They found that the ratio of film area to erythrocyte area was 2:1 for a number of mammalian species. This fact was taken as evidence that the erythrocyte surface was composed of a bimolecular layer of lipids. Recently, Bar et al. (1966) re-examined Gorter and Grendel's work and showed that the ratio of monolayer area to erythrocyte surface area is 2:1 at low surface pressures and 1:1 at collapse pressures. However, it was the earlier monolayer study of Gorter and Grendel that led Davson and Danielli to develop a model of the cell membrane as a lipid bilayer covered with adsorbed proteins (Danielli 1958; Davson 1962). The monolayer approach has provided considerably insight into many phenomena such as the effects of drugs and hormones, hemolysis, formation of lipoproteins, enzymic reactions at interfaces, intestinal absorption of fats, long-range interactions and cell guidance and locomotion, which are summarized as follows.

As discussed above, a monolayer forms one-half of the lipid bilayer in the Davson and Danielli model for membranes. Mono- and divalent cations are known to play a significant role in controlling the permeability properties of membranes. These ions are also associated with phenomena such as nerve propagation and muscle contraction. Using surface pressure, surface potential and surface viscosity techniques, several investigators have studied the interaction of cations with phosphalipid monolayers (Deamer and Cornwell 1966; Shah and Schulman 1965, 1967a–c; Bangham and Papahadjopoulos 1966; Papahadjopoulos 1968; Vilallonga 1968; Vilallonga et al. 1969; Shimojo and Ohnishi 1967). In general, the interaction of cations with phospholipid monolayers depends upon the nature of the polar group, the concentration and valency of cations, the *p*H of the subsolution and the unsaturation of fatty acyl chains. As shown in fig. 5, the molecular area

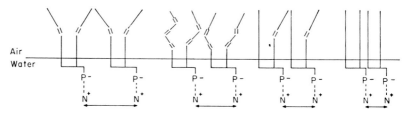

Fig. 5. A schematic representation of the influence of unsaturation on the intermolecular spacing in lecithin monolayers. The lecithin molecules are shown in the order from left to right: dioleoyl lecithin, soybean lecithin, egg lecithin, dipalmitoyl lecithin.

increases with unsaturation of fatty acyl chains and, hence, the interaction with divalent cations decreases due to greater separation between the polar groups (Shah and Schulman 1965, 1967e). The position of metal ions in the dipole lattice of various monolayers is indicated by surface potential measurements (Shah and Schulman 1967e, 1968). The adsorption of cations to various phospholipid monolayers, and the ion-exchange phenomena occurring at phospholipid-water interfaces, have been investigated by measuring the surface radioactivity of subsolutions containing Ca^{45} (Kimizuka and Koketsu 1962; Kimizuka et al. 1967; Rojas and Tobias 1965; Rojas et al. 1966; Hauser and Dawson 1967, 1968). Standish and Pethica (1968) have shown that the presence of metal ions such as Ca^{++}, Al^{+++} and UO_2^{++} in the subsolution strikingly changes the surface pressures and potentials of dipalmitoyl phosphatidylethanolamine monolayers.

The effect of cholesterol on phospholipid monolayers essentially depends upon the nature of the fatty acyl chains (De Bernard 1958; Van Deenen 1965; Chapman et al. 1968, 1969; Demel et al. 1967; Demel and Joos 1968; Demel 1968; Shah and Schulman 1967d; Joos and Demel 1969; Standish and Pethica 1967; Vilallonga et al. 1967). The condensation of phospholipid-cholesterol mixed monolayers has been explained on the basis of the inter-molecular cavities (Shah and Schulman 1967d, 1968), or alternatively on the interaction between cholesterol and fatty acyl chains of phospholipids (Demel 1968; Joos and Demel 1969; De Bernard 1958; Cadenhead and Phillips 1968). The interaction of progesterone, androsterone, aldosterone and deoxycorticosterone with lipid films causes considerable structural change, when changes in surface viscosity and molecular area are used as criteria (Taylor and Haydon 1965; Gershfeld and Pak 1968). It is also interesting that the presence of steroid hormones, or of electrolytes, in the subsolution alters the amount of water carried by a moving monolayer of stearyl alcohol (Pak and Gershfeld 1967). Using Ca^{45} in the subsolution, Kafka and Pak (1969) have shown that insulin (and some analogues of insulin) inhibits the adsorption of Ca^{2+} to monooctadecyl phosphate monolayers.

Monolayer technique has contributed greatly toward understanding the mechanism of cell-lysis. Pethica and Schulman (1953) have stressed the importance of the penetration of surface-active ions into the cholesterol portion of the erythrocyte membrane. Few (1955) has shown that the anti-biotic, polymyxin E, readily penetrates monolayers of cephalin and cardio-lipin, but not those of lecithin. Recently, Salton (1968) has discussed many aspects of lytic agents and their monolayer penetrability.

The monolayer approach provides a suitable model system for studying lipid-protein association at the air-water or oil-water interfaces. The work of Doty and Schulman (1949) demonstrates that the penetration of cardiolipin, cephalin and cholesterol monolayers by albumin, globulin or serum proteins is significantly influenced by the *p*H of the subsolution. Colacicco et al. (1967) have suggested that the mechanism of penetration of a lipid film at low pressures involves interaction between polar groups, rather than simple diffusion of protein into the film. Camejo et al. (1968) have shown that an apoprotein of the high density lipoprotein (HDL) penetrates various lipid monolayers in the following order of penetrability: cholesterol > phospha-tidyl ethanolamine > phosphatidyl choline > stearic acid > sphingomyelin > lysophosphatidyl choline. They suggested that the unusual surface activity of HDL-protein may be intimately related to the mechanism of formation of the lipoprotein. Using surface pressure, potential and bubble stability measurements, Shah (1969b, 1970c) has shown that the helical conformation of poly-L-lysine at *p*H 11 exhibits the maximum interaction with stearic acid monolayers, and decreases the rate of drainage of water in the bubble lamellae.

Several enzymes are known to be associated with membranes and the enzymic reactions occurring at the membrane surface are of importance in realtion to cellular activities. The major difference between the enzymic reactions in the bulk and at the surface is the spatial orientation of enzyme or substrate molecules in the latter. The rates of enzymic hydrolysis of lecithin in bulk solution and in monolayers are strikingly different (Shah and Schulman 1967c). These investigators also have shown that unsaturation of fatty acyl chains increases the intermolecular spacing (fig. 5) and, hence, increases the rate of hydrolysis. The *p*H of the subsolution, the state of compression and the surface charge of the monolayer also influence the rate of enzymic reactions at interfaces (Colacicco and Rapport 1966; Bangham and Dawson 1960).

Although electron microscopy has contributed much to revealing the ultrastructural organization of cells, very little is known about the effect of heavy metals and fixatives on lipids and proteins in biological structures. Monolayers, again, have provided a useful model system for studying the interaction of heavy metal ions with lipids and proteins. It has been shown that both $KMnO_4$ and OsO_4 react with the double bonds of fatty acyl chains, but not with phosphoryl choline groups in lecithin monolayers (Dreher et al. 1967; Shah 1970d). In contrast, uranyl ions react with the phosphate groups and tend to solidify lecithin monolayers (Shah 1969a, 1970a). Surface chemistry of various lipids and their interactions with metal ions and enzymes have been reviewed recently by Shah (1970b).

Cell adhesion and cell-surface interactions have been investigated by using surface chemical techniques. Rosenberg (1962) has shown, by examination of cells separated from the solid surface by multilayers of several hundred Å thickness, that the cells are influenced by the nature of the solid surface; thus, suggesting a long-range cell-surface interaction. Rosenberg (1963) has further shown that cell guidance is strikingly influenced by a change of 60 Å in the height of the contact surface. Rosenberg (1965) has investigated the spreading behavior of cells on monolayers adsorbed at the saline-fluorocarbon interface.

A host of biological phenomena such as lysis, effect of drugs on the permeability of membranes, cell adhesion and cell-surface interactions can be fruitfully investigated using principles and techniques of surface chemistry.

3. Surface phenomena related to membrane formation

3.1. Adsorption

In general, a separation between the hydrophilic and hydrophobic groups in
a molecule causes the molecule to orientate at the air-water or oil-water
interface. The hydrophilic group tends to dissolve in the aqueous phase,
whereas the hydrophobic group tends to stay out of the aqueous phase. This
type of molecule is surface active because it tends to adsorb at the interface
and decrease the interfacial tension. Fig. 6 schematically represents the

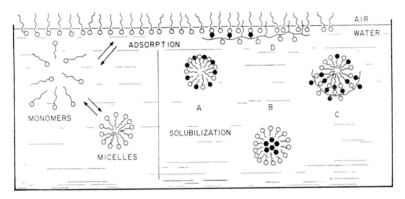

Fig. 6. A schematic representation for adsorption of soluble surface-active molecules at
the air-water interface, and for micelle formation. (a) represents solubilization of polar
molecules in a micelle; (b) represents solubilization of nonpolar molecules within a micelle;
(c) represents adsorption of a polymer at the micellar surface; (d) represents the interaction
of a polymer with the adsorbed film of surface active molecules. ○ and ● represent ionic
and nonionic polar groups respectively.

phenomena of adsorption, micelle formation, solubilization and interactions
at the micellar surface. The polar and the nonpolar parts of surface-active
molecules are shown by 'circles' and 'tails', respectively. In general, if the
hydrocarbon chains ('tails') are small (i.e., less than 12 $-\overset{|}{\underset{|}{C}}-\overset{|}{\underset{|}{C}}-$ bonds), the
molecules will dissolve in the solution; the molecules at the interface and
those in the solution will remain in equilibrium. When adsorption occurs at
the interface, there is a greater concentration of molecules at the interface

than in the solution. If the hydrocarbon chains are longer than 14 $-\overset{|}{\underset{|}{C}}-\overset{|}{\underset{|}{C}}-$ bonds, the surface-active molecules generally form insoluble monolayers. However, a polar group is essential to orientate a molecule at the interface and, hence, will form insoluble monolayers. It has been shown, using labelled surface-active molecules, that under the first adsorbed layer, a further deposition of molecules can occur upon increasing the concentration in the bulk solution (Dixon et al. 1949).

The phenomenon of adsorption results in certain compounds accumulating at an interface in a concentration significantly greater than that in the bulk solution. This process may have caused enrichment of the primitive ocean surface with surface-active molecules. The molecules present in the thin film, formed in the above manner, could have reacted among themselves under the influence of solar radiation. The formation and collapse of such films could have formed the primitive lipoprotein membranes. Goldacre (1958) has suggested a process by which lipoprotein membranes can be formed at the air-water interface (fig. 7). It is likely that lipids could have

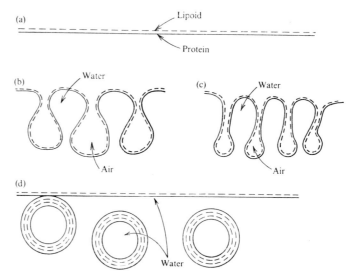

Fig. 7. A proposed mechanism for the formation of lipoprotein vesicles from adsorbed films at the air-water interface. (a) represents an adsorbed film of lipid and protein at the air-water interface; (b) represents the structure of the folded film due to compression by wind or interfacial turbulence; (c) represents the structure of film after diffusion of air from the 'sacs'; (d) represents the structures formed when two adjacent membranes join together. The membrane thus formed resembles the structure shown in fig. 1 (Goldacre 1958).

formed an adsorbed film at the surface of primitive oceans; peptides or proteins having opposite charges to those of lipids or having hydrophobic side chains could have interacted with the lipids and, thus, formed a lipo-protein film at air-sea interface. Wind or turbulence at the surface would tend to produce folds in the film, trapping air and water in small bubbles. The air would gradually leak out of the bubbles, as shown in fig. 7b and 7c. The lamellae formed by collapsed bubbles would combine to form lipo-protein membranes separating the two aqueous phases (fig. 7d). The struc-ture of such membranes would be conceptually the same as that of the Davson and Danielli model for biological membranes (fig. 1).

3.2. Micelle formation

At low concentrations, a soluble surface-active compound is molecularly dispersed (or soluble) in the aqueous solution. However, if the concentra-tion is further increased, the molecules begin to aggregate into micelles as schematically shown in fig. 6. The concentration at which the formation of micelles begins is called critical micelle concentration (or CMC). Nuclear magnetic resonance studies of micellar solutions suggest that molecules in micelles are rapidly exchanging with monomer species in the solution (Naka-gawa and Tori 1964; Eriksson and Gillberg 1965). Hence, micelles and monomers are in equilibrium (fig. 6). Micelle formation is accompanied by an abrupt change in many solution properties such as, surface tension, conductivity, ability to solubilize certain dyes, viscosity, light scattering and osmotic pressure (Preston 1948). These properties may be used to determine the CMC for a particular surface-active compound under specified experi-mental conditions.

A number of factors affect both the CMC for a surface-active compound and the size of the micelles formed. Increasing hydrophobicty of molecules leads to decreasing values of the CMC and increasingly large micelles. Increasing the concentration of electrolytes decreases the CMC and increases the micellar size.

Micelles formed from simple surface-active compounds, such as sodium dodecyl sulfate, are generally roughly spherical and contain 20–100 surfac-tant molecules per micelle, under most conditions. The hydrophilic groups occupy the surface of the micelle and are exposed to the aqueous phase, and the hydrophobic chains occupy the interior. The interior of the micelle has essentially the properties of a liquid hydrocarbon.

Urey (1952) has suggested that the primitive oceans could easily have had a

10% concentration of organic compounds. However, an estimate based on more recent assumptions concerning the rate of abiogenic organic chemical synthesis allows 1% concentration of organic compounds in the primitive oceans (Sagan 1961). This concentration of organic compounds (1–10%) is higher than, or close to, the CMC of most of the common surface-active compounds (Mukerjee 1967). Therefore, it seems very likely that micelles or aggregates of organic matter might have been present in the primitive oceans and that reactions might have taken place at the micellar surface. Selective adsorption of a polypeptide on micellar surface could occur if the polypeptide has a distribution of charges opposite to that of the micellar surface (fig. 6c). If the polypeptide contains nonpolar side chains, the association could occur because of the interaction of these side chains with the nonpolar chains of the surface-active molecules (fig. 6d). Most bond-changing reactions are also catalyzed at the micellar surface (Cordes and Dunlap 1969).

As mentioned previously, the interior of a micelle resembles a liquid hydrocarbon. Thus, it is possible to solubilize significant amounts of nonpolar substances in the interior of micelles (fig. 6b). Micelles are suspended in the aqueous environment because of the interaction of the polar groups of surface-active molecules with water. Several polar compounds present in the aqueous phase penetrate into or are adsorbed at the micellar surface; this results in mixed or swollen micelles (fig. 6a). Therefore, solubilization of a substance within the micelle results in an accumulation of certain type of molecules. This process may have played a significant role in the formation of aggregates of specific organic compounds in the primitive oceans.

3.3. Coacervation

Before we discuss the role of coacervation of organic compounds in the formation of primordial biological systems, it might be worthwhile to discuss briefly the phenomenon of coacervation. The term coacervation is derived from the latin word *acervus*, meaning aggregation, with the prefix co-, signifying the union or aggregation of colloidal particles. A colloidal solution of macromolecules or lipids often separates into two phases: one colloid rich, the other colloid poor, depending upon the temperature, *p*H, and electrolyte concentration of the solution. The colloid rich fraction is called the coacervate and the colloid poor the equilibrium liquid. Often, when two solutions of different macromolecules are mixed, formation of a coacervate and an equilibrium liquid results. Coacervates are colloid-rich liquids which are not spontaneously birefringent, and which exhibit the viscosity characteristics of Newtonian liquids (Bungenberg de Jong 1949). Coacervates are formed as a result of intra- and intermolecular associations between macromolecules. There is a sharp decrease in the specific viscosity (η_{s_p}) preceding coacervation because of the reduction in the amount of occlusion-liquid inside the macromolecule. Further, the coacervate is to be regarded as an association of macromolecules in which the points of contact should not be thought of as static but dynamic, since it is still a typical Newtonian liquid. The coacervation process is completely reversible. The coacervate layer and and the surrounding medium are in thermodynamic equilibrium.

In relation to the origin of life, the interesting coacervates are those which separate out in the form of microscopic droplets rather than as a separate layer. These droplets possess many interesting characteristics. They often contain vacuoles, and take up or absorb various types of organic molecules. They may show a creeping movement and internal circulation when placed in an electrical field. Booij (1967) has reported some very interesting properties of phospholipid coacervates and their behavior in the electric field. Films can form on the surface of coacervate drops in the presence of lipids. However, in the absence of such films, there is always a sharp boundary between the drops and the surrounding equilibrium liquid. It should be stressed that the drops are not isolated from the medium by this boundary, but are capable of interacting with the medium. The role of coacervates in causing a higher level of organization and reactivity on the part of organic compounds present in the primitive oceans will be discussed in the next section.

4. Significance and mechanisms of formation of prebiological membrane systems

It is generally believed that the primitive oceans contained a variety of organic compounds such as amino acids, nucleotides, lipids, hydrocarbons and their polymers synthetized from the primitive atmosphere under the influence of heat, ultraviolet radiation, electrical discharges or ionizing radiations. According to Urey (1952) and Sagan (1961), the concentration of organic matter in the oceans must have reached about 1–10%. As mentioned in the previous section, the phenomena of adsorption, micelle formation and solubilization are likely to occur at this concentration of polar lipids and hydrocarbons. Moreover, it should be pointed out that polymers of amino acids containing hydrophobic side chains would also adsorb at the air-water interface or at the micellar surface (figs. 6c and 6d). The surface activity of polymers, or ability to adsorb at an interface, would also depend upon the conformation of the polymers (Shah 1969b).

Three proposed prebiological systems involving membranes are (1) Goldacre's lipoprotein vesicles (fig. 7d), (2) coacervates, and (3) microspheres made from protenoids. Whether one of these systems is the most probable system for prebiological evolution is a matter for conjecture. It is very likely that all these systems may have existed in different local environments on the primitive earth. These three prebiological systems would be expected to have the following characteristics.

The formation of bilayer vesicles, according to the scheme of Goldare (1958) (fig. 7), is most likely to occur at the air-water interface of the primitive oceans, tidal basins, lakes and lagoons. If the lipids are present in the aqueous phase, they are likely to adsorb at the air-water interface. The charge distribution in this film could introduce a degree of specificity in the lipid-polymer interactions at the interface since, among the polymers present in the aqueous phase, only those having opposite charge distributions will interact with the lipid film. Apart from the ionic interactions, the polymers exhibiting surface activity also would tend to increase the lipid-polymer association (fig. 6d). Thus, polymers having a great number of hydrophobic side chains would tend to adsorb on the lipid layer at the air-water interface. It should be stressed that lipids are not indispensable for the formation of vesicles, since a polymer film is by itself able to form such closed structures. The aqueous phase trapped within the vesicles may contain organic compounds. Reactions occurring within the vesicles may take a selective course, since the transport of various compounds from outside to inside, and of reaction products from inside to outside of the vesicles, may depend upon the surface charge, porosity and surface rheology of the vesicle membrane. Moreover, vesicles containing different reaction systems may coalesce, because of the lability of their membranes to form a more complex structure (vesicle) with many reactions occurring within it. This may evolve into a vesicle containing coupled reactions in which the product of one reaction becomes the substrate for another.

The information content of such vesicle membranes would depend upon the following factors: the intermolecular spacing between lipid molecules, the charge distribution in the lipid film, and upon the polymer molecules as well as the number and character of hydrophobic groups and their distribution on polymers. Shah and Schulman (1965, 1967e, 1968) have shown that unsaturation of hydrocarbon chains increases the intermolecular spacing, and that this markedly influences the interaction of metal ions with phospholipid monolayers. Shah and Schulman (1967c) have also shown that enzymic hydrolysis of phospholipid monolayers is strikingly influenced by the unsaturation of fatty acyl chains and, hence, by the intermolecular spacing in phospholipid monolayers.

The area per molecule of the lipids in a membrane is a very important membrane parameter. In this connection the observations of Meyer and Bloch (1963) on the fatty acid composition of the phospholipids from yeast cells which were grown anaerobically are highly interesting. Under these conditions, the unsaturated fatty acyl chains which are common for aerobically grown cells are replaced by saturated shorter fatty acyl chains. It has been shown by Van Deenen (1965) that both lecithins having either longer unsaturated or shorter saturated fatty acyl chains have the same area per molecule and, hence, the same intermolecular spacing in monolayers. This implies that yeast cells maintain the same area per lipid molecule in both aerobic and anaerobic conditions in order to maintain the structural and functional integrity of the membranes.

The next system of interest in relation to the origin of membranes is coacervate droplets. The process of coacervation is one of the most powerful methods of concentrating highly polymerized substances from very dilute solutions. For example, coacervate drops have been isolated from a solution containing gelatin at a dilution of 0.001 % (Bungenberg de Jong 1947). The polymer concentration in the drops could reach as high as 10%. Although the coacervate drops are liquid in consistency, they are sharply segregated by a boundary layer from the aqueous solution surrounding them. Such separation of coacervate drops must have taken place in the 'primitive soup' where non-specific polymerization of a variety of organic compounds occurred. Coacervate drops represent multimolecular systems, which can easily shift from a static to a dynamic state by their interaction with the environment. Oparin (1965) has incorporated several enzyme systems in such coacervate droplets, which may then increase or decrease in size as a result of the synthesis or hydrolysis of compounds within the droplets. Because of their labile structure, they can coalesce or divide, thereby resulting in the recombination and separation of various chemical systems. This process may have resulted in a gradual evolution of coacervate droplets capable of performing a series of chemical reactions. The lability of their structure is one of the attractive features of coacervates, for it can lead to multi-reactions systems by the recombination of several coacervate drops.

Another system of interest in connection with the origin of membranes is the microspheres which have been reported by Fox et al. (1959). In this case a mixture of amino acids was placed on a piece of lava and heated to 170 °C for several hours. The powder was thereby converted to a light amber-colored viscous liquid which, upon dilution by hot 1 % sodium chloride solution, produced microspheres. These microspheres are stable spherical structures 2–7 μ in diameter which can withstand centrifugation at 3000 rpm. They are stable to the operations involved in the preparation of thin sections for electronmicroscopy. Approximately 20 attributes of the proteins are found in these thermal protenoids (Fox 1965). For instance, in hypotonic and hypertonic solutions the microspheres swell and shrink, respectively, although the responses are less striking than those of biological cells. Fox (1965) has reported that when the micropsheres are placed in aqueous suspension, the resulting pH is about 3; if a buffer of pH 5.5–6.5 is added, the microsphere begins to hollow out, and 'double membranes' form. Such double membranes are evident in the electron micrograph shown in fig. 8. It should be pointed out that the thickness of these layers is 10–20 times greater than those of biological membranes. Recently, Fox et al. (1970) have summarized various properties of protenoids and microspheres.

The author of the present article has proposed a simple and novel method for the formation of prebiological membrane systems as shown in fig. 9. The presence of lipids and proteins (or polypeptides) in the primitive oceans may have resulted in the formation of adsorbed lipid-protein film at the ocean surface. The aerosol droplets formed as a result of collapse of waves, would also possess the lipid-protein film on their surface with the nonpolar portion of lipid molecules pointing out of the aqueous droplets. When such droplets settle down on the ocean surface, they would coat themselves again with a lipid-protein film. This process would produce closed membranous structures. The reactions occurring within these droplets would be influenced by the surface charge and the permeability characteristics of the membrane which is similar in structure to that proposed by Davson and Danielli for red cell membrane (fig. 1). Since a great number and variety of aerosol droplets could be formed at the ocean surface, nature could have experimented with billions of such membranous structures for further evolution. The membrane covered aerosol droplets may have constituted a self-assembled and organized structure in the 'hot dilute soup' of primitive oceans.

This author believes that the sequence of events leading from primitive oceans, which may be described as a 'hot dilute soup' of organic compounds, to the precellular level of organization may have been as follows. As a result

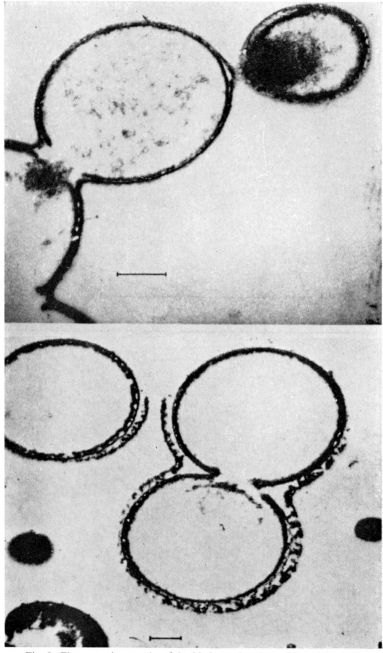

Fig. 8. Electron micrographs of double layers in microspheres (Fox 1965).

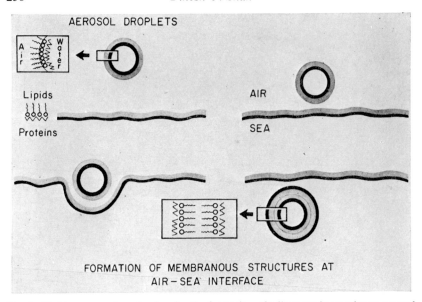

Fig. 9. Shah's proposed mechanism for the formation of a lipoprotein membrane around aerosol droplets. The molecular structure of this membrane is similar to that proposed by Davson and Danielli for erythrocyte membrane (see fig. 1).

of various surface phenomena such as adsorption, micelle formation, solubilization and coacervation, multimolecular aggregates may have come into being. If a protein with a catalytic (or enzymic) activity was present in such an aggregate, the system could carry out a specific reaction. In consequence of the association and dissociation of such aggregates, new systems may have evolved, performing several different catalytic reactions. Such processes of association and dissociation could also increase the complexity of structure of the system of aggregates. Physico-chemical interactions at the surface of an aggregate may have created prebiological membranes, which could then lead to futher evolution by governing the transport processes across the membranes. Processes of association and dissociation could go on more readily between coacervate drops, because of their rather labile structure, than between more rigid structures such as microspheres or vesicles. However, more experimental data are needed on the permeability and dissociation-recombination properties of these three different membrane systems. The conditions of the primitive environment described here point strongly to the inference that the structure and transport properties of membranes of prebiological systems must have contributed significantly to the evolution of such systems that we call 'living systems' today.

Acknowledgements

This article is dedicated to Madhuriben Desai, without whose kind help and encouragement it could not have been written. I would like to thank Dr. R. O. Anderson, Dr. A. A. Benson, Dr. J. F. Danielli, Dr. S. W. Fox, Dr. R. J. Goldacre and Dr. T. E. Thompson for their kind permission to reproduce their diagrams. I would also like to thank Dr. A. Chu and Mr. E. J. Murphy, Senior Research Scientist, for their critical reading and helpful comments; and to Mrs. P. Kilian and Miss P. M. Warwick for their editorial assistance.

References

ADELMAN, W. J. and J. P. SENFT, 1968, J. Gen. Physiol. 51, 1025.

ANDREOLI, T. E., M. TIEFFENBERG and D. C. TOSTESON, 1967, J. Gen. Physiol. 50, 2527.

BANGHAM, A. D. and R. M. C. DAWSON, 1960, Biochem. J. 75, 133.

BANGHAM, A. D. and R. W. HORNE, 1964, J. Mol. Biol. 8, 660.

BANGHAM, A. D. and D. PAPAHADJOPOULOS, 1966, Biochim. Biophys. Acta 126, 181.

BANGHAM, A. D., M. M. STANDISH and N. MILLER, 1965a, Nature 208, 1295.

BANGHAM, A. D., M. M. STANDISH and J. C. WATKINS, 1965b, J. Mol. Biol. 13, 238.

BANGHAM, A. D., M. M. STANDISH and G. WEISSMAN, 1965c, J. Mol. Biol. 13, 252.

BANGHAM, A. D., J. DE GIER and G. D. GREVILLE, 1967, Chem. Phys. Lipids 1, 225.

BAR, R. S., D. W. DEAMER and D. G. CORNWELL, 1966, Science 153, 1010.

BENEDETTI, E. L. and P. EMMELOT, 1968a, J. Cell Biol. 38, 15.

BENEDETTI, E. L. and P. EMMELOT, 1968b, Structure and function of plasma membranes isolated from plasma liver; In: Ultrastructure in biological systems, vol. 4, A. Dalton and F. Haguenau, eds. (Academic Press, New York) p. 33.

BENSON, A. A., 1966, J. Amer. Oil Chem. Soc. 43, 265.

BONTING, S. L. and A. D. BANGHAM, 1967, Exptl. Eye Res. 6, 400.

BOOIJ, H. L., 1967, Physical-chemical aspects of membrane dynamics and model systems; In: Conferences on cellular dynamics: proceedings of the 3rd and 4th inter-disciplinary conference, L. Peachey, ed. (The New York Academy of Sciences, New York) pp. 1–49.

BRANTON, D., 1968, Photophysiology 3, 197.

BROWN, F. and J. F. DANIELLI, 1964, The cell surface and cell physiology; In: Cytology and cell physiology, 3rd ed., G. Bourne, ed. (Academic Press, New York) p. 239.

BUNGENBERG DE JONG, H., 1947, Koninkl. Ned. Akad. Wetenschap., Proc. 50, 707.

BUNGENBERG DE JONG, H., 1949, Crystallization-Coacervation-flocculation; In: Colloid science, vol. 5, II, H. Kruyt, ed. (Elsevier Publishing Company, New York) pp. 232–255.

CADENHEAD, D. A. and M. C. PHILLIPS, 1968, Advan. Chem. Ser. 84, 131.

CAMEJO, G., G. COLACICCO and M. M. RAPPORT, 1968, J. Lipid Res. 9, 562.

CASS, A. and A. FINKELSTEIN, 1967, J. Gen. Physiol. 50, 1765.

CHAPMAN, D., ed., 1968, Biological membranes (Academic Press, New York) pp. 1–423.

CHAPMAN, D., V. B. KAMAT and R. LEVENE, 1968, Science 160, 314.

CHAPMAN, D., N. F. OWENS, M. C. PHILLIPS and D. A. WALKER, 1969, Biochim. Biophys. Acta 183, 458.

CHAPMAN, D. and S. A. PENKETT, 1966, Nature 211, 1304.

COLACICCO, G. and M. M. RAPPORT, 1966, J. Lipid Res. 7, 258.

COLACICCO, G., M.M. RAPPORT and D. SHAPIRO, 1967, J. Colloid Interface Sci. 25, 5.

CORDES, E. H. and R. B. DUNLAP, 1969, Accounts Chem. Res. 2, 329.

COSTER, H. G. L. and R. SIMONS, 1968, Biochim. Biophys. Acta 163, 234.

DAEMEN, F. J. M. and S. L. BONTING, 1969, Biochim. Biophys. Acta 183, 90.

DANIELLI, J. F., 1958, Surface chemistry and cell membranes; In: Surface phenomena in chemistry and biology, J. F. Danielli, K. G. A. Pankhurst and A. C. Riddiford, eds. (Pergamon Press, London) p. 246.

DANIELLI, J. F., 1966, J. Theoret. Biol. 12, 439.

DANIELLI, J. F. and H. DAVSON, 1935, J. Cell. Comp. Physiol. 5, 495.

DAVSON, H., 1962, Circulation 26, 1022.

DEAMER, D. W. and D. G. CORNWELL, 1966, Biochim. Biophys. Acta 116, 555.

DE BERNARD, L., 1958, Bull. Soc. Chim. Biol. 40, 161.

DE GIER, J., J. G. MANDERSLOOT and L. L. M. VAN DEENEN, 1968, Biochim. Biophys. Acta 150, 666.

DE GIER, J., J. G. MANDERSLOOT and L. L. M. VAN DEENEN, 1969, Biochim. Biophys. Acta 173, 143.

DEL CASTILLO, J., A. RODRIQUEZ, C. A. ROMERO and V. SANCHEZ, 1966, Science 153, 185.

DEMEL, R. A., 1968, J. Amer. Oil Chem. Soc. 45, 305.

DEMEL, R. A. and P. JOOS, 1968, Chem. Phys. Lipids 2, 35.

DEMEL, R. A., L. L. M. VAN DEENEN and B. A. PETHICA, 1967, Biochim. Biophys. Acta 135, 11.

DEMEL, R. A., S. C. KINSKY and L. L. M. VAN DEENEN, 1968, Biochim. Biophys. Acta 150, 655.

DIXON, J. K., A. J. WEITH, A. A. ARGYLE and D. J. SALLEY, 1949, Nature 163, 845.

DOTY, P. and J. H. Schulman, 1949, Disc. Faraday Soc. 6, 21.

DREHER, K. D., J. H. SCHULMAN, O. R. ANDERSON and O. A. ROELS, 1967, J. Ultrastruct. Res. 19, 586.

ELBERS, P. F., 1964, The Cell Membrane; In: Recent progress in surface science, vol. 2, J. F. Danielli, K. G. A. Pankhurst and A. C. Riddiford, eds. (Academic Press, New York) pp. 443–503.

Engelman, D. M., 1969, Nature 223, 1279.

ERIKSSON, J. C. and G. GILLBERG, 1965, NMR studies of the solubilization of aromatic compounds in cetyltrimethylammonium bromide solution; In: Surface chemistry, Proceedings 2nd Scandinavian symposium surface activity, P. Ekwall, K. Groth and V. Runnstrom-Reio, eds. (Munksgaard, Copenhagen) pp. 148–156.

EVERITT, C. T. and D. A. HAYDON, 1968, J. Theoret. Biol. 18, 371.

FEW, A. V., 1955, Biochim. Biophys. Acta 16, 137.

FINEAN, J. B., 1966, The molecular organization of cell membranes. In: Prog. Biophys. Mol. Biol. vol. 16, J. A. V. Butler and H. E. Huxley, eds. (Pergamon Press, New York) pp. 143–170.

FINEAN, J. B., 1957, The molecular structure of nerve myelin and its significance in relation to the nerve 'membrane'; In: Metabolism of the nervous system, D. Richter, ed. (Pergamon Press, New York).

Fox, S. W., 1965, Simulated natural experiments in spontaneous organization of morphological units from protenoids; In: The origins of prebiological systems and of their molecular matrices, S. W. Fox, ed. (Academic Press, New York) pp. 361–382.

Fox, S. W., K. HARADA and J. KENDRICK, 1959, Science 129, 1221.

Fox, S. W., 1970, Membrane-like properties in microsystems assembled from synthetic protein-like polymer; In: Ultrastructure in biological systems, vol. 6, F. Snell et al., eds. (Academic Press, New York) pp. 1–30.

GERSHFELD, N. L. and C. Y. C. PAK, 1968, Nature, 219, 495.

GLAUERT, A. M. and J. A. LUCY, 1968, Globular micelles and the organization of membrane lipids; In: Ultrastructure in biological systems, vol. 4, A. J. Dalton and F. Haguenau, eds. (Academic Press, New York) pp. 1–30.

GOLDACRE, R. J., 1958, Surface films, their collapse on compression, the shapes and sizes of cells and the origin of life; In: Surface Phenomena in Chemistry and Biology, J. F. Danielli, K. G. A. Pankhurst and A. C. Riddiford, eds. (Pergamon Press, New York) pp. 278–298.

GOOD, R. J., 1969, J. Colloid Interface Sci. 31, 540.

GORTER, E. and F. GRENDEL, 1925, J. Exptl. Med. 41, 439.

GREEN, D. E. and T. ODA, 1961, J. Biochem. 49, 742.

GREEN, D. E., D. W. ALLMANN, E. BACHMANN, H. BAUM, K. KOPACZYK, E. F. KORMAN, S. LIPTON, D. H. MACLENNAN, D. G. McCONNELL, J. F. PERDUE, J. S. RIESKE and A. TZAGOLOFF, 1967, Arch. Biochem. Biophys. 119, 312.

GULIC-KRZYWICKI, T., E. RIVAS and V. LUZZATI, 1967, J. Mol. Biol. 27, 303.

HALDANE, J. B. S., 1928, The Origin of life. Rationalist Annual p. 148.

HANAI, T. and D. A. HAYDON, 1966, J. Theoret. Biol. 11, 370.

HANAI, T., D. A. HAYDON and J. TAYLOR, 1965a, J. Theoret. Biol. 9, 278.

HANAI, T., D. A. HAYDON and J. TAYLOR, 1965b, J. Theoret. Biol. 9, 422.

HANAI, T., D. A. HAYDON and J. TAYLOR, 1965c, J. Theoret. Biol. 9, 433.

HANAI, T., D. A. HAYDON and J. TAYLOR, 1965d, J. Gen. Physiol. 48, 59.

HASSELBACH, W. and L. G. ELFVIN, 1967, J. Ultrastruct. Res. 17, 598.

HAUSER, H. and R. M. C. DAWSON, 1967, European J. Biochem. 1, 61.

HAUSER, H. and R. M. C. Dawson, 1968, Biochem. J. 109, 909.

HAYDON, D. A., 1968, J. Amer. Oil Chem. Soc. 45, 230.

HAYDON, D. A. and J. TAYLOR, 1963, J. Theoret. Biol. 4, 281.

HENN, F. A., G. L. DECKER, J. W. GREENAWALT and T. E. THOMPSON, 1967, J. Mol. Biol. 24, 51.

HENN, F. A. and T. E. THOMPSON, 1968, J. Mol. Biol. 31, 227.

HENN, F. A. and T. E. THOMPSON, 1969, Ann. Rev. Biochem. 38, 241.

HOFFMAN, J. F., 1962, Circulation 26, 1201.

HUANG, C., L. WHEELDON and T. E. THOMPSON, 1964, J. Mol. Biol. 8, 148.

HUANG, C. and T. E. THOMPSON, 1965, J. Mol. Biol. 13, 183.

HUANG, C. and T. E. THOMPSON, 1966, J. Mol. Biol. 15, 539.

HUBBELL, W. L. and H. M. McCONNELL, 1968, Proc. Nat. Acad. Sci. (U.S.A.) 61, 12.

JENDRASIAK, G. L., 1969, Chem. Phys. Lipids 3, 98.

JOHNSON, S. M. and A. D. BANGHAM, 1969a, Biochim. Biophys. Acta 193, 82.

JOHNSON, S. M. and A. D. BANGHAM, 1969b, Biochim. Biophys. Acta 193, 92.

JOOS, P. and R. A. DEMEL, 1969, Biochim. Biophys. Acta 183, 447.

KAFKA, M. S. and C. Y. C. PAK, 1969, Biochim. Biophys. Acta 193, 117.

KAVANAU, J. L., 1965, Structure and function in biological membranes, vol. I and II (Holden-Day, San Francisco).

KEITH, A., A. S. WAGGONER and O. H. GRIFFITH, 1968, Proc. Nat. Acad. Sci. (U.S.A.) 61, 819.

KIMIZUKA, H. and K. KOKETSU, 1962, Nature 196, 995.

KIMIZUKA, H., T. NAKAHARA, H. UEJO and A. YAMAUCHI, 1967, Biochim. Biophys. Acta 137, 549.

KINSKY, S. C., S. A. LUSE and L. L. M. VAN DEENEN, 1966, Fed. Proc. 25, 1503.

LADBROOKE, B. D., R. M. WILLIAMS and D. CHAPMAN, 1968, Biochim. Biophys. Acta 150, 333.

LANDGRAF, W. C. and G. INESI, 1969, Arch. Biochem. Biophys. 130, 111.

LAUGER, P., W. LESSLAUER, E. MARTI and J. RICHTER, 1967, Biochim. Biophys. Acta 135, 20.

LEHNINGER, A. L., 1959, Rev. Mod. Phys. 31, 136.

LENARD, J. and S. J. SINGER, 1966, Proc. Nat. Acad. Sci. (U.S.A.) 56, 1828.

LESLIE, R. B. and D. CHAPMAN, 1967, Chem. Phys. Lipids 1, 143.

LESSLAUER, W., A. J. SLOTBOOM, N. M. POSTEMA, G. H. DeHAAS and L. L. M. VAN DEENEN, 1968, Biochim. Biophys. Acta 150, 306.

LIPPE, C., 1968, J. Mol. Biol. 35, 635.

LUCY, J. A., 1964, J. Theoret. Biol. 7, 360.

LUZZATI, V., 1968, X-ray diffraction studies of lipid-water; In: Biological membranes, D. Chapman, ed. (Academic Press, New York).

LUZZATI, V. and F. HUSSON, 1962, J. Cell Biol. 12, 207.

MADDY, A. H., 1966, The chemical organization of the plasmamembrane of animal cells; In: Int. Rev. Cyto., vol. 20, G. H. Bourne and J. F. Danielli, eds. (Academic Press, New York) pp. 1–64.

MADDY, A. H., 1967, The organization of protein in the plasma membrane; In: Formation and Fate of Cell Organelles, vol. 6, K. B. Warren, ed. (Academic Press, New York) p. 266.

MEYER, F. and K. BLOCH, 1963, J. Biol. Chem. 238, 2654.

MIYAMOTO, V. K. and T. E. THOMPSON, 1967, J. Colloid Interface Sci. 25, 16.

MUELLER, P. and D. O. RUDIN, 1967a, Nature 213, 603.

MUELLER, P. and D. O. RUDIN, 1967b, Biochem. Biophys. Res. Commun. 26, 398.

MUELLER, P. and D. O. RUDIN, 1968, Nature 217, 713.

MUELLER, P., D. O. RUDIN, H. T. TIEN and W. C. WESCOTT, 1962, Circulation 26, 1167.

MUELLER, P., D. O. RUDIN, H. T. TIEN and W. C. WESCOTT, 1963, J. Phys. Chem. 67, 534.

MUKERJEE, P., 1967, Advan. Colloid Interface Sci. 1, 241.

NAKAGAWA, T. and K. TORI, 1964, Kolloid-Z. 194, 143.

NILSSON, G., 1965, J. Ultrastruct. Res. 12, 207.

OHKI, S., 1969, J. Colloid Interface Sci. 30, 413.

OMURA, T., P. SIEKEVITZ and G. E. PALADE, 1967, J. Biol. Chem. 242, 2389.

OPARIN, A. I., 1924, Proiskhozhdenie zhinzni (Izd. Moskovskiy Rabochiy, Moscow).

OPARIN, A. I., 1965, The pathways of the primary development of metabolism and artificial modeling of this development in coacervate drops; In: The origins of prebiological systems and of their molecular matrices, S. W. Fox, ed. (Academic Press, New York) pp. 331–346.

PAGANO, R. and T. E. THOMPSON, 1967, Biochim. Biophys. Acta 144, 666.

PAGANO, R. and T. E. THOMPSON, 1968, J. Mol. Biol. 38, 41.

PAK, C. Y. C. and N. L. Gershfeld, 1967, Nature 214, 888.

PANT, H. and B. Rosenberg, 1970, Proc. Ann. Meeting Biophys. Soc., p. 70a.

PAPAHADJOPOULOS, D., 1968, Biochim. Biophys. Acta 163, 240.

PAPAHADJOPOULOS, D. and A. D. BANGHAM, 1966, Biochim. Biophys. Acta 126, 185.

PAPAHADJOPOULOS, D. and N. MILLER, 1967, Biochim. Biophys. Acta 135, 624

PAPAHADJOPOULOS, D. and J. C. WATKINS, 1967, Biochim. Biophys. Acta 135, 639.

PETHICA, B. A. and J. H. SCHULMAN, 1953, Biochem. J. 53, 177.

PRESTON, W. C., 1948, J. Phys. and Colloid Chem. 52, 84.

RAND, R. P. and V. LUZZATI, 1968, Biophys. J. 8, 125.

RENDI, R., 1967, Biochim. Biophys. Acta 135, 333.

ROBERTSON, J. D., 1967, Protoplasma 63, 218.

ROBERTSON, J. D., 1969, The Molecular structure and contact relationships of cell membranes; In: Prog. Biophys. Biophys. Chem., vol. 10, J. A. V. Butler and B. Katz, eds. (Pergamon Press, New York) pp. 344–407.

ROGERS, J. and P. A. WINSOR, 1967, Nature 216, 477.

ROJAS, E., J. Y. LETTVIN and W. F. PICKARD, 1966, Nature 209, 886.

ROJAS, E. and J. M. TOBIAS, 1965, Biochim. Biophys. Acta 94, 394.

ROSENBERG, B. and B. B. BHOWMIK, 1969, Chem. Phys. Lipids 3, 109.

ROSENBERG, M. D., 1962, Proc. Nat. Acad. Sci. (U.S.A.) 48, 1342.

ROSENBERG, M. D., 1963, Science 139, 411.

ROSENBERG, M. D., 1965, The culture of cells and tissues at the saline-fluorocarbon interface; In: Tissue culture, C. V. Ramkrishnan, ed. (Junk Publishers, The Hague) p. 93.

ROTHFIELD, L. and A. FINKELSTEIN, 1968, Ann. Rev. Biochem. 37, 463.

SAGAN, C., 1961, Radiation Res. 15, 174.

SALTON, M. R. J., 1968, J. Gen. Physiol. 52, 227.

SCHMITT, F. O., 1959, The Ultrastructure of the Nerve Myelin Sheath; Chapter XVII in: Multiple sclerosis and the demyelinating diseases. Proceedings of the Association for Research in Nervous and Mental Disease, vol. 28, H. W. Worltman, ed. (Williams and Wilkins Company, Baltimore) p. 247.

SESSA, G. and G. WEISSMANN, 1967, Biochim. Biophys. Acta 135, 416.

SESSA, G. and G. WEISSMANN, 1968a, J. Biol. Chem. 243, 4364.

SESSA, G. and G. WEISSMANN, 1968b, Biochim. Biophys. Acta 150, 173.

SHAH, D.O., 1969a, J. Colloid Interface Sci. 29, 210.

SHAH, D. O., 1969b, Biochim. Biophys. Acta 193, 217.

SHAH, D. O., 1970a, Lipid-metal ion interactions in monomolecular films; In: Heavy metals and cells, Proceedings of the 2nd Rochester conference on environmental toxicity, in press.

SHAH, D. O., 1970b, Surface chemistry of lipids; In: Advances in Lipid Research, vol. 8, R. Paoletti and D. Kritschevsky, eds. (Academic Press, New York) pp. 347–431.

SHAH, D. O., 1970c, Lipid-protein association in lung surfactant; In: Advances in experimental Medicine and Biology, vol. 7, M. Blank, ed. (Plenum Press, New York) pp. 101–117.

SHAH, D. O., 1970d, Biochim. Biophys. Acta, 211, 358.

SHAH, D. O. and J. H. SCHULMAN, 1965, J. Lipid Res. 6, 341.

SHAH, D. O. and J. H. SCHULMAN, 1967a, Lipids 2, 21.

SHAH, D. O. and J. H. SCHULMAN, 1967b, Biochim. Biophys. Acta 135, 184.

SHAH, D. O. and J. H. SCHULMAN, 1967c, J. Colloid Interface Sci. 25, 107.

SHAH, D. O. and J. H. SCHULMAN, 1967d, J. Lipid Res. 8, 215.

SHAH, D. O. and J. H. SCHULMAN, 1967e, J. Lipid Res. 8, 227.

SHAH, D. O. and J. H. SCHULMAN, 1968, Advan. Chem. Ser. 84, 189.

SHIMOJO, T. and T. OHNISHI, 1967, J. Biochem. (Japan) 61, 89.

SINGER, I. and I. TASAKI, 1968, Nerve excitability and membrane macromolecules; In: Biological Membranes, D. Chapman, ed. (Academic Press, New York) p. 347.

SJOSTRAND, F. S., 1968, Ultrastructure and function of cellular membranes; In: Ultrastructure in biological systems, vol. 4, A. J. Dalton and F. Haguenau, eds. (Academic Press, New York) p. 151.

SMALL, D. M., 1967, J. Lipid Res. 8, 551.

STANDISH, M. M. and B. A. PETHICA, 1967, Biochim. Biophys. Acta 144, 659.

STANDISH, M. M. and B. A. PETHICA, 1968, Trans. Faraday Soc. 64, 1113.

STEIM, J. M., 1968, Advan. Chem. Ser. 84, 259.

STEIN, W. D., 1967, The movement of molecules across cell membranes (Academic Press, New York).

STOECKENIUS, W. and D. M. ENGELMAN, 1969, J. Cell Biol. 42, 613.

SUTTON, A. M. and D. ROSEN, 1968, Nature 219, 153.

SWEET, C. and J. E. ZULL, 1969, Biochim. Biophys. Acta 173, 94.

TAKAGI, M., K. AZUMA and U. KISHIMOTO, 1965, Ann. Report Biol. Works Fac. Sci. Osaka Univ. (Japan) 13, 107.

TAYLOR, J. L. and D. A. HAYDON, 1965, Biochim. Biophys. Acta 94, 488.

THOMPSON, T. E., 1964, The properties of bimolecular phospholipid membranes; In: Cellular membranes in development, M. Locke, ed. (Academic Press, New York) p. 83.

THOMPSON T. E. and F. A. HENN, 1970, Experimental phospholipid model membranes; In: Structure and function of membranes of mitochondria and chloroplasts, E. Racker, ed. (Reinhold Book Corporation, New York).

TIEN, H. T., 1968a, Advan. Chem. Ser. 84, 104.

TIEN, H. T., 1968b, J. Phys. Chem. 72, 2723.

TIEN, H. T., 1968c, Nature 219, 272.

TIEN, H. T., 1968d, J. Phys. Chem. 72, 4512.

TIEN, H. T. and A. L. DIANA, 1967, J. Colloid Interface Sci. 24, 287.

TIEN, H. T. and A. L. DIANA, 1968, Chem. Phys. Lipids 2, 55.

TSOFINA, L. M., E. A. LIBERMAN and A. V. BABAKOV, 1966, Nature 212, 681.

UREY, H. C., 1952, in: The planets: their origin and development (Yale University Press, New Haven) pp. 152–153.

VAN DEENEN, L. L. M., 1965, Phospholipids and Biomembranes; In: Progress in Chemistry of Fats and Other Lipids, vol. 8, R. T. Holman, ed. (Pergamon Press, New York) p. 60.

VAN DEN BERG, H. J., 1965, J. Mol. Biol. 12, 290.

VAN ZUTPHEN, H., L. L. M. VAN DEENEN and S. C. KINSKY, 1966, Biochem. Biophys. Res. Commun. 22, 393.

VANDENHEUVEL, F. A., 1965, Ann. N.Y. Acad. Sci. 122, 57.

VANDENHEUVEL, F. A., 1966, J. Amer. Oil Chem. Soc. 43, 248.

VILALLONGA, F., M. FERNANDEZ, C. ROTUNNO and M. CEREIJIDO, 1969, Biochim. Biophys. Acta 183, 98.

VILALLONGA, F., 1968, Biochim. Biophys. Acta 163, 290.

VILALLONGA, F., R. ALTSCHUL and M. S. FERNANDEZ, 1967, Biochim. Biophys. Acta 135, 557.

WALLACH, D. and A. Gordon, 1968, Fed. Proc. 27, 1263.

WEIER, T. E. and A. A. BENSON, 1967, Amer. J. Bot. 54, 389.

WEISSMANN, G. and G. SESSA, 1967, J. Biol. Chem. 242, 616.

C. Ponnamperuma (ed.), Exobiology. © North-Holland Publishing Company

CHAPTER 8

Evolution of proteins

M. O. DAYHOFF

1.   Introduction

'Relics' of ancient organisms can be found in the biochemical systems of their living descendants. The exceedingly conservative nature of the evolutionary process has preserved such relics in all living species. Many basic reaction pathways and even many features of complicated polymer structures are derived from extremely remote ancestors living long before the ordinary fossil record was formed. This dynamic preservation of the biochemical components of living cells is often more rigorous than the preservation of sedimentary fossils. We shall see that the biochemical evidence in protein structures gives us a wealth of information from which to construct a phylogenetic tree of all life. Further, it gives us a glimpse of the ancient protein molecules in ancestral organisms and a time scale for measuring the relative antiquity of divergences.

In many ways the biochemical evidence from living species and the fossil evidence complement each other. Unlike fossil evidence, which includes many extinct lines, all of the biochemical information pertains only to direct ancestors. Special features of extinct collateral lines or metabolic pathways that have been completely abandoned in all living groups could not be inferred directly from the biochemical evidence. However, the features of organisms which are surely ancestral to modern lines can be inferred. Particularly clear are the structures which must have existed in the immediate common ancestors of two descendent lines. It is just here that the fossil evidence is weak; often at least one of the descendent lines had very few members at the start and would not be found in the fossil record. The two kinds of evidence are also complementary in estimating time. While the biochemical evidence is useful for precisely ordering phyletic events, it must be related to time in years by reference to one or more points in the fossil record.

The interweaving of the biochemical with the fossil evidence promises to fulfill a long-standing hope, to be able to work out the complete, detailed, quantitative phylogenetic tree – the history of the origin of all living species back to the very beginning of life.

In this chapter, I shall deal mainly with an important portion of the biochemical evidence, the amino acid sequences of proteins from living organisms. Potentially a great quantity of such information can be elicited. There are a small number of generally applicable, theoretically understandable laws that have governed the evolution of proteins; therefore it is possible to apply logical and statistical methods quantitatively to the data. (Dayhoff 1969. Extensive references and results are given here on all protein and nucleic acid sequences.) Fig. 1 shows the amount of phylogenetic organization possible using only the 112 amino acid 'traits' in the sequences of the protein cytochrome c from many species and a few reasonable assumptions about the evolutionary process. The methods used to derive this tree will be discussed below.

The first complete sequence of a protein was published in 1953 by Sanger and his coworkers (Sanger and Thompson 1953). Since then the sequence information has grown exponentially; well over 300 complete sequences have already been established by workers in many fields for a variety of reasons. Their main interests include chemical structure and function, catalysis, polymer chemistry, chemical synthesis, drug manufacture, medicine, genetics, and taxonomy. Due to this broad interest base and recent major improvements in laboratory techniques, sequence information promises to accumulate rapidly for some time to come.

Because of our own research interest in the theoretical aspects of protein chemistry, my laboratory has long maintained a collection of known sequences. In order to integrate the information gathered by diverse groups, to focus attention on this important new area of study and to establish a foundation of readily available, correct data, we decided in 1965 to publish annually all the known sequences in an *Atlas of protein sequence and structure*. Four volumes have already appeared. To aid workers in the correlation and understanding of the protein structures, we have included in the *Atlas* theoretical inferences and the results of computer-aided analyses which are necessary to illuminate such inferences. Many of the ideas discussed in this chapter are taken from the *Atlases* to which many people have

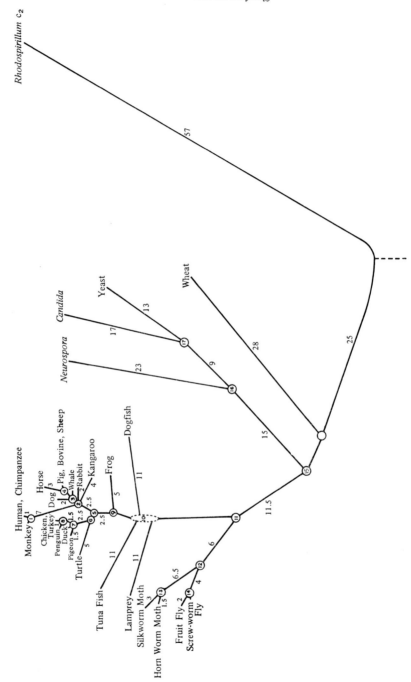

Fig. 1. Phylogenetic tree of cytochrome c. The topology has been derived from the sequences as explained in the text. The numbers of inferred amino acid changes per 100 links are shown on the branches of the tree. The point of earliest time cannot be determined directly from the sequences; we have placed it by assuming that, on the average, species change at the same rate. (From Dayhoff 1969.)

contributed, including C. M. Park, L. T. Hunt, P. J. McLaughlin, W. C. Barker, R. V. Eck and M. R. Sochard.

Each year the number of known sequences has almost doubled, so that in the 1969 *Atlas* there were some 300 sequences of more than 30 links, which have been elucidated by several hundred research workers. The distribution of these known proteins among the biological groups is shown in fig. 2. The area of each circle is proportional to the total number of

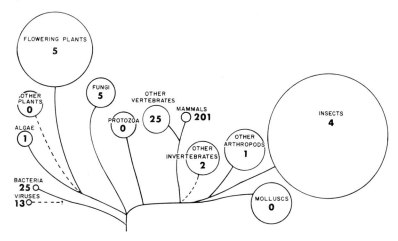

Fig. 2. Distribution of sequence data among biological groups. The area of each circle is proportional to the number of species which have been described for each group (some groups are very incompletely known). The numbers indicate protein sequences more than 30 links long which are presented in the *Atlas of protein sequence and structure* 1969 (From Dayhoff 1969.)

species which have been described for that group. However, some groups are very incompletely known, and there are no generally accepted definitions of 'species' for bacteria or viruses. The integer in the circle represents the number of sequenced proteins. It can readily be seen that the emphasis so far has centered on the mammals. Only rarely have proteins from species of widely different ancestry been considered. Yet a comprehensive approach is important because of the constraining interrelatedness of all living organisms through their common descent; every biological structure, function, and system is largely determined by its evolutionary history. Knowledge of the structure of a cellular component such as a protein in a selection of diverse organisms, of the evolutionary tree of biological species, and of the rules of evolution permits a kind of biological interpolation, the prediction, within ascertainable probability limits, of the nature of the structure in all other living organisms. One can further infer many things about the structure in ancestral organisms. Ultimately a great deal will be learned from living organisms about the protein structures in the immediate ancestor of all living species, which lived over 3 billion years ago, and even in the single line which preceded it.

2. *Structure of proteins*

Proteins are polymers composed of the twenty different kinds of amino acids shown in fig. 3. They are synthesized in the cell, according to the message in the genetic material, by successively adding amino acids to the chain, one molecule of water being eliminated with each addition. As the chain builds up, it begins to coil and adhere weakly to itself to form a three-dimensional structure of characteristic shape, flexibility, and chemical properties (Epstein et al. 1963).

It has become clear that the basic metabolic processes of all living cells are very similar. A number of identical compounds, mechanisms, structures, and reaction pathways are found in all living things so far observed, including such diverse cells as those of bacteria, beans, butterflies, birds, and biochemists (Sallach and McGilvery 1963). Identical proteins are seldom found in different species; however, a number of proteins have identifiable counterparts known as homologues in most living things. Biochemical reactions which involve cooperation between catalytic proteins from widely different species have been carried out in the laboratory. During the past 15 years, homologous proteins have been shown to have very similar amino acid sequences as well as closely similar three-dimensional structures (Kendrew et al. 1960; Perutz et al. 1968; Dickerson 1964; Buse et al. 1969; Huber et al. 1968; Padlan and Love 1968).

The variation in protein structure from one biological group to another is illustrated in fig. 4 by the first halves of the sequences of cytochrome c, a protein of fundamental importance found in mitochondria (Margoliash et al. 1961; Matsubara and Smith 1962; Kreil 1963; Smith and Margoliash 1964; Rothfus and Smith 1965; Stewart and Margoliash 1965; McDowall and Smith 1965; Bahl and Smith 1965; Nakashima et al. 1966; Needleman and Margoliash 1966; Goldstone and Smith 1966, 1967; Nolan and Margoliash 1966; Chan and Margoliash 1966a, b; Chan et al. 1966; Yadi et al. 1966; Heller and Smith 1966; Narita and Titani 1968; Dus et al. 1968; Wojciech and Margoliash 1968; Dayhoff 1969). The information for the

Glycine
G
$$HO-\overset{O}{\overset{\|}{C}}-\overset{NH_2}{\overset{|}{CH}}-H$$

Alanine
A
$$HO-\overset{O}{\overset{\|}{C}}-\overset{NH_2}{\overset{|}{CH}}-CH_3$$

Aspartic acid
D
$$HO-\overset{O}{\overset{\|}{C}}-\overset{NH_2}{\overset{|}{CH}}-CH_2-C-OH$$

Glutamic acid
E
$$HO-\overset{O}{\overset{\|}{C}}-\overset{NH_2}{\overset{|}{CH}}-CH_2-CH_2-\overset{O}{\overset{\|}{C}}-OH$$

Asparagine
N
$$HO-\overset{O}{\overset{\|}{C}}-\overset{NH_2}{\overset{|}{CH}}-CH_2-\overset{O}{\overset{\|}{C}}-NH_2$$

Glutamine
Q
$$HO-\overset{O}{\overset{\|}{C}}-\overset{NH_2}{\overset{|}{CH}}-CH_2-CH_2-\overset{O}{\overset{\|}{C}}-NH_2$$

Proline
P
$$HO-\overset{O}{\overset{\|}{C}}-\overset{NH-CH_2}{\overset{|}{CH}}\quad\underset{CH_2}{\overset{|}{CH_2}}$$

Serine
S
$$HO-\overset{O}{\overset{\|}{C}}-\overset{NH_2}{\overset{|}{CH}}-CH_2-OH$$

Threonine
T
$$HO-\overset{O}{\overset{\|}{C}}-\overset{NH_2}{\overset{|}{CH}}-\overset{CH_3}{\overset{|}{CH}}-OH$$

Cysteine
C
$$HO-\overset{O}{\overset{\|}{C}}-\overset{NH_2}{\overset{|}{CH}}-CH_2-SH$$

Lysine
K
$$HO-\overset{O}{\overset{\|}{C}}-\overset{NH_2}{\overset{|}{CH}}-CH_2-CH_2-CH_2-CH_2-NH_2$$

Arginine
R
$$HO-\overset{O}{\overset{\|}{C}}-\overset{NH_2}{\overset{|}{CH}}-CH_2-CH_2-CH_2-\overset{H}{\overset{|}{N}}-\overset{NH_2}{\overset{\|}{C}}=NH$$

Valine
V
$$HO-\overset{O}{\overset{\|}{C}}-\overset{NH_2}{\overset{|}{CH}}-\overset{CH_3}{\overset{|}{CH}}-CH_3$$

Isoleucine
I
$$HO-\overset{O}{\overset{\|}{C}}-\overset{NH_2}{\overset{|}{CH}}-\overset{CH_3}{\overset{|}{CH}}-CH_2-CH_3$$

Leucine
L
$$HO-\overset{O}{\overset{\|}{C}}-\overset{NH_2}{\overset{|}{CH}}-CH_2-\overset{CH_3}{\overset{|}{CH}}-CH_3$$

Methionine
M
$$HO-\overset{O}{\overset{\|}{C}}-\overset{NH_2}{\overset{|}{CH}}-CH_2-CH_2-S-CH_3$$

Tyrosine
Y
$$HO-\overset{O}{\overset{\|}{C}}-\overset{NH_2}{\overset{|}{CH}}-CH_2-\langle\ \rangle-OH$$

Phenylalanine
F
$$HO-\overset{O}{\overset{\|}{C}}-\overset{NH_2}{\overset{|}{CH}}-CH_2-\langle\ \rangle$$

Tryptophan
W

Histidine
H

Protein
$$\vdots\ \overset{O}{\overset{\|}{C}}-\overset{NH}{\overset{|}{CH}}-R$$
$$\overset{O}{\overset{\|}{C}}-\overset{NH}{\overset{|}{CH}}-R\ \vdots$$

Fig. 3. The twenty amino acids from which proteins are synthesized. These exhibit a variety of shapes and chemical reactivities. The amino acids combine with loss of water to form a protein chain as indicated in the lower right hand corner. The interior of a protein contains mainly the hydrophobic residues, with hydrocarbon side chains packed closely together, while the hydrophilic types are mainly on the outside in contact with the aqueous phase. Two cysteine residues can be combined by oxidation to form a stabilizing disulfide crosslink in the chain. The one-letter abbreviation is shown below the name of each amino acid.

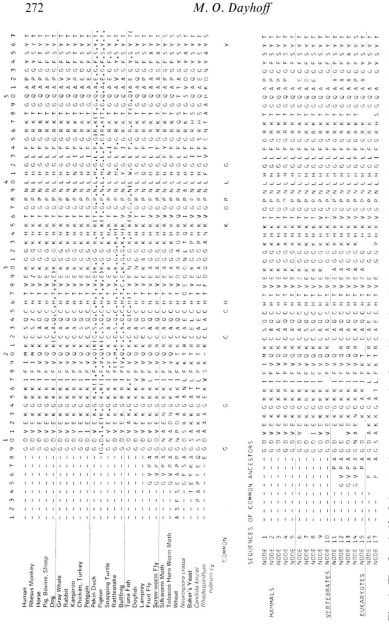

Fig. 4. The first half of the cytochrome c proteins from 29 species. Each amino acid residue is represented by a single letter as shown in fig. 3. A Z represents either E or Q. Sequences of closely related groups are similar. Even though the proteins come from widely different biological groups, only a single amino acid is found in 11 of the 57 positions shown. Very few changes have been accepted in the other positions. The phylogenetic tree of fig. 1, which requires the minimum number of amino acid changes in the ancestral organisms, was derived from the (complete) sequences. The sequences of the ancestors at the nodes of fig. 1, which are derived by the computer methods discussed later, are shown in the lower part of the figure. In the few places that are blank, two or more amino acids are equally likely to have been present. (From Dayhoff 1969.)

synthesis of cytochrome c is located in the cell nucleus. The protein is attached, through the cysteine residues at positions 22 and 25, to a heme group which contains one atom of iron. This iron participates in the transfer of electrons during metabolic oxidation in the cell. A clearly, though very distantly, related protein has also been analysed from a simple photo-synthetic bacterium, *Rhodospirillum rubrum*, where the protein is used in photosynthetic electron transfer.

All the sequences, even though from very diverse organisms, contain so many identities that there is no question about the correct alignment. About $\frac{3}{5}$ of the total number of amino acids at corresponding positions are identical in wheat and human chains, while $\frac{3}{10}$ are identical in human and *R. rubrum*. Certain positions contain the same amino acid in all the sequen-ces; others are filled by any one of several amino acids which are usually of similar shape or chemical properties. Since the proteins differ slightly in length, dashes have been added to maintain a good alignment.

3. Functions of proteins

Many different kinds of proteins are required for the functioning of a living organism; for example, there are several thousand different kinds in the bacterium *Escherichia coli* and possibly a million kinds in a human being. Each protein is delicately adapted to perform important structural, chemical or managerial functions in the cell. Most chemical reactions, such as a step in the synthesis of an amino acid, are catalyzed by proteins. The shape and mechanical properties of subcellular organelles, such as flagella, are largely determined by the shapes and characteristics of their unique structural proteins. The precipitation of the calcium compounds in bone or shell is initiated and stabilized by a protein matrix. Metabolic reactants are transported by proteins; for example, CO_2 and O_2 by the hemoglobin in blood. Foreign substances of particular shapes are removed by protein antibodies which specifically bind to them. The inhibition or activation of enzymes can be accomplished by their reaction with other proteins. Finally, primary control over cell function is exerted by repressor proteins which prohibit the transcription of the genetic message prerequisite to the forma-tion of other proteins. Thus, proteins are very important components of the complex feedback network of cell chemistry. Proteins are the main determinants of the physical and chemical characteristics of a species. Very slight differences in the structures of some of the very many proteins in the human complement lead to individuality.

4. Synthesis of proteins in the cell

The 'library' of a cell, which stores the information necessary for the synthesis of the many structurally unique proteins, is located in very long strands of nucleic acid (DNA) (Srb et al. 1970). The chemical message is written by the linear sequence of four different kinds of 'letters', the nucleotide links of the nucleic acid. Most DNA is found in the chromosomes. The cellular organelles, mitochondria and plastids, also contain DNA.

The chemical processes by which each protein is synthesized, according to the directions in a portion of the chromosome strand called a gene, are quite complex and involve a number of protein and small nucleic acid molecules which uniquely fit other nucleotides and amino acids. Two consecutive information-transfer processes are involved. First, DNA directs the synthesis of ribonucleic acid (RNA) molecules, a simple one-for-one transcription of the nucleotides in the DNA molecules. RNA differs from DNA by the absence of one methyl group in one of the four bases and by the presence of an additional oxygen atom in the sugar which links the nucleotides. The transcription mechanism is chemically and spacially complex, although mathematically simple. Transfer RNA (tRNA) and ribosomal RNA (rRNA), which are metabolically active in the synthesis of

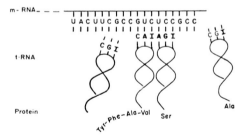

Fig. 5. Information transfer in protein manufacture. Synthesis occurs on the surface of a ribosome. The tRNA molecules have previously been activated, each forming an energetic bond with its specific amino acid by an enzyme-catalyzed process. The activated tRNA molecules line up consecutively along the mRNA, where each tRNA is hydrogen-bonded precisely to its specific codon. As each activated tRNA achieves its proper position, the growing protein chain is attached to the amino acid of this new tRNA. The preceding tRNA is then free to leave. The tRNA must fit next to the other adjacent tRNA molecules as well as to the ribosome. The illustrated shape of the tRNA molecule is diagrammatic only. Messenger RNA contains the following bases: A, adenine; C, cytosine; G, guanine; and U, uracil. Some of the bases in tRNA have been modified by enzymes. Inosine, I, is similar to A and G. (From Dayhoff and Eck 1968.)

proteins, and messenger RNA (mRNA), which carries the information to synthesize a protein, are formed by transcription.

The second, or translation, process is mathematically complex (see fig. 5). The mRNAs determine the sequence of amino acids in the proteins which the cell synthesizes. Nucleotides are read sequentially from the messenger, three at a time. Each such triplet, or 'codon', is translated into one particular amino acid in the growing protein chain. There are 64 possible triplets of the four different kinds of nucleotides, but only 20 amino acids. In most cases there are several codons which specify, or code for, the same amino acid. The nature of this code has been worked out for *E. coli* (Nirenberg et al. 1965). There are indications that it is almost identical in all living organisms.

5. The evolutionary process

The cell routinely forms a copy of its nucleic acids before cell division, passing on one copy of this library to each daughter cell. In multicellular organisms which exhibit sexual reproduction, there is a more complex process, which also involves this copying of the chromosomes; each offspring receives a complete library of chromosomes, half from each parent.

Two forms of chromosomal organization are found in living organisms. In bacteria and blue-green algae there is a single chromosome which is composed of two strands of DNA entwined as a helix. This chromosome is not separated from the rest of the cell by a membrane. In higher organisms, including animals, green algae, plants, and true fungi, the DNA is contained within a nuclear membrane. Typically there are several separate pieces of double-stranded helical DNA, each super-coiled with protein to form more complex chromosomes. The biochemical processes of replication are very similar in the two forms.

The rare errors which occur in passing along the chromosome library to offspring make possible the evolutionary process. Most of these errors are lethal or at least deleterious to the proper functioning of the organism. Some few are beneficial. Direct evidence for the molecular nature of the mistakes is found in the data rapidly accumulating on abnormal forms of proteins or chromosome structures present today in human and other populations (Atkin and Ohno 1967; McKusick 1968; Ohno et al. 1968; Perutz and Lehmann 1968; Schroeder and Jones 1965).

Several kinds of mistakes occur frequently. First there are gross errors in chromosome structure. A duplicated portion may be incorporated in a chromosome resulting in the synthesis of two identical proteins coded from the original and duplicated genes; or a section of a chromosome may be deleted altogether in which case the corresponding proteins would no longer be produced at all. Occasionally, sections of chromosomes are moved so that the order of the genes on the chromosome is altered. In the higher organisms, sections of genetic material can be exchanged between chromosomes, or an extra chromosome may be passed to the offspring as in the case of human mongolism. Even an entire duplicate genetic complement is sometimes found in plants.

There are also errors which occur within single genes. Of these, the changes most commonly accepted by natural selection are changes of one single nucleotide within a gene to one of the other three nucleotides. The resulting protein may then have one amino acid changed. Occasionally an amino acid is inserted or deleted. Most such changes are deleterious, diminishing or destroying the effectiveness of the protein produced. It is these small errors which result in so-called 'point' mutations with which we shall be concerned here. Many examples of the effects of these errors can be seen in the cytochrome sequences of fig. 4. Deletions or insertions are mainly found at the ends of the sequences, although one must have occurred within one sequence at position 20.

Evidence for the occurrence of point mutations and most kinds of chromosomal aberrations is seen in all biological groups. It is reasonable to assume that such changes have been occurring ever since the divergence of the ancestor of all living things into two 'species'. For billions of years, then, biochemical evolution has been a repetitive process producing many billions of generations of organisms, each organism in the succession differing but little from its parent(s) and in turn from its offspring. Many errors have occurred in individuals, some few of which have been selected for perpetuation in the species.

The evolutionary process in biology and biochemistry is extremely conservative, largely because older cell components come to be relied upon by many subsequent additions. The basis of this conservatism is particularly well illustrated by proteins. Each protein has become adapted to perform particular functions; each interacts with many other cell components in several ways, by participating in actual chemical action, through complexes formed with inhibitors and cooperating components, through the rate at which its reactions occur, or by its structural properties. The protein must also fold automatically into the correct three-dimensional structure when it is synthesized, the corresponding genes must replicate and be transcribed normally, and the mRNA must be stable and be translated at the usual rate. All of these properties must be little disturbed by the usual changes in and extremes of the environment that the organism encounters. Almost any change in the protein structure, even though of particular advantage for one function, would coincidentally disturb so many other interactions that it would be extremely disadvantageous to the organism. So severe are these constraints that an identical sequence of each protein is found in most individuals of a species. A given sequence often predominates within a species for several million years. A minor variant, usually differing by only one amino acid, may eventually become preferable because of changes either in the other cell constituents or in the environment. Only infrequently does a more profound structural change prove beneficial. The net effect of this extreme conservatism is the great similarity of protein structures in various species as seen in fig. 4.

Of the almost infinite number of possible combinations of chromosomal aberrations and mutations which might have occurred, only a very insignificant fraction have actually occurred in living organisms. It seems very likely that many which could have occurred, but in fact did not, might have proved beneficial and have been accepted. The evolutionary paths that were actually followed were based on a succession of rare random events, the acceptable mistakes. At any given time, the future of biochemical evolution is largely indeterminate because of these many possible alternatives.

Each diverging species is relatively free to accept a unique series of changes. The preservation of such a series today in different descendent species usually marks them as sharing a common ancestor in which the changes occurred. Only rarely do identical point mutations occur independently in two species. If there is one amino acid change in each of two homologous proteins, the changes occur at the same position only slightly more often than one would expect by chance. If the changes do occur at the same position, then there is about one chance in five that the resultant amino acids will be the same. Because this coincidence rate is so low, we can hope to reconstruct a precise evolutionary history with a reasonable amount of data.

6.   Phylogenetic trees derived from protein sequences

From protein sequence data alone, it is possible to derive a phylogenetic tree which shows in detail the nature of the ancestral relationships of present-day species (Eck and Dayhoff 1966). Fig. 1 shows the tree which we have derived from the cytochrome c data (Dayhoff and Park 1969). This tree, within the limitations of the small quantity of data now available, has the same topology (order of branching without regard for the lengths of branches or for the angles between them) as trees derived from morphological or other biological considerations. Each point on the tree represents a definite time, a particular species, and a definite protein structure within the majority of individuals of this species. There is a 'point of earliest time' on any such tree. Radiating from this point, time increases on all branches. Protein sequences from living organisms all lie at the ends of branches which represent the present time. The pattern of the network of lines in the tree then indicates the relative order in which the proteins became distinct from one another. The location of the point of earliest time, that is, the connection of the trunk to the branching structure, cannot be inferred directly from the sequences, but must be estimated from other considerations. It seems likely that sequences from all organisms have changed with time, even though the morphology of some organisms may appear primitive. In the absence of any evidence to the contrary, we have used a point about equally distant from all the observed sequences as the point of earliest time.

7.   Intuitive considerations

Fig. 4 demonstrates that very few amino acid changes have been accepted in the cytochrome c sequences of all these species – an average of less than two changes per position. Biologically similar organisms tend to have the same amino acid in a given position. In order to illustrate some general considerations in building phylogenetic trees from protein sequence information, let us examine three positions. In position 15 the plants and fungi all have an A, while the animals all have a K. In position 17 the fungi all have an L, while wheat and animal sequences usually have an I. In position 8 the insects and plants have an amino acid, usually A, while the vertebrates lack this portion of the sequence. So seldom do changes occur that the observed changes almost always reflect a mutational event occurring in a single ancestral organism. The evidence at position 15 favors the hypothesis that there was a single mutation in a single lineage connecting the animal group with the plant group. The mutation at position 17 indicates that there is a single lineage between fungi and the other species, while that at position 8 indicates a similar single lineage between vertebrates on the one hand and insects and plants on the other. We combine this evidence to make a diagram of these lineages, shown in fig. 6a. This topology accommodates the information from the three sites, requiring the occurrence of only three changes in three ancestral organisms. Either of the other two possible topological configurations (figs. 6b and 6c) requires the occurrence of at

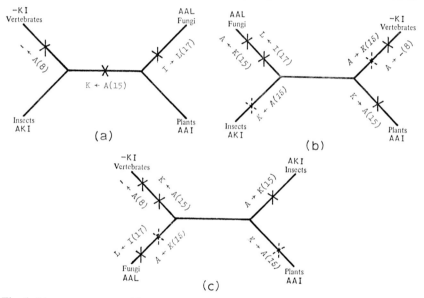

Fig. 6. Lineages constructed from the evidence in positions 8, 15, and 17. The most likely topology is shown in (a). Three mutations would have occurred on the branches as indicated. The two alternative topological configurations, (b) and (c), are less likely to be correct. At least four mutations must have occurred in either case. In both, the two mutations in position 15 could have occurred in alternative ways shown by solid and dashed X's. (From Dayhoff 1969.)

least four changes. Because changes in sequences are so rare in this system, configuration 6a is very likely to be correct.

In building a tree, we must consider all the evidence, not just that from three selected positions. There are a number of other amino acid positions that confirm the correctness of the topology in fig. 6a. Occasionally there is conflicting information; for example, there is evidence in position 17 that the fruit fly belongs with the fungi. Since this conclusion is contradicted by the weight of all the other evidence, we must presume that in this position there have been two separate mutations in two separate organisms, which by coincidence resulted in the same amino acid. This is an example of parallel evolution on a molecular scale.

8.   *Computer methods*

The quantity of data that must be considered in constructing a phylogenetic tree is so great, and objectivity is so important, that processing of the information is clearly an appropriate task for the computer. We have three computer programs which we use to determine that topology which requires the smallest number of mutations, the 'best' topology.

The first program considers a series of present-day sequences in a preassigned topology. It computes the ancestral sequences at all the branch points or nodes and counts the total number of mutations which occurred throughout the tree. This number is a measure of the 'size' of the topology. The second and third programs automatically generate a large number of topologies which are then measured by this first program. We seek that topology which has a minimum size.

To begin with, the first program infers the sequences in ancestral organisms from the given topology. Positions along the chain are considered one at a time. Where only one amino acid is found in all the sequences, it was almost certainly found also in all the ancestors at all times. Further, a number of reasonable conjectures can be made regarding the ancestral sequences at the nodes. Fig. 7 shows two special cases of the way in which amino acids might occur at one position in five proteins that are linked by a definite topology. In fig. 7a, A is most likely to have been the amino acid in the

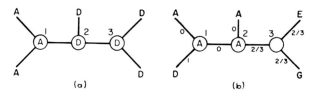

(a)   (b)

Fig. 7. Procedure for inferring ancestral sequences. Positions are considered one at a time. Amino acids which might occur at one position in different cytochrome c sequences are shown by letters. The amino acids which most probably existed in the ancestors at the three nodes are placed within circles. Where no one choice is clearly preferable, the node is left blank. The number of changes inferred to have occurred on each branch is shown in (b). (From Dayhoff 1969.)

ancestor at node 1, while D was present in ancestors at nodes 2 and 3. There was one mutation from D to A at some time between nodes 2 and 1. Any other assignment of ancestral structures would necessitate an assumption of 2 or more mutations. In the second figure, A is the best choice for nodes 1 and 2, but there is no single best choice for node 3. In this case we simply leave a blank in the ancestral structure, to reflect a considerable measure of uncertainty.

The computer must be instructed so that it will treat all possible topological configurations, not merely one particular case. For this purpose, any tree can be thought of as being made up entirely of nodes connecting three branches. More complex branching simply involves two or more such nodes with zero distance between them.

The general procedure for finding the size of the tree is as follows. For each amino acid position of each node, a list of amino acids occurring on each branch is made. For node 2 of fig. 7b, the lists would read:

Branch 1	Branch 2	Branch 3
A	A	E
D		G

The amino acid which occurs on more branches than any other, in this case A, is selected for the node. If no single first choice exists, the position is left blank.

There are a number of cases for which this simple program gives an equivocal assignment when it need not. Consider the case in fig. 8. The procedure described above would have assigned blanks to nodes 1 and 3. The choice of A as the amino acid at all three nodes, necessitating two independent mutations from A to B, is best; the choice of any other combination would require at least three mutations.

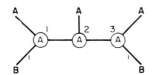

Fig. 8. A particular topology, with the amino acids at one position, from five proteins. (From Dayhoff 1969.)

We have added some further steps to enable the computer to reconsider its initial decisions. The first assignment of nodal structures is examined. If at least two of the three neighbors to a blank node contain the same amino acid, this acid is assigned to the node. The resulting configuration is examined again and again until no more changes occur. Finally, any node which does not agree in structure with two of its neighbors is changed to a blank. This results in the following nodal amino acids for the structure of fig. 8.

	Node 1	Node 2	Node 3
First pass	Blank	A	Blank
Second pass	A	A	A
Final pass	A	A	A

This procedure produces a definite assignment of ancestral amino acids in positions where one choice is clearly preferable. It leaves blanks where there is a reasonable doubt. On rare occasions one of two equally likely amino acids may be picked, because of the great length and complexity of topological trees and of the shortsightedness of the computer. By the application of these procedures to each position, the ancestral sequences are produced for each node.

Once the assignments of the ancestral sequences have been made, the amino acid changes along each of the branches of the tree are counted. Each sequence is compared with each adjacent sequence, position by position, and the number of amino acid changes is totaled. Even when we do not assign an amino acid to a position, we still can estimate the total number of changes which must have occurred in that position. A blank in a nodal sequence, as in node 3 of fig. 7b, indicates that several amino acids are equally likely as common ancestors. The minimum number of mutations near node 3 is two. We distribute these over the three branches about the node, counting 0.67 changes for each of the branches. Where two adjacent nodes are both blank, we count 0.33 changes for the branch length between them. For the special case of positions involving parallel mutations, as in fig. 9, it is likely that both nodal sequences contained either A or B and that only two mutations occurred. It is quite unlikely that one contained A and the other B as three mutations would have been required. In this special case, we assign the contributions to branch lengths from this position as shown in fig. 9.

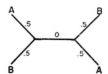

Fig. 9. Branch lengths accumulated for the case of a parallel mutation.

A sum of the contributions from all positions is formed by the first program. This gives a single number for the topology which is the total number of changes. This number from program 1 is used by programs 2 and 3 to evaluate the many alternative topologies.

The second of the three programs derives a good approximation to the minimal tree. We start with three sequences, which can have only the topology shown in fig. 10a. The complete order of events, showing the point of earliest time, as represented in fig. 10b, cannot be reconstructed from the sequence information alone. The program next considers the topologies produced by adding a fourth sequence to each of the three branches, as in figs. 10c, d, and e. The total length of each of these alternative trees is computed, and the shortest configuration is retained as best. Other sequences are evaluated and placed one at a time. The position of each sequence must be decided without regard for the sequences to be considered later. Inevitably one makes at least one decision wrong in retrospect, leading to a tree similar to but not identical with the optimal one. It is an approximation to the correct tree. Such an approximation may alternatively be derived from biological considerations.

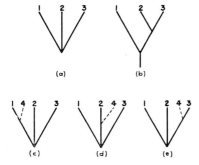

Fig. 10. Topological connections. The only topological connection for three sequences is shown in (a). The complete history, suggested by (b), cannot be determined from sequences alone. The three possible alternative topologies for four sequences are shown in (c), (d) and (e). (From Dayhoff 1969.)

The third program shifts each of the branches on the approximate tree to all alternative locations. There are millions of possible configurations for the cytochrome c tree. A manageable subclass of these, among which the best answer is almost surely to be found, is obtained by the following procedure. For every branch of the approximate tree, a cut is made and each resultant part is grafted onto all the branches of the other part (see fig. 11). Each such topology is evaluated by the first program, and the best one is

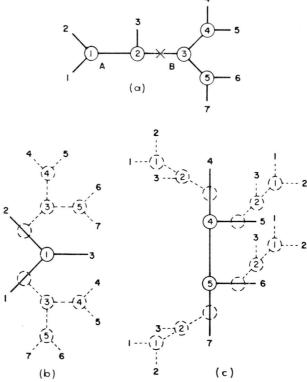

Fig. 11. Examples of alternative topologies systematically considered by program 3 (Dayhoff and Park 1969). One representative complex topology is shown in (a). The program cuts each of the 11 branches of this topology in turn. Each piece is grafted onto each branch of the other piece. Consider one such cut between nodes 2 and 3 dividing the figure into two pieces, A and B. The two new structures resulting from grafting B onto A are indicated in (b). The four new structures to be considered as a result of grafting A onto B are given in (c). (From Dayhoff 1969.)

chosen as the new approximation to the tree. This one is used as the next starting tree and alternatives of it considered. Finally, a tree is found that is superior to all alternatives.

9. Ancestral sequences

The sequences of the common-ancestor proteins at the nodes indicated in fig. 1, as calculated by the computer, are displayed at the bottom of fig. 4. The very small number of blanks show how few positions are doubtful.

These ancestral sequences take on a very real meaning in view of the present possibility of synthesizing the sequences in the laboratory and measuring their chemical activities (Hirschmann et al. 1969; Gutte and Merrifield 1969). A great deal about the chemical capabilities of ancient organisms may thus one day be known.

10. The cytochrome c phylogenetic tree

Fig. 1 shows the best tree for the cytochromes obtained by these procedures. The branch lengths are expressed in PAMs (accepted point mutations per 100 links). This unit includes a correction for the superimposed mutations estimated to have occurred (Dayhoff and Eck 1968) (see table 1). The major

TABLE 1

Correspondence between the observed number of amino acid differences per 100 links of two sequences and the actual total number of accepted point mutations (some of them superimposed) that must have occurred, i.e. the evolutionary distance. This number in PAM units is derived from sequence data and inferred ancestral sequences in the families which have been studied. It takes into account the unequal frequency and mutability of the amino acids in estimating the superimposed and back mutations to be expected on the basis of random changes. This table is used to correct the branch lengths of the phylogenetic trees.

Observed differences in 100 links	Evolutionary distance in PAMs
1	1
5	5
10	11
15	17
20	23
25	31
30	39
35	48
40	58
45	70
50	83
55	98
60	117
65	140
70	170
75	208
80	260
85	370

(From Dayhoff and Eck 1968.)

groups fall clearly into the topology shown. In the time interval during which the lines to the kangaroo, the rabbit, the ungulates, and the primates diverged, either there were no mutations in cytochrome c or the evidence of these mutations was obliterated by subsequent changes. Similarly, the order of divergence of the lines to lamprey, cartilaginous fishes, and bony fishes cannot be fixed; these lines probably diverged over a short period of time. For some species, such as the dog and the whale, the optimum topology depends on a single amino acid position. There is perhaps one chance in 5 or 10 that the position of the branch point of such a sequence deviates by one PAM unit from the true topology. Other protein sequences from these animals should settle these uncertainties.

One aspect of the topology needs special consideration – where to place the bacterial sequence. The evidence found in positions 13, 29, and 57 indicates a connection to the wheat branch. Position 55 alone gives weak conflicting evidence for a connection to the fungus branch. If the preponderance of evidence is true, then the fungi and animals must have diverged from each other after their divergence from the wheat line. In the resulting tree, the bacterial sequence is at the end of a very long branch.

11. *Geological time scale*

We are interested in knowing the relative geological times associated with the events of the phylogenetic tree of cytochrome c. We estimate these times from the number of mutations on the branches. For each family of proteins, the rate of accumulation of mutations is different (see table 2). However, the risk of mutation within a family such as cytochrome c is evidently rather constant, at least for nucleated organisms, over the interval during which the species under consideration diverged. The variation of branch lengths which correspond to the same time interval is about what would be expected for a random process of mutation. The proportionality constant of time per mutation must be fixed from considerations such as the divergence point of the fish and mammalian lines. Paleontological evidence (Young 1962; Romer 1966) places this particular divergence at about 400 million years ago. The cytochrome c tree places it at 11.5 PAMs, on the average. Therefore, 400 million years correspond to 11.5 PAMs. Because of the random nature of mutational events, any branch length B (for short branches), as an estimate of time, has a standard deviation of $\pm \sqrt{100B/L}$ where L is the

TABLE 2

The rates of mutation of selected proteins (McLaughlin and Dayhoff 1969). The figures are given in PAMs, which we have estimated would occur in 100 million years of evolution. For most of the proteins, the rate was based on the time of divergence of the mammalian orders at 75 million years ago.

Proteins	Mutations per 100 million years
Fibrinopeptides	90
Growth hormones	60
Immunoglobulins	34
Kappa C region	40
Kappa V region	34
Heavy C region	28
Ribonucleases	30
Hemoglobins	12
Beta	13
Alpha	11
Myoglobins	9
Gastrins	9
Adenohypophyseal hormones	9
Encephalitogenic proteins	7
Insulins	4
Cytochromes c	3
Glyceraldehyde 3-PO_4 dehydrogenases	2
Histones	0.06

(From Dayhoff 1969.)

length of the protein. Thus comparison using 400 links allows the evolutionary distance to be estimated twice as precisely as one using only 100 links. As more proteins become available, the precision of the time estimates for species divergences will improve.

It may well turn out that the mutation rate varies in major groups and that there is an occasional species which undergoes a sudden, large change. However, there is sufficient potential information within living species to establish the average mutation rates in various phyla and to develop a useful relative time scale of phylogenetic events. Using the constant rate assumption, let us consider the implications of the bacterial cytochrome sequence.

The information in fig. 1 is redrawn in fig. 12 with the meaning of the branch lengths altered. Now the vertical coordinate is proportional to time, with the present day at the top. The position of nodes is derived from the average number of mutations on the branches above each node. The con-

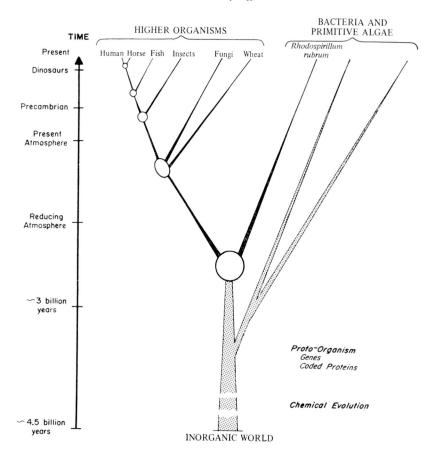

Fig. 12. Temporal aspects of the cytochrome phylogenetic tree. Details of the upper part of the tree, shown in black, are derived from information in the cytochrome c sequences. The size of any node reflects the uncertainty in its position. Other parts of the tree, shown in gray, are inferred from the details of biochemical and biological structures and from geological and fossil evidence. Fossils resembling present-day bacteria and blue-green algae have been found in cherts more then three billion years old (Barghoorn and Schopf 1966; Barghoorn and Tyler 1965). All living phyla seem to be derived from a proto-organism which possessed many metabolic capacities, nucleic acid genes, and protein enzymes (produced by the genetic code). Preceding the time of the proto-organism, the genetic code and the metabolism of nucleic acids and small molecules evolved (Jukes 1966; Eck and Dayhoff 1966; Crick 1968). The time scale is derived from the radioactive dating of rocks. The oxidation level of the atmosphere is inferred from the oxidation states of ancient sediments (Rutten 1962; Calvin 1969). (From Dayhoff 1969.)

nection of the branches to the trunk was presumed to be equidistant from the bacterial and the other sequences. The meaning of the earliest divergence shown is not entirely settled. Probably it is a divergence of the bacterial line and that of the eukaryote nucleus, where the gene for cytochrome c is found. However, cytochrome c is used in the mitochondria, which are subcellular organelles that may be present as a result of an ancient symbiotic relationship of a bacterium and a eukaryote. It is possible that the cytochrome c gene originally came into the cell with the mitochondrial genome. In this case, the earliest divergence would represent the mitochondrial–bacterial divergence. The dimensions of the tree would be the same for either interpretation. From this one family of clearly related proteins, estimates of the time relationships on an extensive phylogenetic tree of life are derived.

12. Ferredoxin

The evolutionary tree derived from the protein of ferredoxin is particularly interesting in connection with the study of the early evolution of life. Ferredoxin is found in anaerobic and photosynthetic bacteria, in blue-green and green algae, and in the higher plants. The protein is bound to iron and inorganic sulfur through some of its cysteine residues. Ferredoxin is the most electronegative metabolic enzyme known, with a potential close to that of molecular hydrogen. It participates in a wide variety of biochemical processes fundamental to life, including photosynthesis, nitrogen fixation, sulfate reduction, and other oxidation-reduction reactions. It may have achieved these functions very early in the differentiation of the biological kingdoms, and its structure been conserved strongly ever since. The evidence from the plastids of higher plants indicates a rate of change of the protein comparable to that of cytochrome c.

The overall topology of the evolutionary tree of ferredoxin, shown in fig. 13 (Dayhoff and McLaughlin 1969) is clearly established by the sequences. Within the groups of plants and the group of nonphotosynthetic bacteria, the relative branch lengths have been estimated from the point mutations as explained above.

The sequences of ferredoxin from the plant plastids are almost twice as long as those from anaerobic bacteria, whereas the one from *Chromatium*, a photosynthetic bacterium, is intermediate. There has clearly been a duplication of genetic material in the bacterial line. The two halves of the molecule are closely similar (Matsubara et al. 1969; Eck and Dayhoff 1966;

Jukes 1966b). The plant line also appears to have incorporated duplications of genetic material, possibly independently of the one occurring in bacteria. The number of iron and sulfur atoms attached to the protein differs in the three biological groups. This variation would indicate a somewhat different function and environment for the three ferredoxin types. Therefore, it is to be expected that the rate of change in the three groups of ferredoxins would be different. A quantitative estimate of the evolutionary distance between the plant plastids and the bacteria in terms of point mutations is impossible from these sequences because of the many changes in length. It must be a very long distance.

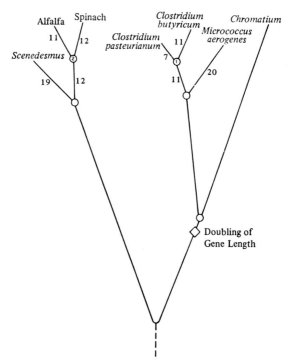

Fig. 13. Evolutionary tree of ferredoxin. The sequences from bacteria and from plants are clearly related, but very distantly so. The doubling of gene length in the bacterial line probably occurred as shown here. The details of duplications in the plant line are obscured by the many subsequent changes. (From Dayhoff 1969.)

The biological meaning of the first divergence on the tree is not clear. It is likely that the plastids are symbionts acquired by a primitive higher organism about the time of the divergence of the plant and animal lines. The gene for plant ferredoxin may have derived either from the nuclear line or the plastid line. If it is a plastid gene, the earliest branching point on the tree refers to the plastid–bacterium divergence rather than to the plant-nucleus–bacterium divergence.

13. Globins

Extensive sequence work has been done on hemoglobin, the principal oxygen-transporting substance in vertebrate blood. These studies illuminate the evolution of species and the duplication of genetic material in the last billion years (Braunitzer et al. 1964; Hill and Buettner-Janusch 1964; Zuckerkandl and Pauling 1965). The main events in the globin family are shown in fig. 14 (Dayhoff and Eck 1968).

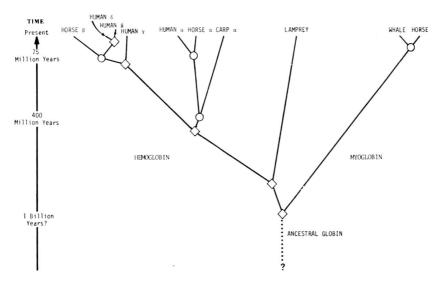

Fig. 14. Evolutionary tree of the globins. The lengths of the branches are proportional to PAM units. The scale at the left shows two geological time markers. The time of the ancestral globin is estimated by extrapolation. Circles indicate species divergence, and diamonds, gene duplications. In general, it is impossible to construct a tree that shows both the number of changes on each branch and time levels. In this simple case it happens to be possible and is instructive. The insect sequence diverged very early, possibly even before the myoglobin-hemoglobin duplication. (From Dayhoff and Eck 1968.)

In higher vertebrates, hemoglobin typically has a tetrameric structure consisting of two alpha-type and two beta-type protein chains; a heme group is attached to each chain. Two atoms of oxygen can be transported by each heme group.

Genes coding for alpha- and beta-type chains, presently located on different chromosomes, were derived originally from a single ancestral gene by an ancient duplication of genetic material. More recently, another duplication gave rise to separate genes coding for the human beta chain, which is synthesized only after birth, and to the human gamma chain, a beta-type chain which is synthesized before birth. This separation in time of the production of the chains indicates that not only the gene for the original protein structure, but also the mechanism which controlled protein synthesis, was duplicated in each case. Very recent duplications have given rise to genes which lie close to each other on the same chromosome, for example, the human delta and beta chain genes.

Myoglobin, a heme protein found in mammalian muscle, has a sequence still recognizably similar to that of the hemoglobins, evidently the result of an ancient duplication. It occurs as a monomer. Other similar proteins which have diverged from the mammalian hemoglobins a long time ago, have been examined from the lamprey, a jawless fish, and from an insect larva.

Not only are the sequences of the globin family similar, but the three-dimensional structures of myoglobin, hemoglobin, and the insect globin as determined by X-ray crystallography, are also similar (Kendrew at al. 1960; Dickerson 1964; Huber et al. 1968; Padlan and Love 1968; Perutz et al. 1968; Buse et al. 1969). This conservation of the structure makes it probable that the sequences of some of the heme proteins of more distant organisms such as protozoa or even plants may prove to be recognizably similar to the known globins.

There have been a great many changes in the hemoglobins and myoglobins since they diverged from a common ancestor. These mutations cannot be followed in detail from the small number of sequences now known, but their approximate number can be inferred from the number of differences that are observed between two sequences (see table 1).

The time scale of fig. 14 has been drawn proportional to the inferred number of accepted point mutations between major protein or species types. The topology has been derived using the detailed methods described above. Myoglobin represents a very old divergence, preceding the origin of the lamprey hemoglobin. The lamprey protein diverged before the chromosome duplication giving rise to the alpha and beta chains of hemoglobin. The insect sequence branched off before the lamprey, probably around the time of divergence of hemoglobin and myoglobin. There is not sufficient evidence to place it precisely; further, we cannot yet decide if the divergences of insect and lamprey represent species divergences or gene duplications which preceded the species divergences. Both of these animals produce several kinds of hemoglobin.

The sequence from carp clearly belongs on the a branch after the divergence of the a and β chains. From the information in the sequences alone, therefore, we would expect the bony fishes to have an a-β tetramer system similar to that of mammals. On the evidence of the β-γ-δ branch, the hoofed mammals diverged later than the human γ chain. The implication is that the separation of the gene which has subsequently evolved into human fetal hemoglobin γ is more ancient than the mammalian radiation.

Extrapolating from the estimated geological times of divergence of some species on the globin tree (400 million years for mammals and bony fish), we could expect that the divergence of hemoglobin and myoglobin took place in the Precambrian, perhaps a billion years ago.

14.   *Information potentially available*

Evolutionary studies based on protein sequence 'traits' have so far utilized only a few small proteins such as cytochrome c, hemoglobin, insulin, fibrinopeptides, trypsin, and ferredoxin. Potentially there is available a great deal more material on which precise estimates of evolutionary history can be based. An upper limit of the information can be calculated from the total quantity of DNA found in the major groups (McLaughlin and Dayhoff 1969; Dayhoff 1969; Shapiro 1968) (see table 3). If all of this genetic material codes for proteins which are, on the average, 500 amino acids long, then the numbers of proteins shown in the last column would be produced. These numbers might be too large by even a factor of ten, since some of the genome does not code for structural proteins, but functions as regulatory or structural DNA or to produce metabolically useful RNA. However, the nucleic acid sequences which have been examined such as tRNA and rRNA are highly conserved. Like the protein sequences, they contain information on which a phylogeny can be based. As measures of information, the numbers in the last column are suggestive of the potential of the genome.

Not all proteins will be useful in phylogenetic studies based on amino acid traits, only those whose sequences have been highly conserved over the time span of interest can be used. Recognizable homologues of similar

TABLE 3

Genome size in biological groups.

Group	Haploid number of nucleotides	Maximum number of average proteins which can be coded
Mammals	3.2×10^9	2.2×10^6
Bony fish	0.9×10^9	6×10^5
Fruit fly	0.1×10^9	7×10^4
Sponges	0.06×10^9	4×10^4
Fungi (ascomycetes)	0.03×10^9	2×10^4
Bacteria	0.013×10^9	9×10^3
Plant Viruses	0.3×10^4	2
Mitochondria	1.7×10^4 to 16×10^4	11 to 107
Cytochrome c	330	1/5

(From Dayhoff 1969.)

length must be located whose amino acid traits can be correlated. As yet there is little information on the relations of extremely distant groups. Only in the cytochrome family have complete sequences been done for homologues in both bacteria and the kingdoms of the higher organisms. From exploratory experiments and sequence work on the active centers of homologous proteins of very distantly related groups, there is every reason to believe that ultimately at least hundreds of the proteins of *E. coli* will be found to be usefully similar to homologues in plants or animals. In table 4 are shown the active centers which have so far been elucidated (Dayhoff and Eck 1968; Groskopf et al. 1969; Möckel and Barnard 1969; Jurasek et al. 1969; Hofmann 1971; Shotton 1971). Bacterial sequences of the digestive enzymes trypsin and pepsinogen and of the metabolic enzymes glyceraldehyde 3-phosphate dehydrogenase and phosphoglucomutase are very similar to those of higher organisms. Sequences with such extensive similarity in their active sites usually show homology throughout the chains. This has been observed in the case of trypsin, for which some additional information is available; the many scattered portions of the bacterial sequence are similar to portions of bovine trypsin. We would predict that the many other enzymes whose functions were fixed long ago, for example, those which are involved in basic metabolic reactions, or with the replication of the genome or with the coding of proteins, would similarly be highly conserved. From the larger and more closely related genomes of plants, animals, protozoa, and fungi, there will surely be produced a great many proteins which are very similar.

The precision with which the topological connections can be resolved and with which time intervals can be estimated is proportional to the square root of the number of amino acid traits studied. Thus the inclusion of 25 families of sequences the size of cytochrome c will give five times the present precision. From such a quantity of data, further gains in accuracy may be made from an improved understanding of the evolutionary process. There are already at least 25 kinds of vertebrate sequences known which can be used in studying the phylogeny of kingdoms. Unfortunately, sequences homologous to most of these proteins from the other kingdoms are presently lacking.

The usefulness of different protein families for estimating different areas of phylogenetic relations can be judged from another aspect of the present information. This is based on the observation, made on a number of families, that the rate of change of a protein whose function is well established is surprisingly constant over the time during which the different kingdoms of higher organisms have diverged. Evidence from the tRNA molecules even suggests a similar overall rate of accepting changes in bacterial as well as in plant and animal sequences (McLaughlin and Dayhoff 1970). Table 2 shows the rates of change of all proteins so far examined in more than one higher species. A wide range is displayed for this heterogeneous collection. Presumably the other vertebrate proteins would also fall into such a range. On the basis of constant rates, we would predict that the sequences of the bacterial homologues of proteins which change as slowly as cytochrome c, such as glyceraldehyde 3-phosphate dehydrogenase (GPDH) would also be recognizably related to the sequences in higher organisms. This prediction is supported by the active site sequences of GPDH shown in table 4 which are almost identical. It is likely that slowly changing proteins like GPDH and cytochrome will help unravel the obscure phylogenetic history of the bacteria.

From table 2 we see that the histones, which function in the control of the transcription of the genome prerequisite to protein synthesis in the higher organisms, are much more rigidly conserved than cytochrome c. If bacteria are found to contain homologous compounds, this type of protein should provide valuable insight into early phylogeny.

Proteins which change as rapidly as the hemoglobins are useful in studying the relationships among the invertebrates and the vertebrates, while the more rapidly changing proteins, such as ribonuclease or growth hormone, will be helpful in distinguishing closely spaced phyletic events among the vertebrate groups.

It will also prove possible to learn about the evolution of the genome in

TABLE 4

Active site peptides, the portions of proteins intimately involved in the chemical reactions which they catalyze (Dayhoff et al. 1969; Groskopf et al. 1969; Jurasek et al. 1969; Möckel and Barnard 1969; Hofmann 1971; Shotton 1971). The sequences of homologous proteins are very similar even though they are derived from widely diverse biological groups. The exact order of the amino acids enclosed in brackets has not been experimentally determined. The abbreviation B represents either D or N; Z represents either E or Q.

Proteolytic enzymes

Serine

```
Trypsin – bovine                          D S C Q G D S G G P V V C S G K
Trypsin – pig                             N S C Q G D S G G P V V C G Q Q L
Chymotrypsin A – bovine                   S S C M G D S G G P L V C K K N
Chymotrypsin B – bovine                   S S C M G D S G G P L V C Q K N
Elastase – pig                            S G C Q G D S G G P L H C L V N
Thrombin – bovine                         D A C E G D S G G P F V M K S P
Protease – bacterium (Streptomyces griseus)   T C Q G D S G G P M F
Plasmin – human                           S C Q G D S G G P L V C F E K
```

Histidine

```
Chymotrypsin A – bovine                   V V T A A H C G V T T S D
Chymotrypsin A – pig                          A A H C G V T T S D
Chymotrypsin B – bovine                   V V T A A H C G V T T S D
Chymotrypsin C – pig                          A A H C I N S G T S R T
Chymotrypsin – snapping turtle                A (A, H, C) G V T T S
Trypsin – bovine                          V V S A A H C Y K S G I Q
Elastase – pig                            V M T A A H C V D R E L T F R
Protease – bacterium (Streptomyces griseus)   T A A H C V (N, S, G, S, N, G)
```

Aspartic acid

Pepsinogen – pig	I V D T G T S
Pepsinogen – fungus (*Penicillium janthinellum*)	I A D T G T T L

Cysteine

Dehydrogenases

Alcohol dehydrogenase – yeast	Y S G V C H T D L H A W H G D
Alcohol dehydrogenase – horse	A T G I C R S D D H V T S G L
GPDH – pig	Y D N S L K M V I V S N A S C T T N C L A P L A K
GPDH – rabbit, chicken, ostrich, sturgeon, honey bee and yeast	I V S N A S C T T N C L A P L A K
GPDH – human	I H I V S N A S C T T N C L A P L A K
GPDH – halibut	V V V S N A S C T T N C L A P L A K
GPDH – *E. coli*	Y(Z,B,G) – I V S N A S C T T N C L A P V A K
GPDH – lobster	Y S K D M T V V V S N A S C T T N C L A P V A K
GPDH – blue crab	V V V S N A S C T T N C L A P V A K
LDH – pig and bovine (heart)	V I G S G C N L D S A R
LDH – pig, chicken and frog (muscle)	V I G S G C N L D S A R
LDH – dogfish (muscle)	(I,G,S,G,C,B,L,B,S,A)R

Serine

Prosphorylating enzymes

Phosphoglucomutase – *E. coli*	A I G G I I L T A S H B
Phosphoglucomutase – rabbit	T A S H B P G G P(B.B.G)F G I K
Phosphoglucomutase – flounder	T A S H D P G G P D D G F

GPDH = glyceraldehyde 3-PO$_4$ dehydrogenase; LDH = lactate dehydrogenase. (From Dayhoff 1969.)

the primordial single cell line which led to the proto-organism, the most immediate common ancestor of all living species. Just as mammals today have several related genes coding for the various types of globins, so the proto-organism already had families of related genes which are still preserved. The tRNA molecules constitute such a family which must have arisen through gene duplication and subsequent adaptation to the specialized recognition of a codon and reaction with a specific amino acid (Eck and Dayhoff 1966; Jukes 1966a; Crick 1968). The protein enzymes which catalyze the attachment of the amino acids to the tRNAs may also constitute such a system. The enzymes which control the replication of DNA and the transcription of RNA may be similar. No sequences are presently known for these ancient proteins.

There is a great amount of evolutionary information implicit in those macromolecules which have a common ancestry of their own within that primordial single cell line. Eventually it should be possible to extrapolate far back into this primeval era, using the knowledge of macromolecular structures and the principles of the evolutionary process. It has long been the dream of biologists to work out the history of all the living species back to the very beginning. The many traits obtainable from biochemistry promise to make this a reality.

References

ATKIN, N. B. and S. OHNO, 1967, Chromosoma 23, 10.

BAHL, O. P. and E. L. SMITH, 1965, J. Biol. Chem. 240, 3585.

BARGHOORN, E. S. and J. W. SCHOPF, 1966, Science 152, 758.

BARGHOORN, E. S. and S. A. TYLER, 1965, Science 147, 563.

BRAUNITZER, G., K. HILSE, V. RUDLOFF and N. HILSCHMANN, 1964, Adv. Protein Chem. 19, 1.

BUSE, G., S. BRAIG and G. BRAUNITZER, 1969, Z. Physiol. Chemie 350, 1686.

CALVIN, M., 1969, Chemical evolution. (Oxford University Press, New York) 278 pp.

CHAN, S. K. and E. MARGOLIASH, 1966a, J. Biol. Chem. 241, 507.

CHAN, S. K. and E. MARGOLIASH, 1966b, J. Biol. Chem. 241, 335.

CHAN, S. K., I. TULLOSS and E. MARGOLIASH, 1966, Biochem. 5, 2586.

CRICK, F. H. C., 1968, J. Mol. Biol. 38, 367.

DAYHOFF, M. O., ed. 1969, Atlas of protein sequence and structure 1969. (National Biomedical Research Foundation, Silver Spring, Md.) 365 pp.

DAYHOFF, M. O. and R. V. ECK, 1968, Atlas of protein sequence and structure, 1967–68. (National Biomedical Research Foundation, Silver Spring, Md.) 365 pp.

DAYHOFF, M. O., L. T. HUNT, J. K. HARDMAN and W. C. BARKER, 1969, In: Atlas of protein sequence and structure 1969, ed. M. O. Dayhoff (National Biomedical Research Foundation, Silver Spring, Maryland) 47.

DAYHOFF, M. O. and P. J. McLAUGHLIN, 1969, In: Atlas of protein sequence and structure 1969, ed. M. O. Dayhoff (National Biomedical Research Foundation, Silver Spring, Maryland) 33.

DAYHOFF, M. O. and C. M. PARK, 1969, In: Atlas of protein sequence and structure 1969, ed. M. O. Dayhoff (National Biomedical Research Foundation, Silver Spring, Maryland) 7.

DICKERSON, R. E., 1964, In: The proteins, Vol. 2, ed. H. Neurath. (Academic Press, New York) 603.

DICKERSON, R.E. and I. GEIS, 1969, The structure and action of proteins. (Harper and Row, New York) 128 pp.

DUS, K., K. SLETTEN and M. D. KAMEN, 1968, J. Biol. Chem. 243, 5507.

ECK, R. V. and M. O. DAYHOFF, 1966, Atlas of protein sequence and structure 1966. (National Biomedical Research Foundation, Silver Spring, Md.) 215 pp.

ECK, R. V. and M. O. DAYHOFF, 1966, Science 152, 363.

EPSTEIN, C. J., R. F. GOLDBERGER and C. B. ANFINSEN, 1963, Cold Spring Harbor Symp. Quant. Biol. 28, 439.

GOLDSTONE, A. and E. L. SMITH, 1966, J. Biol. Chem. 241, 4480.

GOLDSTONE, A. and E. L. SMITH, 1967, J. Biol. Chem. 242, 4702.

GROSKOPF, W. R., L. SUMMARIA and K. C. ROBBINS, 1969, J. Biol. Chem. 244, 3590.

GUTTE, B. and R. B. MERRIFIELD, 1969, J. Amer. Chem. Soc. 91, 501.

HELLER, J. and E. L. SMITH, 1966, J. Biol. Chem. 241, 3158.

HILL, R. L. and J. BUETTNER-JANUSCH, 1964, Fed. Proc. 23, 1236.

HIRSCHMANN, R., R. F. NUTT, D. F. VEBER, R. A. VITALI, S. L. VARGA, T. A. JACOB, F. W. Holly and R. G. DENKEWALTER, 1969, J. Amer. Chem. Soc. 91, 507.

HOFMANN, T., 1971, Atlas of protein sequence and structure 1971, ed. M. O. Dayhoff (National Biomedical Research Foundation, Silver Spring, Md.).

HUBER, R., H. FORMANEK and O. EPP, 1968, Naturwissenschaften 2, 75.

JUKES, T. H., 1966a, Molecules and evolution. (Columbia Univ. Press, New York) p. 229.

JUKES, T. H., 1966b, Biochem. Biophys. Res. Commun. 24, 744.

JURASEK, L., D. FACKRE and L. B. SMILLIE, 1969, Biochem. Biophys. Res. Commun. 37, 99.

KENDREW, J. C., R. E. DICKERSON, B. E. STRANDBERG, R. G. HART, D. R. DAVIES, D. C. PHILLIPS and V. C. SHORE, 1960, Nature 185, 422.

KOSHLAND, D. E., Jr. and K. E. NEET, 1968, Ann. Rev. Biochem. 37, 359.

KREIL, G., 1963, Z. Physiol. Chemie 334, 154.

MARGOLIASH, E. and E. L. SMITH, 1965, In: Evolving genes and proteins, ed. V. Bryson and H. J. Vogel. (Academic Press, New York) 221.

MARGOLIASH, E., E. L. SMITH, G. KREIL and H. TUPPY, 1961, Nature 192, 1125.

MATSUBARA, H., T. H. JUKES and C. R. CANTOR, 1969, Structure, function and evolution in proteins, Brookhaven Symposium in Biology: No. 21 (1968). (Brookhaven National Laboratory, Upton, New York) 201.

MATSUBARA, H. and E. L. SMITH, 1962 ,J. Biol. Chem. 237, PC3575.

MAYR, E., 1966, Animal species and evolution. (Harvard Univ. Press, Cambridge) 797 pp.

McDOWALL, M. A. and E. L. SMITH, 1965, J. Biol. Chem. 240, 4635.

McKUSICK, V. A., 1968, Mendelian inheritance in man. 2nd ed. (Johns Hopkins Press, Baltimore) 521 pp.

McLaughlin, P. J. and M. O. Dayhoff, 1969, In: Atlas of protein sequence and structure, ed. M. O. Dayhoff (National Biomedical Research Foundation, Silver Spring, Maryland) 39.

McLaughlin, P. J. and M. O. Dayhoff, 1970, Science 168, 1469.

Möckel, W. and E. A. Barnard, 1969, Biochim. Biophys. Acta 194, 622.

Nakashima, T., H. Higa, H. Matsubara, A. Benson and K. T. Yasunobu, 1966, J. Biol. Chem. 241, 1166.

Narita, K. and K. Titani, 1968, J. Biochem. 63, 226.

Needleman, S. B. and E. Margoliash, 1966, J. Biol. Chem. 241, 853.

Nirenberg, M., P. Leder, M. Bernfield, R. Brimacomb, J. Trupin, F. Rottman and C. O'Neal, 1965, Proc. Natl. Acad. Sci. U.S. 53, 1161.

Nolan, C. and E. Margoliash, 1966, J. Biol. Chem. 241, 1049.

Ohno, S., U. Wolfe and N. B. Atkin, 1968, Hereditas 59, 169.

Padlan, E. A. and W. E. Love, 1968, Nature 220, 376.

Perutz, M. F. and H. Lehmann, 1968, Nature 219, 902.

Perutz, M. F., H. Muirhead, J. M. Cox and L. C. G. Goaman, 1968, Nature 219, 131.

Romer, A. S., 1966, Vertebrate paleontology. 3rd ed. (Univ. Chicago Press, Chicago) 468 pp.

Rothfus, J. A. and E. L. Smith, 1965, J. Biol. Chem. 240, 4277.

Rutten, M. G., 1962, The geological aspects of the origin of life on earth. (Elsevier, New York) 146 pp.

Sallach, H. J. and R. W. McGilvery, 1963, Chart of intermediary metabolism. (Gilson Medical Electronics, Middleton, Wis.).

Sanger, F. and E. O. P. Thompson, 1953, Biochem. J. 53, 366.

Sanger, F. and H. Tuppy, 1951, Biochem. J. 49, 481.

Schroeder, W. A. and R. T. Jones, 1965, Prog. Chem. Organic Natural Prod. 23, 113.

Shapiro, H. S., 1968, In: Handbook of biochemistry, ed. H. A. Sober. (Chemical Rubber Co., Cleveland) H52.

Shotton, D. M., 1971. Atlas of protein sequence and structure 1971, ed. M. O. Dayhoff (National Biomedical Research Foundation, Silver Spring, Md.).

Smith, E. L. and E. Margoliash, 1964, Fed. Proc. 23, 1243.

Srb, A. M., R. D. Owen and R. S. Edgar, eds., 1970, Facets of genetics, Readings from the Scientific American. (W. H. Freeman and Company, San Francisco).

Stevens, F. C., A. N. Glazer and E. L. Smith, 1967, J. Biol. Chem. 242, 2764.

Stewart, J. W. and E. Margoliash, 1965, Canad. J. Biochem. 43, 1187.

Wojciech, R. and E. Margoliash, 1968, In: Handbook of biochemistry, ed. H. A. Sober. (Chemical Rubber Co., Cleveland) C158.

Yadi, Y., K. Titani and K. Narita, 1966, J. Biochem. 59, 247.

Young, J. Z., 1962, The life of vertebrates. 2nd ed. (Oxford Univ. Press, New York) 820 pp.

Zuckerkandl, E. and L. Pauling, 1965, In: Evolving genes and proteins, ed. V. Bryson and H. J. Vogel. (Academic Press, New York) 97.

C. Ponnamperuma (ed.), Exobiology. © North-Holland Publishing Company

CHAPTER 9

The emergence of genetic organization

CARL R. WOESE

1. Introduction

The fundamental questions in biology are basically evolutionary. We can lay no claim to a true understanding of biology until we can discuss in specific terms how cells came into being and what general types of biological systems emerge under various evolutionary conditions.

In the past our concept of evolution was confined of necessity to what can be called 'post-cellular evolution' – that which occurred subsequent to the emergence of the prototype of the present-day cell. Over the last two decades, however, we have come to a profound understanding of macro-molecular structure and cellular function in general. This knowledge provides a conceptual background specific enough to cope now with the problems of cellular evolution and the basic evolutionary principles.

At the heart of today's cell are the information transferring processes – the capacities to map nucleic acid primary structure into the primary structure of nucleic acid or protein. And the evolution of these processes, particularly that of translation, is the most important single facet of cellular evolution. In the present article, then, we will examine the problems of cellular evolution concerned with emergence of cellular 'tape reading' processes – emphasizing the evolution of translation.

2. Dynamics of the evolution of macromolecules: the role of quaternary structure

To start we ask what general principles governed the cell's evolution.

Perhaps the most general and fundamental that can be formulated at present is the principle of increasing complexity: Living systems tend with time to assume increasingly complex configurations. One can point to probable examples of this principle at all stages of biological organization and throughout all evolutionary time. It was blatantly manifested on the intercellular level over the last 500 million years or so; viz. the spectacular burgeoning of the metazoans. It is seemingly still at work, on the inter-organismal level, as human social evolution shows. During an ancient era, the principle was applied on an intracellular level – in generation of organelles and the emergence of progressively larger genomes.* Before that even, it must have operated (perhaps during the evolution of translation and immediately thereafter) on the intramolecular level – giving rise to increasingly sophisticated classes of enzymes, and so forth. And, of course, it had ultimately to apply to the abiotic earth in producing the units from which the very first self-replicating entities were constructed.

The complexity principle so stated has no specific predictive value. Much of the study of evolution henceforth will be devoted to formulating the principle in ways that permit prediction of the outcome of evolutions having certain starting and boundary conditions. At present we are mainly concerned with the evolution of complexity immediately preceding, during, and succeeding the establishment of a translation capacity – complexity both intramolecular and on the molecular aggregate level.

The first problem is to define 'complexity'. Clearly, equating it to 'information content' is insufficient – particularly if information content itself is defined in the usual way (e.g. assigning 4.3 bits to every amino acid in a polypeptide). It is intuitively obvious that any enzyme is more complex than a polypeptide having a 'random' sequence of the same length. Yet complexity is a function of information content in that the latter places an upper bound on the former. Complexity is more akin to that elusive quality, information 'meaning' – something that is a function of the nature of the system for which the entity is information, but something that we cannot yet define well enough to quantitate (MacKay 1954; Ashby 1960). It is useful then to distinguish two aspects of macromolecular complexity, an 'information content' – determined by the number of residues in its primary structure – and an 'intricacy' – some measure of the complexity of interactions between

---

* The latter can be deduced from the recent evidence on the primary structure of proteins indicating that these and, therefore their corresponding genes, tend to be grouped into families.

a molecule and other components in the system. We shall consider this in further detail below.

Given a rudimentary genetic system, the problem of evolving macro-molecular complexity is ostensibly straightforward. The modern geneticist's armamentarium includes all manner of techniques for joining and otherwise 'tailoring' genetic segments. Thus the question faced is not 'Is there a means to evolve macromolecular complexity', but rather, 'Of the genetic mechanisms that potentially generate macromolecular complexity, which one or ones did the cell employ'?

Viewed on a genetic level only, the above question seems unanswerable and perhaps trivial. Viewed in its entirety, however, the question takes on a different dimension. There may be but one, or one major, way to evolve macromolecular complexity.

In terms of my argument the single most important property of proteins is their capacity to form quaternary structure. The conventional wisdom here assumes quaternary structure to be merely the empirical basis for allo-steric properties. Yet its greatest significance may lie in its evolutionary function. The answers to two questions concerning quaternary structure lead one to such a conclusion. The first is why quaternary structure appears always to involve either identical subunits or genetically related subunits. The second is why quaternary structure exists at all – the alternative being larger, more complicated tertiary structures.

Use of identical (or related) subunits in constructing quaternary structure may reflect the fact that the forces (interactions) creating quaternary structure are essentially those involved in tertiary structure. In other words, the 'complementary' interactions that cause a portion of a protein chain to fold back on itself could under certain circumstances, be utilized to associate two identical protein chains. Experimental support for this notion is found, for example, in the dimerization of pancreatic RNAse and particular properties of such dimers (Crestfield and Fruchter 1968). A corollary to this reasoning is that evolution of quaternary structure involving *unrelated* subunits is virtually impossible (when the subunits are as complex as those found in the cell today).

Why quaternary structure should exist in the first place has been rational-ized by any of several arguments. For example, quaternary structure is a means to circumvent inaccurate protein synthesis; or quaternary structure is the only feasible way to obtain the correct sorts of weak interactions that give rise to allosteric properties. Such explanations tend not to bear up under close scrutiny, however.

Let us attempt, then, to understand quaternary structure in terms of evolution – that is, it permits a maximizing of *intricacy*; in other words, quaternary structures can become more intricate than tertiary structures of comparable informational complexity (Woese 1971a).

The above arguments lead one to the cyclic mechanism depicted in fig. 1

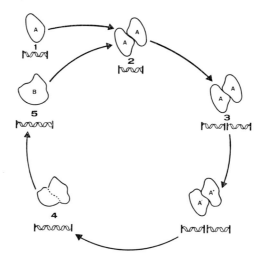

Fig. 1. Co-dimerization cycle for the evolution of macromolecular complexity. A monomeric molecule (stage 1) evolves to become a dimer (stage 2). After duplication of the underlying gene (stage 3), the dimer can evolve to contain non-identical (but related) subunits. Joining of the two genes (stage 4) results in a 'co-dimer', a single polypeptide chain retaining the essential properties of the dimer immediately preceeding it. Further evolution of the co-dimer leads to a molecule in which the two 'halves' are no longer similar, i.e. to a new molecule, B, approximately twice the size of the starting molecule, A. This cycle then operates on B as the starting monomer, and so on. It is hypothesized that this cycle is a necessary and sufficient condition for evolution of macromolecular complexity, at the level of a gene based genetic system (Woese 1971a).

for the evolution of intramolecular complexity. The cycle begins with a monomeric polypeptide (stage 1). This then becomes a dimer (stage 2) – an evolution rationalized on the grounds that the dimeric form can become more intricate than otherwise possible, as stated above. Next there would be a chance duplication of the underlying gene – a relatively common evolutionary occurrence – which however, potentially *doubles* the informational content of the system (see stage 3). Sooner or later a joining of the two genes may occur (stage 4). The result is a covalently linked 'dimer', or *co-dimer*. Again genetic precedent permits this assumption. Certain genetic 'tailoring' may be required in the linking in order that the essential features of the original quaternary structure be retained. Ever since step 3, the two halves of the dimer or the co-dimer have been able to evolve to some extent independently of one another. Thus the selective advantage that will eventually come from creation of a co-dimer may not be immediately apparent. Ultimately the simple peptide link joining both sections of a co-dimer will prove less of a constraint on the structure's evolution than the more complicated interactions necessary to maintain integrity of a true dimer.* In the final step (stage 5), the two portions of the new molecule have evolved in different ways – one half of the molecule perhaps 'fine-tuning' or otherwise controlling the function of the other. A point is eventually reached where the two halves of the structure cease to resemble one another, and therefore, the molecule must be defined as a totally new entity, B in the figure. The cycle then begins again, using this more complex protein B as its starting point, and so on (Woese 1971a).

Among the protein primary structures catalogued to date several examples consistent with such a cyclic genesis occur. The bacterial ferridoxin molecule for example, gives clear evidence of what could be its most recent 'co-dimerization' step (Tsunoda et al. 1968). Hopefully, more extensive studies of protein primary structure will reveal cases in which not only the most recent, but the preceding co-dimerization step can also be detected. It has already been suggested for cytochrome c that the molecule for some reason comprises a number of related polypeptide units of about twelve amino acid residues each (Dus et al. 1968). If true, this may mean that cytochrome c has undergone three cycles of co-dimerization in its evolution – i.e. $12 \times 2^3 \sim 100$.

Discussion has been confined so far to a system possessing the rudiments at least, of a genetic apparatus. The above evolutionary cycle could have had its beginnings in a pretranslational era, however, for the central principle – i.e. that formation of regular aggregates (of which a dimer is a special case) leads to increased intricacy – does not necessarily depend upon translation or genetic phenomena.

---

* Picture an evolution in which this covalent joining step is omitted. In this case proteins would eventually arise that were composed of subunits, which themselves were composed of smaller subunits, and so on. The resultant hierarchy of intersubunit associations would clearly create rather severe constraints on – and certain other disadvantages for – the super-aggregate's function. Note too that perhaps the most commonly employed intermolecular interaction in quaternary structure today *is* a simple *covalent* bond, the -S-S- bridge; it is not likely, however, that such a bond could be utilized in a primitive, i.e. reducing, environment.

If, as is likely, nontranslational peptide synthesis predated translation, the primitive earth may have contained aggregates in which peptides were regularly arrayed about what could be called a 'nucleation center'. Examples would be coordination complexes involving particular peptides and a transition metal ion, or peptides of a basic character regularly arrayed in the major groove of a nucleic acid double helix, or certain peptides arranged on surfaces, etc. As translation evolved, however, these nontranslationally produced polypeptides would tend to be replaced by peptides of translational origin – the latter having perhaps a somewhat narrower range of primary structures and the capacity to become fused at the genetic level, as seen above. The larger peptides, resulting from such fusion, might invest complexes with greater stability and/or more specific reactivity.

Starting thus, with peptides juxtaposed by nucleation centers, the above cycle would eventually generate peptides complex enough that some of them would undergo dimerization in the absence of a nucleation center, and so on.

Although the concept of nucleation centers in evolution is presently untestable, it has the value of suggesting that the geneology of any protein may eventually be traced to one of a relatively small number of prototypes, each defined in terms of some simple complex involving certain peptides and a given nucleation center in the primitive environment (Woese 1971a).

The genesis of complexity through the above co-dimerization cycle need not be restricted to proteins. As will be seen below, RNA evolution may proceed similarly. Furthermore, the codimerization cycle itself seems a special case of a universal mode of 'becoming' – of creating 'macro-order' from 'elements' in a 'microuniverse' (Bohm 1969; Woese 1971b). We will consider below the nature of the 'elements' that constitute macromolecules for the case of the translation system.

3. The evolution of translation

The barest knowledge of the translation process is sufficient to convince one that its evolution is no simple matter. The varied kinds of components and the number of components of any kind required, surely make evolution of the whole machine a most protracted affair, one undoubtedly involving a series of stages some of which even differed in quality. Thus in approaching translation's evolution we must attempt to reconstruct those stages, arriving ultimately at an understanding of the archetype translation machine. In the process we will encounter conceptual as well as factual problems. Although several interesting experimental approaches to this matter have recently opened up, the yield of data therefrom is still too meagre to make the present discussion anything more than a conceptual exercise.

In one respect at least, translation differs markedly from its companion tape reading processes, nucleic acid replication and transcription. In the latter cases, mere knowledge of the physical interactions that produce double helical structure provide immediate insight into the input–output relationship (i.e. $G \rightarrow C$, $C \rightarrow G$, etc.), into the molecular mechanism involved in the processes, and into the bases for their evolutions (at least in part). No comparable monolithic principle seems to pervade translation. Though we know the input–output relationship for translation (i.e. the set of codon assignments) we have no physical basis for these (e.g. a 'codon–amino acid pairing' interaction analogous to Watson–Crick base pairing). Furthermore, we are still uncertain of the molecular mechanics that bring about translation. And, of course, we have no real reason for suspecting why a translation relationship had to evolve in the first place – if indeed there be any reason.

Perhaps the principal clue at present to translations' evolution lies in what has been termed the 'central paradox' of the process: a number of complex protein-mediated functions are critical to translation; yet these functions (proteins) could not have come into being unless the cell already possessed the capacity to translate (Woese 1967). This 'paradox' can be resolved by assuming all the protein-mediated functions of the machine today to be 'derived functions' – ones evolving after the emergence of the archetype translation machine. This assumption generates a number of conceptual problems in itself, however.

First, the archetype machine would have to be composed essentially of nucleic acid. (It could, of course, contain proteins that were produced non-translationally. But certainly, such proteins, if they existed, would not have a complexity even approaching that of modern proteins. Thus the basic features of the machine would have to be properties of its nucleic acid components.) But what kinds of nucleic acid interactions can (1) provide the mechanical movement of translation, and (2) give the process what kinds and degrees of 'specificity'?

Although it is not unreasonable to invest nucleic acids with some degree of biological specificity, it does seem unreasonable to give them a degree of specificity characteristic of modern proteins. tRNA is a good example of a nucleic acid having biological specificity. Yet we notice that the precision of tRNA's function is probably brought about through post-transcriptional modification of particular bases (again protein mediated), not through evolution of a 'pure' nucleic acid structure. Furthermore, tRNA always functions in a context that includes quite sophisticated proteins.

Thus, though we cannot rule out the capacity of primitive nucleic acids to recognize certain amino acids (in fact, as we shall see, evidence favors such a notion), we must content ourselves with a recognition that is preferential, not an all-or-none discrimination. At the very least, this means that primitive translation would have been, by modern standards, a very error-ridden process; in all probability it would have been worse – a highly ambiguous process (Woese 1967, 1970a).

In considering the nature of primitive proteins we then must abandon three of our cherished preconceptions: (1) the concept of protein having a unique primary structure, (2) certain of our prejudices concerning protein function – biological specificity, etc., and (3) our ideas of protein size and perhaps certain other physical properties as well. If nucleic acid components of an archetype translation mechanism cannot distinguish clearly among amino acids, then 'group codon assignments' will result. In other words, certain groups of 'related' amino acids will be assigned (as a whole) to certain groups of codons. The proteins synthesized on the basis of such mapping rules will then be 'statistical proteins'; a statistical protein, by definition, is a group of proteins many of which have very similar physical properties, and all of which are related by being an approximate translation (by today's standards) of a given genetic segment (Woese 1965). Since biological specificity is in essence dependent upon unique primary structure, a statistical protein cannot manifest what we would call biological specificity. There is no reason however, to preclude their functioning as catalysts for certain general reactions.

Inaccuracy in protein synthesis is of several kinds – e.g. that associated with placement of an amino acid in response to a codon, and that associated with maintenance of a constant correct reading frame. Inaccuracy of either sort would tend to prevent evolution of large proteins, and of very complicated tertiary structures. Thus, the primitive proteins should be small and have simple or ill-defined tertiary structures – their function relying mainly upon the properties of particular primary structures per se and secondary structure.

3.1. The molecular mechanics of the translation process

A major obstacle to understanding the evolution of translation may be that we do not know how translation really works. What little understanding we have of the process is in terms of 'templating', 'tRNA adaptors', the 'A-site–P-site' model, etc. According to the last of these, the ribosome is considered to possess two translation sites, an A site accommodating the incoming, aminoacylated tRNA, and a P site, wherein the peptidylated tRNA resides (Watson 1964; Erb et al. 1969). Following the transfer of the nacent peptide from the peptidyl tRNA (in the P site) to the aminoacyl tRNA (in the A site), it is thought that a 'translocation' of the newly peptidylated tRNA from A site to P site occurs – with a concomitant ejection of the previously peptidylated tRNA (now discharged) from the P site (Erb et al. 1969; Kaji et al. 1969).

Three main features characterize the A-site–P-site model: (1) a ribosome having two *kinds* of sites (one of each kind); (2) the idea that tRNA and its codon are *moved* from one site to the other; and (3) the implicit notion that the ribosome and associated protein factors play the active roles in a process in which tRNA and mRNA are but static, passive entities.

The A-site–P-site model has been presented largely for orientation and as a point of departure for a rather different conception of translation, one arising out of the frustration generated of trying to reconcile the A-site–P-site model with the evolution of translation. Clearly the A-site–P-site model is worse than useless for this purpose. It is too general to provide any insight into features of a primitive translation machine. Moreover, it may be detrimental to the gaining of such insight, in that the kind of mechanism it does suggest is so complex, so highly evolved, that one cannot imagine how translation could ever come into being if it had to start with even a simplified version of such a machine.

I feel it necessary, therefore, to divest one's thinking of this sort of model. It helps to realize that there actually exists *no* evidence proving the basic features of the model. Of course, myriad facts are in *accord* with it. For example, the fact that isolated 30S and 50S ribosomal subunits manifest very different biological properties, is often taken as a proof that the A site resides on one subunit, the P site on the other (Pestka and Nirenberg 1966; Gilbert 1963). The facts here can equally well be explained on the assumption that separation of the 30S and 50S subunits distorts subunit structure to the point that the properties of sites on either subunit are altered. (Note too that precedent exists for assuming any ribosomal site to comprise elements from both the 30S *and* the 50S subunit, in which case splitting the subunits apart would yield two *partial* sites, the properties of which obviously need not even remotely resemble each other.)

With a *tabula rasa* we can try to understand the workings of today's translation machine on the assumption that the basic mechanical features of the mechanism today are common even to the most primitive versions thereof.

For largely intuitive reasons attention will center on the tRNA molecule. This may be because tRNA really is central to translation today, and nobody has been able to suggest any biological alternative to a translation utilizing an 'adaptor' type of mechanism. Or, it may be because tRNA seems a molecular anachronism, a nucleic acid, as has often been said, 'that is trying to become a protein'. Certainly the cell does go to considerable expense to modify this molecule after its transcription. Could it be, therefore, that tRNA was an integral part of translation from the very beginning, and the cell was forced to carry it along – modifying it to improve it – merely because there never could occur a stage in evolution when tRNA was dispensable?

The most intriguing feature of the tRNA molecule uncovered to date is the so-called anticodon arm. In all cases this segment of the molecule comprises 17 nucleotides, the first and last five of which are complementary – giving the arm a double helical stalk of five base pairs (Madison 1968). What then is the configuration of the remaining middle seven nucleotides? Fuller and Hodgson suggested that these would arrange themselves so as to maximize the base stacking interactions over the entire arm (Fuller and Hodgson 1967). Models show that the double helical stalk of the arm can be extended in a single stranded fashion for five additional nucleotides, the extension having the same pitch and axis as the stalk itself. In this arrangement, the remaining two nucleotides of the middle seven can connect the top of the extension to the top of the double stranded stalk proper (see fig. 2). Notice also that symmetry of the anticodon arm permits *two* structures of this type – one through extension of the $3' \rightarrow 5'$ chain of the stalk, the other through extension of the $5' \rightarrow 3'$ chain thereof (Fuller and Hodgson 1967). Either conformation places the anticodon triplet at the top of the single stranded extension. In one conformation the III* anticodon base (i.e. the one pairing with the III codon base) is topmost; in the other it is the I* anticodon base. Fuller and Hogdson argue that tRNA actually exists in the form with the $3' \rightarrow 5'$ chain extended: in this conformation (here called FH) strong stacking purines (which generally occur in the number 6 and number 7 positions) are included in the helical extension (Fuller and Hodgson 1967). In contrast, weak stacking pyrimidines (which always occupy the number 11 and number 12 positions) must be included in the helical extension in the alternate (hf) conformation (see fig. 2).

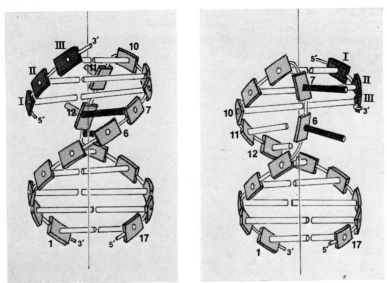

Fig. 2. Alternative structures for the tRNA anticodon arm. The first and last five of the 17 nucleotides in the arm are complementary and form a double helical stalk. Extension of the 5′–3′ chain of the stalk for five additional nucleotides in a single stranded fashion yields the so-called hf conformation; extension of the 3′–5′ chain in this manner yields the so-called FH conformation. In both cases the anticodon triplet constitutes the topmost three nucleotides in the single stranded extension. In the hf conformation (right) the I* anticodon base (number 8) is topmost; in the FH (left) conformation, however, it is the III* anticodon base (number 10) that is topmost. Numbering of the bases is that employed by Fuller and Hogdson (1967).

The basic Fuller–Hodgson argument that base stacking interactions in the anticodon arm be maximized sounds reasonable. The assumption that the maximization be done on an anticodon arm in isolation does not. Better to maximize stacking in an anticodon arm in the act of translation.*

In that latter case, the fact that two tRNAs and their respective codons simultaneously occupy adjacent sites on the ribosome must be considered. The two codons and their anticodons potentially form a double helical segment of six base pairs length. Were this to form, only *one* of the anti-

---

* Also, their question as to which *one* of the two structures is the *correct* one should prove historically interesting. The more general question: Why there are two structurally equivalent forms of the anticodon arm, was not asked – by *any* molecular biologist at the time. This is a clear example of the implicit prejudices introduced by dogmatic concepts – in this case the *static* tRNA inherent in the adaptor-template view of translation.

codon arms (that from the peptidyl tRNA) could also exist in the FH conformation. However, the other anticodon arm (that from the aminoacyl tRNA) could exist in the hf conformation. The two anticodon arms and their codons could thus form a joint double helical structure equivalent to twenty base pairs in length, as fig. 3 illustrates (Woese 1970b).

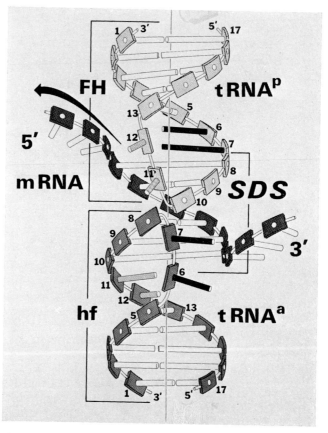

Fig. 3. The basic translation complex. The complex (in essence an RNA double helix the equivalent of twenty base pairs in total length) comprises anticodon arms of both a peptidyl tRNA (tRNA^p) in the FH conformation and an aminoacyl tRNA (tRNA^a) in the hf conformation, plus their respective codons (combined with the anticodons proper in a central 'sextuplet duplex structure' or SDS). mRNA as drawn would move from right to left in translation (indicated by the large arrow). The axis of two-fold rotational symmetry relating the anticodon stalks is formed by the intersection of two planes: one is perpendicular to the axis of the complex at its midpoint; the other contains that axis and is perpendicular to the bars (base pairs) linking the no. 5–no. 13 positions in *either* anticodon stalk.

What this means is that peptide bond synthesis is conceivably accompanied by a configurational change in the tRNA molecule. So tRNA may not be the static entity it is generally taken to be. Rather than having one configuration, tRNA may have any of several configurations, and transitions among these may play an active role in the mechanics of translation.

hf → FH transitions are not inconsistent with the A-site–P-site model; of course it must be refined so that the A site contains the bottom half of the translation complex as drawn in fig. 3, and the P site its top half. The symmetry of the translation complex begins to suggest that all of translation may be a more symmetrical process, however, than it is generally envisioned to be; and a specific allosteric type of translation mechanism can be designed around the hf → FH transitions. This mechanism, which does not need to invoke translocation, is depicted in fig. 4 (Woese 1970b).

The ribosome is taken to have two sites, R and R', which play *equivalent* roles in translation. When a peptidyl tRNA is in one of the sites, e.g. R, an aminoacyl tRNA can enter the other site, in this case R' (see fig. 4a). The former tRNA is in an FH conformation, the latter in the hf. Peptidyl transfer is accompanied by an in situ hf → FH transition involving what was the aminoacyl tRNA. The transition automatically expels what was the peptidyl tRNA (now deacylated) along with its accompanying codon from the R site, and simultaneously brings the next codon in line into that site, as fig. 4b shows. Then an aminoacyl tRNA (hf conformation) can enter the R site to pair with the newest codon (fig. 4c). In connection with the subsequent peptidyl transfer event the tRNA now in the R site undergoes an hf → FH transition, causing expulsion of the opposite tRNA (now deacylated) with its codon from site R', thus concluding one cycle in the process and setting the stage for the start of the next – (fig. 4d) (Woese 1970b).

Mechanically such a translation mechanism would work in the manner of a *double reciprocating ratchet*, alternate codons being translated in alternate sites.

The mechanism described demands a functional two-fold rotational symmetry for a portion at least of the 70S ribosome. At present there is no certain indication of the orientation the symmetry axis would take relative to the ribosome. It is reasonable to put it perpendicular to the 'plane' of the 50S–30S join, which permits each R site to contain elements from both the 50S and the 30S ribosomal subunit. However, the possibility that one R site resides on the 50S, the other on the 30S subunit cannot be ruled out. As in the case of the A-site–P-site model, no evidence proving or disproving the Ratchet model now exists, although predictions of the latter are, of course, being tested.

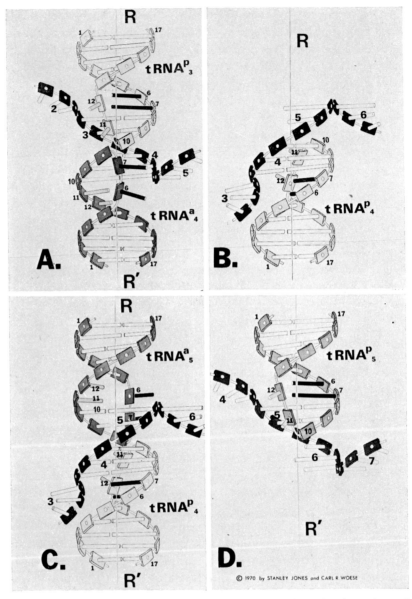

Fig. 4. The molecular mechanics of translation. Small numerical subscripts refer to the
time order of appearance of tRNAs in the translation process; large numbers refer to the
codons in their order.

Translation is seen to occur as follows: A depicts an arbitrary point in the process, e.g.
where the third tRNA is peptidylated and the fourth aminoacylated. The former (in FH

3.2. *Stages in the evolution of translation*

I have argued above that a primitive translation apparatus cannot possess those precise functions of the modern apparatus mediated solely by the evolved proteins. The most profound implication of this concerns the codon-amino acid relationship. Unique, unambiguous codon assignments are not a part of primitive translation, which produces 'statistical proteins' only. Consequently, translation should have evolved in three major stages at least. Stage I is the establishment of the archetype translation mechanism – one having the basic mechanics of the process, but using ambiguous or 'group' codon assignments. Stage II produces the evolution of the modern, unique codon assignments. And since translation is becoming more and more precise as this evolution proceeds, the third and final stage would be the logical extension of Stage II – that is, the evolution of mechanisms that reduce the noise level of the process to its present day values, and 'fine tune' the machine in other ways as well.

Stage I. *Evolution of the archetype translation machine*

Here we deal with the question of what interactions among the primitive biopolymers gave rise to the initial form of translation. Our concern is with both the origin of the machine that gives mRNA tape movement, and the establishment of the primitive nucleic acid → protein mapping rules, the aboriginal codon catalog.

---

conformation) is shown in the ribosomal R site, the latter (in hf conformation) in the R′ site. When peptidyl transfer occurs, tRNA$_4$ will change from the hf to the FH conformation (for reasons here unspecified). This transition has four major consequences: (1) it necessarily breaks the existing SDS duplex, (2) displacing codon number 3 from the axis of the translation complex in the process, which (3) causes tRNA$_3$ to vacate the R site, and, moreover (4) it automatically places codon number 5 in the R site – as illustrated in B. In C tRNA$_5$ Paminoacyl, hf conformation) has entered the R site to pair with codon number 5, re-establishing a translation complex. The subsequent peptidyl transfer event causes an hf →FH transition in the anticodon arm of tRNA$_5$. As before, this breaks the SDS duplex, now displacing codon number 4 and its accompanying tRNA from site R′, and bringing codon number 6 into that site – as D shows. When tRNA$_6$ (aminoacyl, hf conformation) enters site R′, to pair with codon number 6, the cycle is completed, so that a translation complex exactly as shown in A again forms.

The hf →FH transitions, occurring first in one site, then in the other, form a cyclic process that pulls the RNA message (from right to left as shown) through the ribosome in a manner reminiscent of a *double reciprocating ratchet*. tRNAs designated by odd numbers are 'decoded' in one site, those by even numbers, in the other site.

(a) The basic molecular mechanics.   The above reciprocating ratchet model for translation contains several important evolutionary implications. Since the essential mechanism is a tRNA–mRNA interaction, this suggests: (a) that translation *always* employed some sort of tRNA molecule, and (b) that the primitive versions of the ribosome can, in principle, be made static and simple; indeed no ribosome whatever seems needed to begin with. Also the model's basic ratchet mechanism per se is sufficiently simple that it could well occur in a primitive environment containing nucleic acids.

The modern tRNA molecule, on the other hand, is not sufficiently simple that it can be invoked with confidence in a primitive system. Let us, therefore, take the tRNA prototype to be more or less equivalent to the anticodon arm of tRNA today – i.e. proto-tRNA was a simple self-complementary nucleic acid molecule containing a central non-paired segment of about seven nucleotides. The existence of self-complementary nucleic acids in a primitive environment can be rationalized in a number of ways – for example, being 'selected' for reason of their relative resistance to degradation.

The price one pays for this simplification is having to account for the point and manner of amino acid attachment to such a proto-tRNA, in addition to the above-mentioned specificity problem (which would obtain in any case). For reasons that will become clearer below, I feel that tRNA evolutionarily is a collection of self-complementary loop segments (arms); even the -CCA arm is taken to have originated as such a loop, that is now broken. On this basis the aboriginal amino acid attachment site would be *within* a loop. One particularly unusual feature of today's tRNA molecule may provide a clue to the position of this site and manner of amino acid attachment. An adenine residue (almost always highly modified) occurs immediately adjacent to every anticodon triplet (Madison 1968). This modification is the key to the properties of the anticodon arm (perhaps the entire tRNA molecule). If no modification is permitted to occur, the hydrophobic–hydrophilic partitioning of the molecule changes drastically, and it looses its capacity to interact with some or all of its codons (Gefter and Russell 1969; Ghosh and Ghosh 1970). (It remains changeable, though! (Gefter and Russell 1969).)

The modifications in question are accomplished today by highly evolved enzymes, of course. The reason for such a modification evolving in the first place, however, could lie in the fact that certain alterations of adenine residues in this same position in proto-tRNA molecules were essential to function. We will take this supposed aboriginal modification to have been an adenine–amino acid complex, the amino acid bound in an amide link through its COOH group to the NH_2 group of the adenine – similar to the adenine modification reported for $tRNA_{ilu}$ (Schweizer et al. 1969).

To complete this chain of conjecture, I propose the following mechanism for archetype translation (Woese 1970a). Two proto-tRNAs (and a primitive mRNA) form the type of translation complex shown in fig. 3. One tRNA, carrying an adenine-bound amino acid, is in the hf conformation; the other, carrying an adenine-bound peptide, is in the FH conformation. These conformations (hf vs. FH) are somehow determined by differences between the properties of the bound amino acid vs. those of the bound peptide. (For example, the peptide could stabilize an FH conformation by lying in the major groove of the structure.) Peptidyl transfer in this instance is taken to be a spontaneous event (being energetically favored) occurring when the a-NH_2 group of the incoming amino acid contacts the (first) carbonyl carbon of the peptide – an event pictured in detail as follows: thermal fluctuations cause the hf structure (in the aminoacyl tRNA) to undergo movements similar to the start of the above-described hf → FH transition (see fig. 4 caption), the axis of the *stalk* of the hf structure moving relative to the axis of the remainder of the complex (which stays intact). A CPK model of this hypothetical interaction demonstrates that when such perturbation is permitted in the translation complex, the a-NH_2 group of the amino acid (adenine bound) can swing into juxtaposition with the carbonyl carbon of the peptide (even though the adenine residue binding the peptide remains a part of a double helical structure throughout) (Woese, unpublished).

Upon peptidyl transfer, the peptide chain now becomes the determining factor for the conformation of what had been the aminoacylated tRNA, causing it to complete its transition into an FH conformation. Similarly, the fact that the peptide no longer resides on the original FH structure, will facilitate the breakup of the translation complex. And breakup of the old translation complex sets the stage for formation of a new one, in the manner described in fig. 4 (Woese 1970a).

(b) The problem of amino acid recognition in primitive translation. A primitive translation process would have little value unless there were some relationship between the primary structure of the polypeptide produced and the primary structure of the nucleic acid message translated. What interactions involving nucleic acids would produce this? What gives rise to the aboriginal codon assignments?

One can imagine a link arising between amino acids of certain compositions and nucleic acids of 'corresponding' compositions in either of two ways. By analogy with enzyme specificity and the base pairing interaction, a physical 'recognition' of amino acids (or derivatives thereof) by nucleotides (or derivatives thereof) is a possibility. It might be called 'codonamino acid pairing'. Since nucleic acids do not appear to manifest the biological specificity of proteins – at least not in degree – one would not expect codon–amino acid pairing to be so highly selective as an enzyme-substrate interaction.

Chemical reactivity provides an alternative genesis for the amino acid–codon link. The side groups of certain amino acids may turn out to be especially reactive with corresponding purines or pyrimidines. In this way a given base and a given type of amino acid could become associated – a mechanism quite dissimilar from that pictured as occurring in codon–amino acid pairing.

The idea of codon–amino acid pairing has received experimental support of an indirect nature. A strong correlation exists between the ranking of amino acids according to their chromatographic partitioning in heterocyclic base solvent systems, and their ranking according to codon assignment (Woese et al. 1966). All pairs of amino acids defined by having codons that differ only in the III position show almost identical 'polar requirements' (a measure of their chromatographic partitioning) for the amino acids within each pair. Furthermore, all amino acids having codons with C in the II position have a characteristic common 'polar requirement', and so on (Woese et al. 1966).

The interactions that differentiate among amino acids in this case are very weak, and distinction can by no means be all-or-none. Yet asking interactions between amino acids and heterocyclic bases to be rather specific is like asking that amino acids manifest the biological specificity characteristic of enzymes. Hopefully, though, certain oligonucleotides might exhibit somewhat better discrimination among the amino acids than do individual bases. If so, the problem of amino acid discrimination in primitive translation would be solved by having the proto-tRNAs distinguish amino acids physically (into crude groups), as a prelude to their chemical attachment to the anti-codon-adjacent adenine residue discussed above. (Either the amino acid or the adenine amino group would have to be activated for this attachment to occur.)

Stage II. Emergence of the modern codon assignments

The configuration of the codon catalog today is highly ordered – as table 1

TABLE 1

The catalog of codon assignments.

First position		Second position U	Second position C	Second position A	Second position G	Third position
	U	PHE	SER	TYR	CYS	U
	U	PHE	SER	TYR	CYS	C
	U	LEU	SER	CT-1	CT-3	A
	U	LEU	SER	CT-2	TRY	G
	C	LEU	PRO	HIS	ARG	U
	C	LEU	PRO	HIS	ARG	C
	C	LEU	PRO	GLN	ARG	A
	C	LEU	PRO	GLN	ARG	G
	A	ILU	THR	ASN	SER	U
	A	ILU	THR	ASN	SER	C
	A	ILU	THR	LYS	ARG	A
	A	MET	THR	LYS	ARG	G
	G	VAL	ALA	ASP	GLY	U
	G	VAL	ALA	ASP	GLY	C
	G	VAL	ALA	GLU	GLY	A
	G	VAL	ALA	GLU	GLY	G

shows. There can be no possibility whatever that this order arose by chance alone, which makes elucidation of the constraints shaping this ultimate configuration a crucial issue. Corollary issues are the relationship of the ultimate to the aboriginal configuration of the codon catalog, and the reason for its (apparent) universality today. Furthermore, as the aboriginal codon catalog cannot be identical to that of today, a means must exist for altering codon assignments during the cell's evolution.

Constraints that may have shaped the ultimate configuration of the codon catalog are of two general sorts – (a) those generated by the translation process itself, and (b) those *not* directly associated with that process. Most attempts to explain the origin of today's codon assignments have hinged upon the constraint of type b; for example, a codon catalog evolved to minimize the phenotypic consequences of mutation (Sonneborn 1965; Goldberg and Wittes 1966). The question is not, however, how plausible the latter type of constraint is: the question is how certain properties of the primitive translations mechanism could possibly have failed to exert an effect on the ultimate configuration of the codon catalog. As we have seen, there is every reason to suspect that primitive translation was highly inaccurate, and early on, even ambiguous. Any improvement that lowers the translation noise level would have a strong selective advantage. (Woese 1965). Therefore, the only ways one can *avoid* concluding that properties of the translation apparatus (i.e. its transmission noise pattern) shaped the ultimate configuration of the codon catalog are by assuming (1) that the noise generated in primitive translation was unstructured (completely random), and so bore no relationship to particular configurations of the codon catalog, (2) that the noise level of the machine had become so low by the time the ultimate assignments were determined that the noise no longer was a significant factor, or (3) that certain kinds of stringent constraints, not a function of the translation process per se, exerted selective pressures strong enough to override the effects of the translation noise pattern.

The first of these explanations is unlikely in view of the fact that even today the noise pattern in translation is highly structured. Moreover, this noise pattern even correlates to some extent with the present configuration of the codon catalog.*

The second explanation I tend to reject on the grounds that once the noise level of translation became low, the cell would soon evolve to a state of great complexity, making further evolution of codon assignments impossible, as it is today. In other words, I am making explicit the notion implied above, that codon assignments had to evolve *during* the evolution of the translation apparatus itself.

The third explanation might be valid if one could point to some strong specific constraints not themselves properties of the translation apparatus. So far none have been suggested. Readjusting a codon assignment to reduce the deleterious effects of a mutation is not a strong constraint when compared to readjusting a codon assignment to reduce the level of error in translation. The former would apply to the occasional isolated mutational event; the latter could apply nearly every time a particular codon was translated (Woese 1967).

---

* For example, translation noise is a function of codon position, the III base being the least accurately translated, the II base the most accurately translated. Furthermore, noise is a function of what bases are involved; a pyrimidine is most easily mistaken for another base; G is difficult to mistake. The III position is also the one manifesting *degeneracy* in codon assignment and in tRNA recognition patterns, etc. (Woese 1965; Davies et al. 1964; Davies 1969).

The ultimate configuration of the codon catalog is often taken to be largely or solely a historical accident, in the sense that its order is imprecisely determined and purely relative (Crick 1968). Were this ultimate configuration to evolve independently more than once, no two of these would bear any relationship to one another, although they could be ordered to comparable extents. Also a historical accident ordering is considered to be unrelated to the aboriginal ordering(s) (Crick 1968). In the extreme, this view applies not only to the amino acids now encoded, but to the choice of *which* amino acids become encoded as well (Crick 1968). Opposing schools of thought here then range from this one, a feeling that so long as the codon catalog develops to a moderately organized state, it makes little difference what its configuration is, to its opposite, whereby the configuration of the codon catalog today is a unique and predetermined happening (Woese 1969).

Confining attention solely to selective advantages of certain configurations of the codon catalog and the nature of constraints shaping these configurations, neglects the important matter of how the codon assignments become altered during evolution of the catalog. It is generally agreed that the configuration of the codon catalog today is immutable; it is to all intents and purposes locked in. Clearly, were a codon assignment to be changed in a cell today, it would effect the translation of every occurrence of that codon throughout the genome, something tantamount to creating myriad mutations simultaneously, and clearly a lethal event. Then what kind of cell *could* alter codon assignments, and under what conditions? Considerable thought has gone into designing types of primitive cells and conditions whereby codon reassignments would become feasible (Woese 1967, 1969).

Perhaps drastically overhauling an existing, but unproven concept is not the answer. We might do better to discard it completely, and ask whether an alternative to codon reassignment exists. The possibility is not so absurd as it sounds initially, for the above conception of primitive translation permits an evolution of the codon catalog that does not strictly require reassignment of codons (Woese 1965, 1967).

Given aboriginal 'group' codon assignments, their subsequent evolution could proceed by splitting a group assignment into two distinguishable subgroups. This would not reassign codons, though it would *refine* the original group assignment, increasing the complexity of the codon catalog. A small number of (rather large) aboriginal group codon assignments could be refined in a series of steps to the point that the unique type of codon assignment of today eventually emerges. A codon *refinement* evolution is preferable to reassignment, because refinement of a group into two subgroups does not put the cell at any selective disadvantage – the cell being unable to distinguish among those amino acids within the group to begin with (Woese 1970a).

A refinement evolution of the codon catalog has to be a function of the properties of the evolving translation machine. What refinements can occur at which stages is limited by the noise-ambiguity pattern of the machine at any given instant and the possible changes therein that benefit the cell.

The extent to which evolution of the codon catalog by codon refinement would produce a catalog whose configuration was unique, was predetermined, is moot at present (Woese 1970a). My own prejudice is that a configuration so evolved would tend to be highly predetermined. The reasons for feeling so are that the aboriginal group assignments themselves may have been predetermined by the (low) biological specificity manifested by the nucleic acid components of the primitive machine. And refinement of a group codon assignment may be nothing more than emergence of a protein function that 'fine-tunes' the distinctions inherent (and partially manifested) in nucleic acid–amino acid interactions. As long as evolution of translation continues merely to manifest properties inherent in simple nucleic acids, evolution of the codon catalog must continue largely along a predetermined track.

It is not now possible to explain or to place significance on the fact that the configuration of the codon catalog today is universal. Four basic types of explanation will account for the fact. One is that the ultimate configuration uniquely reflects some characteristic(s) – e.g. codon–amino acid pairing interactions – of the archetype translation process. Alternatively, one particular configuration of the catalog is selected for the reason that it is far superior to all others – in terms of the modern cell type. This is highly unlikely, however. The remaining two are basically historical accident explanations. At some point in evolution, one cell line is selected over all others; and that cell line just happens to possess the present configuration of the codon catalog. Similarly, were the evolution of translation a snow-balling affair, the *first* cell line to evolve a slightly better translation apparatus than the archetype machine would be the ancestor of the next improved version, and so on. (In other words, it is a race in which the runners continually increase in speed as a function of distance run, so that the one leading at the start must win. Since one cell line only wins such a race, only one configuration of the codon catalog emerges.)

Evolution of the translation apparatus during stage II. The main point of the

previous section is that one should not consider the evolution of the configuration of the codon catalog as independent of the evolution of all other
aspects of the translation process. Neither should one think of the evolution
of the assignments as being merely a matter of reassigning codons until a
'proper' configuration is achieved; the situation is not analogous to a poker
game with unlimited draws.

Evolution of all aspects of translation will be in accord with the Principle
of Increasing Complexity. Therefore, one starts with a mechanism whose
properties are not only simple but imprecisely defined. With time the mechanism becomes increasingly complicated, acquiring complex properties that
are at the same time precisely defined ('predictable'). Codon assignments
become refined from a few simple, ill-defined groupings to more complex,
well defined orderings. Improvements in machine design and changes in the
character of the codon assignments are interrelated.

We have seen above one possible model for an archetype translation
process – a simple reciprocating ratchet mechanism employing a stripped
down ('one armed') version of tRNA. Clearly, so simple a machine will be
imprecise (ill-defined) in its functioning – e.g. in amino acid-codon matching,
in maintenance of reading frame, in preventing chance termination of its
peptide output. Such a machine is at the mercy of fluctuations in its environment. We must now account for the evolutionary growth of tRNAs, the
appearance of the ribosome, etc. – we must find the principles by which this
crude version of the translation process becomes the elegant refined mechanism it is today.

It may seem that a discussion of this sort at this point in time is either an
exercise in sheer futility or an escape into phantasy. Now can one possible
discuss ribosome evolution, for example, unless one understands the ribosome first? To this it might be added that the above view of translation only
further muddies the waters, for the ratchet mechanism assigns *both* of the
major functions in translation (codon recognition *and* the mechanics of 'tape'
movement) to the tRNA molecule – ostensibly leaving no major function in
the process to be performed by the ribosome.

Not only is it proper to consider such evolutionary problems at this point,
but past failure to consider the translation process in the light of its evolution
may be largely what is responsible for the fact that we understand the basic
nature of translation today little better than we did a decade ago.* Further-

---

* Historically, the decade of the 1960's in Molecular Biology may be noted for the
great effort that went into attempts to elucidate the nature of the translation mechanism, in

more, in the evolving ribosome we come face to face with the general and
fundamental problem of creating macro-order out of a 'microuniverse'
through an organization of the 'elements' in the microuniverse into 'hierarchical patterns' (Bohm 1969; Woese 1971b).

In considering evolution of the ribosome I feel one should take the implications of the Ratchet mechanism at face value: accepting that tRNA alone is
the basis for the process, and that the archetype translation mechanism may
have operated without the aid of a ribosome. The ribosome *per se* then has
no function in translation. On the other hand, we must keep in mind that the
aboriginal process *is* inexact, ill-defined. Thus ribosomal evolution should
be a part of the evolution that refines translation, that brings it to the point
that its character, its functioning, are immune to fluctuations in its environment ('microuniverse'). In other words, if we were to take the tRNA translation complex (fig. 3) as the 'active site' of an enzyme, the ribosome would
then represent the 'rest of the enzyme'.

Let us ask then what relationship the Stage II evolution of the ribosome
bore to tRNA and to tRNA evolution. The structure of tRNA today suggests
certain features of its evolution. In particular its multiple arm structure
speaks of an evolution by gene duplication and joining. It is increasingly
apparent however, that tRNA's functions do not bear a *simple* relationship
to its structure. For example, codon recognition is not just a property of the
anticodon arm, for alterations in the dihydrouracil arm can effect the process
as well (Abelson et al. 1970). The emerging picture of tRNA is that of a
molecule whose functions are the products of coupled interactions among the
arms, not a collection of independent arms each with a separate function.

In evolutionary terms then, addition of some of the arms to the tRNA
molecule could have been to improve an existing function, not to acquire a
new one. For example, it is entirely possible that the so-called 'common'
arm arose to stabilize a tRNA–tRNA interaction. This arm contains the
TψCGA sequence (as a single stranded extension of an hf structure) and is
generally thought to bind to the ribosome (Zamir et al. 1965; Ofengand and
Henes 1969). However, in that ψCGA is a *self*-complementary sequence, two
tRNAs could easily bind to one another (Woese 1970b).

Perhaps the most important feature of the arm structures found in tRNA is
that they appear capable of existing in at least two 'stable' configurations –

---

spite of which little or no progress occurred in understanding the basic nature of the
process. I think it will also be recorded that the fault here lay not in the complexity of the
problem, but in the way in which it was conceptualized.

e.g. FH vs. hf. This bistability is precisely what is needed in order to use them in building larger stable structures. Since these 'arms' are ubiquitous among the *ribosomal* RNAs as well, I would therefore suggest that *they* are the functional units, the 'elements' from which one must evolve both tRNA and the ribosome (more properly the tRNA-ribosome complex). To a first approximation then, the tRNA-ribosome complex can be treated as an ensemble of coupled bistable elements. By increasing the number of (properly coupled) elements in the ensemble, Nature has constructed a series of mechanisms of increasing versatility and precision (Woese 1971b).

'Randomly' connected ensembles of bi- or multi-stable elements have interesting properties. [In brief such ensembles have generally been constructed by randomly connecting two inputs of an elements to outputs of other elements in the set. The logic of an element (its input-output relationship) can be randomly assigned as well.] More often than not such randomly connected ensembles exhibit stable (cyclic) behavior, sometimes of a very complicated nature (Walker and Ashby 1965; Kauffman 1969).

These sorts of ensembles are not quite what we need here, however. In that they are randomly constructed (and 'construction' here is an evolution of sorts) these ensembles are an evolutionary absurdity – they are not subject to the evolutionary 'principles of becoming'.

Therefore, in developing an ensemble view of the ribosome – i.e. in evolving the ribosome – the elements should be interconnected (coupled) in a somewhat ordered way, according to the hierarchical building principle discussed in section 2, I would claim. In other words, the first stage in building biologically is to couple two (or a few) bistable elements in such a way that certain combinations of their states become more stable as a result of their association. Next, such a combined unit is coupled to a similar unit – the coupling being to a first approximation between units as a whole, not between the individual elements therein. Again certain groups of states of the individual elements are further stabilized thereby. In its turn the new larger unit is coupled to another like itself – again 'as a whole' – to further stabilize particular subsets of the set of all possible microstates of the system, and so on. In this way, one starts with a simple system (one or a few bistable elements), transitions between whose states (i.e. the functions) are at the mercy of fluctuations in its microuniverse (are unpredictable and imprecise). Through hierarchical coupling one builds a complex system having two (or a few) 'macrostates' (restricted classes, subgroups, of its 'microstates' – which ultimately become enormous in number). As the structure builds, macrostate transitions become progressively insensitive to random perturbations, and ultimately occur 'only on command' (when information is supplied). Function then becomes precise (Woese 1971b). This I take to be a general Principle of Nature, one applying wherever Nature defined multi-stable elements.

In the more specific terms of a translation apparatus, what such an ensemble structure would mean is that the anticodon arm-ratchet becomes an element in a tightly coupled ensemble that comprises the other 'arms' in the tRNAs, arm structures in the ribosomal RNAs, etc. And the hf → FH transition (which is responsible not only for translation movement, but perhaps for the accuracy of codon-anticodon 'recognition' as well) now occurs only as part of a *macrostate* transition in the complex *as a whole* (Woese 1971b). In this way translation has become precise – insensitive to thermal perturbations, etc. I would contend that the information $(-\Delta S)$ needed to effect macrostate transitions for this system is in the form of the so-called 'G'- and 'T'-factors (Woese 1971b).

Structures built of hierarchically coupled pairs of bistable elements, pairs of this pairs, and so on, would tend themselves to be bistable (or quadristable, etc.) entities (Woese 1971b). Therefore, one should attempt to relate the bistable nature and two-fold symmetry of the tRNA–mRNA translation complex (fig. 3) to comparable macroproperties of the ribosome. Clearly the ribosome is a bipartite entity (30S and 50S subunits): Occam's razor would therefore suggest that we place one of the ratchet sites on one subunit, one the other. In its functioning the ribosome passes through a cycle comprising at least two shape transitions. The above ratchet mechanism suggests a cycle of four transitions, which is still consistent with the data at present (Schreier and Noll 1970).

We have no evidence for the above type of evolutionary model for the ribosome as yet. However, the model can readily be experimentally refined and/or tested. I would note several lines of evidence suggestive of a ribosome evolution via a codimerization cycle. The 50S subunit physically does resemble a dimer of 30S-like units, in the number of proteins it contains, in the size of its RNA, etc. Furthermore, the 23S rRNA is reported to contain segments present in *two* copies per molecule (Fellner 1970). A comprehensive analysis of the primary structures of the 50S and 30S ribosomes should rapidly reveal whether the ribosome did indeed arise by a cyclic doubling mechanism.

Protein has been conspicuously absent from the discussion so far. As stated above the bases for primitive translation must reasonably lie in nucleic acid: and contrary to the conventional view I would contend as well that nucleic acids (particularly the arm structures) are the basis for ribosome function today. I do not intend this to mean that proteins were totally absent from a primitive translation mechanism. If the primitive ribosome did not contain protein (basic protein of non-translational origin) to begin with, it must have become a ribonucleoprotein particle soon after the onset of translation. (A large polyanion would tend to bind basic proteins strongly in any case, and it is better that these be evolved for the purpose than suffer the perturbations that can result from binding non-specific polyamines.)

It is preferable to derive the protein components of the translation apparatus from as restricted an ancestry as possible. Given the nature of the interaction between nucleic acid and basic proteins, it is reasonable to derive them all from 'structural' proteins – i.e. proteins binding tightly to (major grooves, etc. of) nucleic acid. Initially such proteins would serve to stabilize nucleic acid structures, and thereby contribute accuracy to various nucleic acid functions. Gradually the proteins *per se* could come to modify refine, and extend these functions.

The T- and G-factors, which I take to be signals for state transitions, would not be required until translation became sufficiently stabilized that thermal (or chemical) perturbations could not longer drive it rapidly. Possibly the various initiation and termination proteins also came from this family, in that their functions are all similar.

The activating enzymes might have arisen to stabilize proto-tRNAs in their amino acid recognition and/or codon recognition functions. Their evolution (through dimerization) may have been a factor in moving the amino acid attachment site from its postulated anticodon-adjacent position to its present location.

Stage III. Final refinements of the translation process

By the time that evolution of translation reaches stage III, all the basic features of the modern translation process have been fashioned, by definition. Subsequent evolution merely 'fine-tunes' this machine. Two general considerations dominate here: to make translation so precise a process that its inaccuracy ceases to be a factor in evolution; and, to make translation a rapid process. We have not considered speed – one of the usual evolutionary considerations – as a dominant factor in the previous phases of the evolution of translation, for the advantage of more rapid translation at a primitive stage is far outweighed by advantages stemming from improvement in machine design. Thus, although selection for speed of translation may have operated at earlier stages, speed can be only a peripheral consideration at best during those stages.

Translation can be made precise through the use of precise functioning components whose states are narrowly defined – i.e. through error prevention – or through detection and correction of the errors that do occur. The errors the machine deals with concern the exact correspondence of amino acid and its codon, the precise maintenance of reading frame, and the correct placement of the 'discontinuities', the beginnings and ends of polypeptides.

A series of reactions establish the amino acid–codon correspondence. Each amino acid is 'chosen' by its own activating enzyme (here called a 'discriminase' for obvious reasons) in the initial, activation step. The activated amino acid is subsequently placed on a corresponding tRNA, the charging step; and the tRNA goes on to 'recognize' the codon (Berg 1961). An error in amino acid recognition at the first step in the process would be propagated into an error in decoding, unless it were somehow detected and corrected.

Since the mistake rate in protein synthesis is at least as low as a few parts in 10^4, and the mistake rate in amino acid recognition (activation) by the discriminases can be as high as a few parts in 10^2, it appears that error detection–correction features must operate within this chain of events (Loftfield 1963; Bergmann 1961). Indeed, it has been shown that such activation mistakes (e.g. formation of an enzyme-bound val-AMP complex by the isoleucine discriminase) are *not* propagated to the tRNA charging step. Some aspect of the discriminase–tRNA interaction detects and eliminates these mistakes – in the present example, the val-AMP is released from the isoleucine discriminase in the form of AMP and free valine, in the presence of $tRNA_{ilu}$ (Baldwin and Berg 1966). However, were an amino acid recognition error to get beyond this point it *would* become an error in protein synthesis (Chapeville et al. 1961).

The companion recognition problem here is that of codon with anticodon. Base pairing between nucleotide triplets does not seem sufficiently specific to yield the requisite accuracy, a point reinforced by the finding of what appear to be 'wobble pairs' (i.e. G...U pairs) in double stranded regions of the tRNA molecule (Loftfield 1963; Madison 1968). Error in codon recognition may be detected and corrected by the cell. This is suggested by the fact that $tRNA_{asp}$ (GAU and GAC codons) will bind well to ribosomes in the presence of an AGAGAG... message; yet no aspartic acid appears in the protein product of such a translation (Ghosh et al. 1967).

The mechanisms that underlie precision and the correction of errors are unknown at present. Error correction at the tRNA charging step could reflect either some direct recognition of the val-AMP (on the enzyme) or a recognition by the tRNA of an abnormal configuration in the discriminase that carries an incorrect amino acid.

If the ribosome (ribosome-tRNA complex) is indeed the ensemble machine described in the previous section, it will of course, be responsible for the bulk of the accuracy in the decoding steps of translation. The major in-accuracies that concern us here are mismatching amino acid and codon, 'slipping' the reading frame, and chance termination of the peptide output (and improper initiation of reading). It is easy to see that the ribosome might play a role in reading frame maintenance and preventing chance termination of the output (which latter can result from a 'backward' movement of the ratchet mechanism). The main issue, however, is the ribosome's possible role in codon 'recognition'; more specifically: (1) how inherently accurate is codon recognition (e.g. in a primitive translation system), (2) what does 'recognition' amount to in this case (and in general), and (3) how can the ribosome specifically affect recognition?

Evidence strongly implicates the ribosome in codon recognition – e.g. a ribosomal component is involved in streptomycin's effect on translation, and certain ribosomal mutants suppress Su tRNAs, etc. (Traub and Nomura 1968; Apirion and Schlessinger 1969).

Codon recognition has been conceived as a simple (and accurate) 'tem-plating' of codon by anticodon. However, triplet binding itself cannot be accurate, particularly in view of the G...U pairing possibility. In any case the matter is just not this simple, for bases not a part of the anticodon proper, some not even a part of the anticodon arm, can affect this 'recognition' (Abelson et al. 1970). Thus recognition must turn on some larger structure, e.g. the translation complex of fig. 3, where the equivalent of twenty base pairs could be involved. Note here also that suppression of codons is a function of 'context' – i.e. the composition of neighboring codons (Salser et al. 1969).

At this point however, one needs to question our basic notions of 'recog-nition' – i.e. a 'fit' between 'template' and the thing 'templated'. I would contend that this picture is now too simple, and always has been too static. The essence of 'recognition' may not lie in static 'fit', but in a *process*. Incorrect tRNAs do *bind* to codons on the ribosome, *but* the transitions that would then produce faulty protein do *not* follow.

It is difficult to see specifically how the ribosome (tRNA complex) could increase accuracy of codon 'recognition'. However, the ensemble machine described seems isomorphic with various electronic circuitry, and in these cases, one *can* obtain a sharpened response by properly increasing the number of 'elements' – as any radio buff knows.

Consider now the matter of reading frame maintenance. Uhlenbeck et al. (1970) have shown tRNAs to bind certain 'tetracodons' – codons with an additional base, generally at the 3' end – *better* than their normal triplet counterparts. Why then is tetracodon binding not a source of reading frame slippage? Inherent in the ratchet mechanism (figs. 3 and 4) is a device for correcting such errors. The critical base involved in tetracodon binding is number 11 (a uracil residue adjacent to the 5' end of the anticodon proper). This base can form a pair in the hf conformation (only), but the pair then is broken by the hf → FH transition.

It is in Stage III that one can expect protein to assume a prominent role in translation. At a more primitive stage, chain termination, for example, could have been accomplished merely by leaving certain codons unassigned. Sooner or later the ratchet mechanism would move 'backwards' in waiting for a tRNA to 'read' these codons, and termination would result. Initial setting of correct reading frame becomes a problem whenever the cell can no longer tolerate 'junk' protein. Crude initiation might be accomplished through a tRNA mechanism, for in forming the initial translation complex (two tRNAs) the first one has to *enter* in the FH conformation. Only a specially evolved tRNA may be capable of doing this.

Once the ensemble machine evolved to become immune to thermal fluctuations, these mechanisms could no longer operate.

On the present view the main function of ribosomal proteins is 'structural' – acting as very precise 'struts' to help in holding the rRNA in precise configurations. These proteins too become more crucial the more accurate translation becomes.

Speed and accuracy of information transfer are not independent parameters. A good case in point is the relatively long time required to transmit accurately a picture of the Martian surface from a space probe back to Earth. Therefore, many occasions may arise in evolution when a cell can have speed in translation or accuracy, but not both. Certainly the requirements for translation speed vs. translation accuracy are a function of the type of protein produced. On the one hand the cell may need many copies of certain proteins that are rather small and have very simple functions (e.g. structural proteins). In addition some inaccuracy in their synthesis may be tolerable. On the other hand, complicated, large proteins, proteins having sophisticated control functions, may have to be produced with negligible mistakes in order to function precisely. It is unlikely the latter proteins would often be needed in high amounts. In general, the more complex the organism the more demands it makes upon the accuracy of its information transfer processes (up to a point). Therefore, in opting for speed of information transfer at some juncture in evolution rather than accuracy, a cell may be forced to remain simple. By choosing accuracy it could have become more complex.

The difference between the prokaryotic and the eukaryotic cell may be precisely what we are discussing here. Translation systems in the two cases differ considerably – i.e. in terms of size and composition of the components, in terms of antibiotic sensitivity spectra, and (at least for the metazoans vs. prokaryotes) in terms of the speed of translation (Loening 1968; Lin et al. 1966). Accuracy of DNA → DNA information transfer also differs in the two cell types (Drake 1970).

Do considerations of this sort underlie some of the organellar evolutions as well? Of late, organellar evolution is discussed largely in terms of the capacity of a cell to 'capture' another cell, etc. (Stanier 1970). This I feel to be a mistake, for in any case organellar evolutions must turn on the *nature* of the original symbiotic relationship, not on how the two partners came together. From the above we might have predicted a cell to evolve at some stage with *two* protein synthesizing systems, one to produce proteins at a fast rate with little concern for accuracy, another to produce them as accurately as was possible at that time. A prokaryote–'proto-eukaryote' symbiosis may be the feasible way to achieve this. Note in this connection that *all* the membrane protein of Neurospora is reported to be controlled by cytoplasmic inheritance, and so presumably, is of mitochondrial origin (Woodward and Munkres 1966).

4. *Nontranslational protein synthesis: pretranslational evolution*

Little is accomplished by discussing nontranslational mechanisms of protein biosynthesis per se. We have no indication that such mechanisms ever existed. Yet the concept is intimately bound up with the evolution of complexity in a pretranslational–pregenic era, and in considering it one encounters the central evolutionary questions.

A number of general modes of nontranslational protein biosynthesis are conceivable. For purposes of this discussion we will examine two specific possibilities – a *positional templating* mechanism, suggested by the structure of certain nucleic acid–protein complexes, and a second mechanism that would produce proteins of repeating sequence, suggested by the above primitive ratchet translation mechanism.

Double helical nucleic acid will form a stoichiometric complex with poly-lysine, the polypeptide lying in the major groove of the structure with the phosphates of each polynucleotide chain binding to alternate amino groups of the polylysine (Olins 1967). Interresidue spacings in both polymers are such that the fit of polypeptide to polynucleotide is exact – involving little distortion of the double helical nucleic acid structure. In an environment containing double stranded nucleic acid and small basic peptides (or even the basic amino acids themselves), it is difficult to see why the latter would not become aligned along nucleic acid major grooves, and thereby auto-matically become positioned in a manner conducive to their condensation into larger peptides. *Positional templating* would then permit the primitive living system to synthesize one general class of proteins (Woese 1970a).

Proteins made in this way would not manifest the point–point primary structural correspondence to nucleic acid (as proteins of today do). To a first approximation at least, the protein primary structure in this case would be independent of the nucleic acid primary structure. However, a loose coupling between the two (a 'region–region' correspondence) is conceivable in view of the fact that A–T-rich DNA preferentially binds polylysine (rather than polyargine), while the reverse is true for poly G–C (Leng and Felsenfeld 1966). Furthermore, lysine-rich 'proteinoids' preferentially bind poly-pyrimidines, whereas arginine-rich 'proteinoids' prefer polypurines (Yuki and Fox 1969).

Let us turn again to the archetype translation mechanism. Its central structure, the translation complex, is in essence a double helix that juxtaposes two proto-tRNAs and with them, a peptide and an amino acid. Messenger RNA per se does not seem essential to the function of that complex; any nucleic acid–nucleic acid interaction bringing peptide and amino acid into a favorable orientation should suffice. If a primitive environment contained such proto-tRNAs, then it could well contain similar self-complementary nucleic acids that accomplished the juxtaposition of amino acid and peptide in the absence of mRNA. To be more specific, two self-complementary nucleic acid chains both in the *same* conformation (either FH or hf) sterically can form a joint double helical structure, if sequences in their single stranded extensions are complementary one to the other. (The 'common' arms of the peptidyl- and the aminoacyl-tRNAs today are thought to do precisely this during translation – see above.) Were these primitive self-complementary nucleic acids then carrying, one an amino acid, the other a growing peptide, peptidyl transfer might be effected. (The relative positioning of the amino acid and the peptide would be a function of the size of the complex – i.e. the number of base pairs made between the two nucleic acid chains and the size of their single stranded 'loop' regions to begin with.) CPK models show that the problem of amino acid–peptide juxtaposition is sterically as simple, or simpler, as in the case of the archetype translation mechanism discussed above (Woese, unpublished). The peptides synthesized in such a way would have peculiar compositions; they would be repeating dipeptide sequences, or in special cases homopolymers – raising the question of what role, if any, such a peptide might have in a primitive biological system.

These hypothetical nontranslational protein syntheses differ from translation in two important, related respects. For one, no point–point correspondence, no conventional type of information transfer between a nucleic acid and the protein, exists. In the *positional templating* example the protein primary structure is independent of that of the nucleic acid, at least to a first approximation. In the synthesis of repeating dipeptides, the possibility of a colinear relationship between protein and nucleic acid did not exist to begin with. Thus one is led to question whether these are in fact information transferring processes, to ask what forms information transfer in a pretranslational system could take, and to question biological information transfer in general. The second major difference is that, lacking a point mapping relationship to nucleic acid, the nontranslationally produced proteins cannot evolve through a genetic mutational mechanism, and so presumably cannot evolve at all. In this way one is led to question what precisely *is* the role of the gene and its mutation in evolution, particularly early evolution, and what produces evolution in the absence of a genetic system (i.e. absence of the capacity to read out a gene).

The amount of information in, e.g., a protein sequence is defined in terms of a universe of all possible sequences of that sort. Such a universe is neither absolutely nor easily defined. For example, it is not the totality of sequences of that length that can be constructed from twenty different amino acids. That definition overestimates the amount of information in a protein, particularly a primitive protein, by an enormous amount. It tends, too, to give a false impression of evolution, which becomes an 'exercise in miracles' – i.e. an extremely small fraction of the informational universe is explored during evolution.

The definition of the informational universe to be useful must be in terms of the capabilities of the 'transmitters' and the 'receivers' that exist. The universe should not be defined to include messages the transmitters cannot transmit, nor messages the receivers cannot detect. Similarly, it cannot distinguish among messages indistinguishable to the receivers. In our terms the 'transmitter' is a protein synthesizing system, the 'receivers' the interactions of the protein produced with the system as a whole and the effects produced thereby.

Thus, in a primitive environment, where biological specificity is low, where distinctions among certain amino acids and many amino acid sequences are not possible, and where perhaps only short polypeptides can be produced, the informational universe would be relatively small. The same protein that has a high information content in the cell today, would have a far smaller (or even no) information content in a primitive world. (The biological system may differ from more conventional information processing systems in one respect. In that 'receiver' is equated to 'effect on the system', the frequency of occurrence of a biological 'message' will affect its 'reception'. Proteins with very special functions – e.g. a specific β-galactosidase function – might occur from time to time in a primitive environment by chance, but each with a frequency too low to perturb the systems. Such a rare occurrence protein would not then be 'received'.) In this sense evolution becomes an expansion of an informational universe. It also follows that early evolution, far from being an 'exercise in miracles', may in fact be a process whereby (nearly) *all* of the informational universe is explored at any given stage.

It is apparent that any comprehensive discussion of the informational properties of primitive macromolecules requires that one define aspects of the complexity of the system of which the molecule is a part – in which the molecule has 'meaning'. The above concept of the ensemble machine, I think, begins to show us how these effects, the molecule's meaning, might be measured. A macrostate transition in an ensemble of N bistable elements requires some particular fraction, aN, of the elements to undergo individual state transitions. Any molecule that can effect transitions in these particular aN elements of the ensemble would then have, say, aN bits of *meaning* for the ensemble.

5. *Evolution of nucleic acid → nucleic acid tape reading systems: the origin of the gene*

By comparison to translation our understanding of the remaining cellular 'tape reading' processes – those involving nucleic acid → nucleic acid – seems near complete. Base pairing and stacking simply 'template' the incoming monomer nucleotides. And it is easy to picture such a templating interaction (utilizing either mono- or small oligo-nucleotide units) occurring in a primitive environment to begin with. In so far as a static templating picture does fit these phenomena, our understanding *is* virtually complete.

What we tend to overlook once again is the process aspect, the matter of 'polymerases' moving along nucleic acid tapes. Generally a polymerase is viewed as some late evolutionary adjunct to a process that works satisfactorily, but not perfectly, in its absence. But is this so? Are base pairing interactions *sufficient* to effect the mechanical movement in nucleic acid replication? I do not pretend to answer this, but I would emphasize the question by posing alternatives. Yet this is not done for purely pedantic reasons.

Evidence suggests already that simple polynucleotide-mononucleotide interactions are not a sufficient condition for primitive nucleic acid replication. Mononucleotides, for example, will form 'monomer-polymer' double (or triple) helicies with preexisting polynucleotides – in which the monomer 'chain' is stacked, but has no covalent backbone. *But*, the polymer species has in all cases to be a polypyrimidine; polynucleotides of other compositions will not template monomers in this fashion (Howard et al. 1966; Huang and T'so 1966). Furthermore, attempts to polymerize the monomers so aligned, through the use of 'primitive' condensing agents, have yielded less than spectacular results; the resultant oligonucleotides are very small and contain linkages other than the expected 3'–5' phosphodiester bond (Sulston et al. 1968). It is possible therefore, that primitive nucleic acid replication may *not* be solely a matter of preexisting nucleic acids templating monomer subunits into a 'monomer-polymer' double helix, in preparation for their nonspecific condensation.

As discussed above, primitive protein-nucleic acid associations were likely to have involved extensive regions of their respective structures (rather than more or less localized interaction 'sites'). Two types of these complexes are of possible interest in the present context.

One has been introduced above – a basic polypeptide produced through *positional templating* by nucleic acid. Such a polypeptide might catalyze nucleic acid replication by interacting with one of the above monomer-polymer double helicies to form a triple stranded complex (the peptide lying in the major groove of the nucleic acid double helix). Formation of such a tightly bound complex could be a necessary condition for the existence of the general monomer-polymer double helix. In any case, in stabilizing such a helix the basic polypeptide could make the base pairs more accurate, increase discrimination between 2′–5′ vs. 3′–5′ phophodiester links, etc.

Peptides of repeating *di*peptide sequence – produced by a variation of the ratcheting mechanism (see section 4) – are also candidates for a role in primitive nucleic acid replication. Note that a polylysine complex with double stranded nucleic acid is arranged so that alternate ε-NH_2 groups bind to opposite polynucleotide chains. Thus, were a repeating dipeptide sequence of the form $[lys \cdot X]_n$ (where X is a neutral amino acid for example) to bind to a nucleic acid double helix, it should bind to *one* of the chains only (locally). In other words, it is possible that such an alternating peptide would bind to and hold *single* stranded nucleic acid in a helical configuration resembling the configuration it has in an actual double helix. Such an alternating peptide may also have the property of moving rather freely (one dimensional diffusion) along nucleic acid. One or both of these properties could serve as the basis for a primitive nucleic acid replicating (or transcribing) machine.

Over and above the mechanics of a primitive transcription process is the broader question of why such a process should evolve in any case. *A priori* there seems no reason why the genome cannot be translated directly – and this indeed seems to occur for RNA viruses today. Thus one possibility is that transcription arose as some 'late' refinement of cellular dynamics to minimize the handling (and so perturbation) of the genes. Alternatively, transcription could have arisen *before* translation – the reason being that its evolution was, in fact, the evolution of the Gene.

The older framing of this evolutionary question, in terms of whether the gene or the gene product (protein) came first, is clearly naive. The gene did not suddenly appear, nor did it exist in the absence of the gene product. In accordance with general evolutionary principles, the essence of the Gene – the gene-gene product relationship – undoubtedly emerged through a gradual process of refinement, of definition, starting from something that would hardly be recognized as such by today's standards. Thus, the forerunner of the gene may not have been the very stable, unreactive entity it is today. Not only would this primitive 'gene' be responsive to some extent to its environment, but the distinction between gene and gene product may not always have been the sharp one it is now.

For these reasons it seems preferable theoretically that transcription, not translation, be the relationship that originally defined, that evolved, the Gene. Nucleic acid appears capable of existing in *both* a 'gene' form and a 'gene-product' form. The former, of course, is the highly stable, unreactive double helix; the latter, I would claim, is the class of nucleic acid 'arm' structures, structures 'responsive' to their environments by virtue of their bistable character and general geometry. If we do not view biological specificity as a matter of static 'fit', but rather as one of process (i.e. state transitions in 'ensemble machines'), then one no longer need deny that perfectly reasonable (but perhaps somewhat crude) enzymatic functions might be performed by properly evolved nucleic acids. And so an early and complex stage of 'nucleic acid life' may have occurred on the evolving earth, a stage that gave rise to that relationship we call the Gene.

6. Conclusion

One initially tends to view the evolution of the primitive cell as a collection of separate, special evolutionary problems – e.g. obtaining a translating machine, 'deciding upon' a genetic code, evolving an enzyme to read out nucleic acid, evolving this or that enzymatic pathway, etc. However, it gradually becomes apparent that many of these may not be separate problems at all. The evolution of the genetic code may be a part of the evolution of the translation machine, etc. But more importantly, one comes to realize. that in one sense there are no such things as 'special' evolutionary problems. Evolution is not a mixed bag of various 'historical accidents' – happenings that are totally unlike one another and are little interconnected, happenings that in any case would undoubtedly not occur again were the evolutionary process repeated. All evolutionary problems appear to have important common aspects – whether they are intracellular problems or problems on higher levels of biological organization. Thus in ostensibly addressing ourselves to evolutions of genetic codes, translating mechanisms, etc., we are actually discovering and defining the basic principles for all evolution – the laws by which macroorder emerges from a microuniverse. The nature of the 'elements' involved changes from one example to another, but the patterning of events in time, the 'principles of evolutionary construction', seem to remain invariant. A major lesson to be learned from the study of the evolution of the translation apparatus, etc., is that evolution should not be viewed as a card game with unlimited draws, or as a 'walk' from one point to another in some evolutionary phase space (comprising all possible potential outcomes) until some optimal point is found. Evolution has an important aspect of *refinement* lacking in these analogies. Not only do entities become more complex with evolutionary time, but they become more refined, more definite in their properties, as the process proceeds. This quality of refinement, of creating, defining, or expanding some 'universe' (phase space) as a result of evolution, seems to epitomize the process.

Its interaction with the concepts of Mendelian (and later Molecular) genetics has given the study of evolution a strong genetic bias. It is too easy to think of evolution solely in terms of gene mutations, flow in gene pools, and, of course, some vague 'selection' parameters. Too much emphasis is placed on the micro-changes at the expense of the macro-ordering. The doctrine of increasing complexity, however, is not framed in terms of genes and gene mutations – and, at least on a metaphysical level, one feels it should not be. Fortunately, the problems to be encountered in evolution of the cell should bridge the gap between evolution in gene-based systems and that in non-genic systems.

The evolutionary importance of the Gene is not that it permits evolution to occur. [If anything, evolution is in spite of the gene!] Rather, the precise mapping relationship that links the gene (a simple, stable, 'unresponsive' entity that is not evolvable in its own right) to the gene product (an intricate, 'responsive', evolvable, but necessarily 'unstable' entity) thereby permits the existence of incredibly complicated systems – systems too unstable to evolve or even exist in their own right, i.e. in the absence of the capacity to be generated or repaired through reference to a less intricate 'transform' of themselves.

The complexity and homeostatic properties of the randomly constructed networks of bistable elements (see above) certainly provide a good starting point for conceptualizing the nature of pre-genic systems, although the former admittedly are a long way from modeling evolution or replication. With regard to this last point we should recognize that some of the implicit prejudices we gain from experience with modern living systems may be narrowing our perspective regarding 'replication' and evolution in pre-genic systems. How much of the exactness and apparent simplicity of, say, bacterial replication should we demand of a pre-genic system? Is replication in this latter case merely a steady increase in the number of all components in the system, which eventually 'divides' into several parts because it has become large? A conceptually instructive alternative is to pattern the primitive process here on metazoan development (which I and others would claim manifests some degree of extra-genic inheritance and expression). In other words, from some starting point let a primitive system grow in *intricacy* to the point that it ultimately becomes unstable, but as a necessary consequence of which, two or more systems with the 'macroproperties' of the starting system emerge from it at some stage.

Studies on cellular evolution will eventually force an active consideration of a deeper issue, one which has languished so far in the realm of metaphysics for the most part. If there are indeed certain principles of Evolution, how fundamental are these? A 'dimerization' process seems a basic mechanism in macromolecular evolution. But, are its resemblances to certain phenomena in the domain of atomic physics purely superficial? Or are the Principles of structuring uncovered in the study of evolution actually fundamental Laws of Nature? [There is an important corollary issue involved here – viz. whether biology is in principle – as it is treated in practice – a highly specialized branch of the fundamental science, physics, or whether biology is a fundamental science in its own right. On the answer to this turns the attitude of society toward biology and therefore, biology's ultimate impact on the nature of the society.]

References

ABELSON, J. N., M. L. GEFTER, L. BARNETT, A. LANDY, R. L. RUSSELL and J. S. SMITH, 1970, J. Mol. Biol. 47, 15.

APIRION, D. and D. SCHLESSINGER, 1969, The effect of ribosome alterations on ribosome function and on expression of ribosome and non-ribosome mutations CIBA Found. Symp. on Mutations as Cellular Processes.

ASHBY, W. R., 1960, Nature 187, 532.

BALDWIN, A. N. and P. BERG, 1966, J. Biol., Chem. 241, 831.

BARNETT, W. E., 1965, Proc. Natl. Acad. Sci. U.S. 53, 1462.

BARNETT, W. E., D. H. BROWN and J. L. EPLER, 1967, Proc. Natl. Acad. Sci. U.S. 57, 1775.

BERG, P., 1961, Ann. Rev. Biochem. 30, 293.

BERGMANN, F. H., P. BERG and M. BIECKMANN, 1961, J. Biol. Chem. 236, 1735.

BOHM, O., in "Towards a Theoretical Biology 2", p. 41, ed. C. H. Waddington, Edinburgh University Press 1969.

BROWNLEE, G. G., F. SANGER and B. G. BARRELL, 1968, J. Mol. Biol. 34, 379.

CAPECCHI, M. R., 1967, Proc. Natl. Acad. Sci. U.S. 58, 1145.

CHAPEVILLE, F., F. LIPMANN, G. VON EHRENSTEIN, B. WEISBLUM, W. RAY and S. BENZER, 1966, Proc. Natl. Acad. Sci. U.S. 48, 1086.

CHUGUEV, I. I., V. S. AXELROD and A. A. BAYEV, 1969, Biochem. Biophys, Res. Commun. 34, 348.

CRESTFIELD, A. M. and R. G. FRUCHTER, 1968, J. Biol. Chem. 242, 3279.

CRICK, F. H. C., 1968, J. Mol. Biol. 38, 367.

DAVIES, J., 1969, Errors in translation; In: Progress in molecular and subcellular biology. (Springer-Verlag, Berlin) 47.

DAVIES, J., W. GILBERT and L. GORINI, 1964, Proc. Natl. Acad. Sci. U.S. 51, 883.

DOCTOR, B. P., W. FULLER and N. L. W. WEBB, 1970, Nature 225, 508.

DRAKE, J. W., "The Molecular Basis of Mutation", Holden-Day, San Francisco, 1970.

DUS, K., K. SLETTEN and M. KAMEN, 1968, J. Biol. Chem. 243, 5507.

ERB, R. W., M. M. NAU and P. LEDER, 1969, J. Mol. Biol. 38, 441.

FELLNER, P. 1969, Euro. J. Biochem. 11, 12.

FULLER, W. and A. HODGSON, 1967, Nature 215, 817.

GEFTER, M. L. and R. L. RUSSELL, 1969, J. Mol. Biol. 39, 145.

GHOSH, K. and H. P. GHOSH, 1970, Biophys. Res. Comm. 0000.

GHOSH, H. P., S. SOLL and H. G. KHORANA, 1967, J. Mol. Biol. 25, 275.

GILBERT, W., 1963, J. Mol. Biol. 6, 389.

GOLDBERG, A. L. and R. E. WITTE, 1966, Science 153, 420.

HOWARD, F. B., J. FRAZIER, M. F. SINGER and H. R. MILES, 1966, J. Mol. Biol. 16, 415.

HUANG, W. M. and P. O. P. T'SO, 1966, J. Mol. Biol. 16, 523.

KAJI, A., K. IGARASHI and H. ISHITSUKA, 1969, Cold Spring Harb. Symp. Quant. Biol. 34, 67.

KAUFFMAN, S. A., 1969, J. Theoret. Biol. 22, 437.

KONDO, M., 1967, Studies on transfer RNA. PhD thesis. (University of Illinois).

LENG, M. and G. FELSENFELF, 1966, Proc. Natl. Acad. Aci. U.S. 56, 1324.

LIN, S., R. D. MOSTELLER and B. HARDESTY, 1966, J. Mol. Biol. 21, 51.

LOENING, U. E., 1968, J. Mol. Biol. 38, 365.

LOFTFIELD, R. B., 1963, Biochem. J. 89, 82.

MACKAY, D. M., 1954, Synthese 9, 182.

MADISON, J. T., 1968, Ann. Rev. Biochem. 37, 131.

MIZUSHIMA, S. and M. NOMURA, 1970, Nature 226, 1214.

NIRENBERG, M. W., P. LEDER, M. BIRNFIELD, R. BRIMACOMBE, J. TRUPIN, F. ROTTMAN and C. O'NEAL, 1965, Proc. Natl. Acad. Sci. U.S. 53, 1161.

NISHIZUKA, Y. and F. LIPMANN, 1956, Proc. Natl. Acad. Sci. U.S. 55, 212.

NOMURA, M. and C. B. LOWRY, 1967, Proc. Natl. Acad. Sci. U.S. 58, 946.

OFENGAND, J. and C. HENES, 1969, J. Biol. Chem. 244, 62411.

OLINS, D. E., A. L. OLINS and P. VON HIPPEL, 1967, J. Mol. Biol. 24, 257.

PESTKA, S. and M. W. NIRENBERG, 1966, J. Mol. Biol. 21, 145.

SALAS, M., M. B. HILLE, J. A. LAST, A. J. WAHBA, and S. OCHOA, 1967, Proc. Natl. Acad. Sci. U.S. 57, 387.

SALSER, W., M. FLUCK and R. EPSTEIN, 1969, Cold Spring Harb. Symp. Quant. Biol. 34, 513.

SCHREIER, M. H. and H. NOLL, 1970, Nature 227, 128.

SCHWEIGER, M. P., G. B. CHHEDA, L. BACZYNSKYJ and R. H. HALL, 1969, Biochemistry 8, 3283.

SOLL, D., E. OHTSUKA, D. S. JONES, R. LOHRMANN, H. HAYATSU, S. NISHIMURAN and H. G. KHORANA, 1965, Proc. Natl. Acad. Sci. U.S. 54, 1378.

SONNEBORN, T. M., 1965, Degeneracy of the genetic code: Extent, nature, and genetic implications; In: Evolving genes and proteins, V. Bryson and H. Vogel eds. (Academic Press, New York) 377.

STANIER, R. Y., 1970, Some aspects of the biology of cells and their possible evolutionary significance. 20th Symp. of the Soc. for Gen. Microbiol, H. P. Charles and B. C. J. G. Knight eds. (Cambridge Univ. Press, New York) 1.

STULBERG, M. P. and K. R. ISHAM, 1967, Proc. Natl. Acad. Sci. U.S. 57, 1311.

SULSTON, J., R. LOHRMANN, L. E. ORGEL and H. T. MILES, 1968, Procl. Natl. Acad. Sci. U.S. 60, 409.

THIEBE and H. G. ZACHAU, 1968, Europ. J. Biochem. 5, 546.

TRAUB, P. and M. NOMURA, 1968, Proc. Natl. Acad. Sci. U.S. 58, 777.

TSUNODA, J. M., K. T. YASUNOBU and H. R. WHITELEY, 1966, J. Biol. Chem. 243, 6262.

UHLENBECK, O. C., J. BALLER and P. DOTY, 1970, Nature 225, 508.

WALKER, C. C. and R. W. ASHBY, 1965, Kybernetics 3, 100.

WATSON, J. D., 1964, Bull. Soc. Chim. Biol. 46, 1399.

WOESE, C. R., 1965, Proc. Natl. Acad. Sci. U.S. 54, 1546.

WOESE, C. R., 1967, The genetic code: the molecular basis for genetic expression. (Harper and Row, New York).

WOESE, C., 1968, Proc. Natl. Acad. Sci. U.S. 59, 110.

WOESE, C. R., 1969, J. Mol. Biol. 43, 235.

WOESE, C. R., 1970a, Bioscience 20, 471.

WOESE, C. R., 1970b, Nature 226, 817.

WOESE, C. R., 1971a, J. Theor. Biol. 32.

WOESE, C. R., 1971b, Mss in preparation.

WOESE, C. R., D. DUGRE, W. C. SAXINGER and S. A. DUGRE, 1966, Proc. Natl. Acad. Sci. U.S. 55, 966.

WOODWARD, D. O. and K. D. MUNKRES, 1966, Proc. Natl. Acad. Sci. U.S. 55, 872.

YUKI, A. and S. FOX, 1969, Biochem. and Biophys. Res. Comm. 36, 657.

ZAMIR A., R. W. HOLLEY and M. MARQUISSE, 1965, J. Biol. Chem. 240, 1267.

C. Ponnamperuma (ed.), Exobiology. © North-Holland Publishing Company

CHAPTER 10

Early cellular evolution

LYNN MARGULIS*

1. Introduction**

Theories of evolution of microbes on the early earth can only be plausible and consistent with modern knowledge; they are, at best, explanations of what already has irreversibly occurred and therefore can never be confirmed as 'correct'. Recently, astronomers who amused themselves deducing lunar conditions from indirect evidence have suffered the jolt of direct confrontation with samples of the moon. Fortunately, we evolutionary biologists will never be embarrassed by direct observations of Precambrian microbial evolution. Accordingly, we may imagine what occurred in the earliest stages of terrestrial evolution unrestrained by definitive information. I make no pretentions to bring anything more to bear on this problem than a profound interest.

Probably the major concept emerging from recent work related to problems of cellular evolution, of relevance to 'exobiology', is that of the antiquity of our particular sort of cellular life. The question: 'What was the origin of terrestrial life?' has recently become much more circumscribed and specific: how did triplet-coded anaerobic heterotrophic prokaryote cells orginate? (Echlin 1969; Horowitz 1970 in Margulis 1970b).

Ancestral cells, admittedly complex, probably evolved and became stabil-

* I acknowledge with gratitude aid and comments on this manuscript given by Drs. M. G. Rutten, T. D. Ford, M. Muir, D. T. Rickard, and John Olson. I thank the NASA (NGR-004-002) and the National Science Foundation for support.

** Many of the problems mentioned in this paper have been discussed in greater detail in my book, *Origin of eukaryotic cells*, Yale University Press (Margulis, 1970). The tables and figures appear therein.

ized in their major modern aspects before the Fig Tree sediments of South Africa were laid down about 3 billion years ago. The recent work reviewed in this chapter suggests that from about 3 billion years to the present the accepted processes of neodarwinian evolution were already entirely in play. This leaves less of the terrestrial record to be explained by prebiotic processes presumably involved in the origin of the ancestral cells. The reader is referred to chapters 2 and 4 of the present volume for recent thought concerning the 1500 million opening years of earth's history (from about 4.5 to 3 billion years ago) from which there are nearly no sedimentary rocks. This chapter is concerned only with evolutionary developments that occurred subsequent to the origin of these first cells.

What types of relevant scientific information can be useful in reconstructing the earliest evolutionary history of life on the primitive earth? Fundamentally there are two avenues that hopefully converge, each with attendant difficulties: modern microbial physiology, taxonomy and ecology, and direct examination of the fossil record.

Microbial physiology and ecology are potential sharp tools for shaping our concepts of microbial evolution. Recent data reinforces the feeling: many sedimentary deposits must have had an intimate relationship to microbial floras known (or suspected) to have flourished at the time the sediments were formed (Gutstadt and Schopf 1969; La Berge 1967). Louisiana sulfur domes may be products of marine microbial activity; significant amounts of reduced carbon are associated with sediments older than 3 billion years in South Africa (Barghoorn 1970); the banded iron formations of North America are presumably related to iron bacteria (La Berge 1967, Cloud 1968). Although the bacterial fossils associated with iron pyrite (Ehlers et al. 1965) were only Pennsylvanian in age, the applications of their electron microscopic techniques to far more ancient rocks have already revealed systematic associations of microbial floras and sedimentary iron (Schopf et al. 1965).

Before we can assess the historical contribution of microbes to the sediment we must understand well the ecological role of the extant microbes (Ferguson Wood 1967). Although it has long been realized that microbes are intimately involved with recycling elements indispensable to life such as carbon, sulfur, nitrogen, oxygen and phosphorus and that they are responsible for massive deposition of sediment today, except with regard to sulfide production few scholars have tried to apply modern microbial ecological knowledge to sedimentary rocks of the Precambrian. This is in part attributable to lack of information ... the study of microorganisms 'in the sea have received far too little attention when we consider the vast potentiality of the sea both as an adjunct to man's survival and as the place of deposition of the major part of the world's sedimentary rocks and the minerals and ores contained therein' (Ferguson Wood 1958, page 1). Principles of microbial ecology have not been applied to analyses of the Precambrian environment because it is only recently been realized that the entire Precambrian is not 'unfossiliferous' as previously thought, but is 'microfossiliferous' (see Schopf's contribution, this volume). Much of the difficulty, too, comes from well-recognized problems of reconstructing microbial phylogenies (Mandel 1969; De Ley 1968). Yet, this still is only part of the story. In my opinion, the fundamental reason that the Precambrian fossil record has not been successfully related to microbial evolution is conceptual; it stems from certain erroneous, often unstated assumptions, found to varying degrees in the literature.

There is no outline of evolution or set of reliable taxonomic criteria universally accepted by active microbiologists (Ainsworth and Sneath 1962). Therefore it is extremely difficult, even impossible, to extrapolate widely accepted 'primitive' to 'advanced' characters from living microbial forms. This is in sharp contrast to the situation in mammals or flowering plants, organisms that can be acceptably related both to extant and fossil forms on morphological criteria. Successful interpretation of the Phanerozoic fossil record rests upon firm constructs of animal and plant evolution; eventual successful interpretation of the Precambrian record depends intimately upon our concepts of microbial evolutionary relationships. But these are very vague (De Ley 1969; Mandel 1969).

Although microbiologists agree now that their ancient, primarily asexual organisms are products of neodarwinian evolution, there is substantial disagreement at even the highest levels of systematic taxa (kingdom and phylum levels: Copeland 1956; Whittaker 1968; Margulis 1971; Klein and Cronquist 1969) that merely reflects our ignorance of microbial relationships. Therefore it is not possible to simply review this field without explicit cognizance of unstated assumptions some of which, I personally consider to be misconceptions; and it therefore becomes necessary to state, defend and refer to a set of alternative assumptions (Margulis 1968) that provide the framework for the rest of this discussion.

2. Popular unjustified assumptions

2.1. Molecules evolve

Implicit in recent work on the origin of the genetic code is the concept that nothing less than a reproducing, mutating system can evolve (Crick 1968; Orgel 1968). Although the possible evolutionary sequence is only recently the subject of some thought (Orgel 1968), the necessity for the very early coupling of nucleic acid and protein synthesis has been realized by many authors (Horowitz and Miller 1962; Rich 1962; Bernal 1968). That the earliest 'sloppily' replicating entity was a coacervate-type metabolising 'proto-biont' (Oparin 1968) not a more-or-less free polynucleotide-polypeptide synthesizing system has been argued by Oparin. However, Calvin in considering the problem of the origin of photosynthesis, wants us still to 'devise some way of evolving this particular molecule, chlorophyll, belonging to the general class of tetrapyrrolic substances known as porphyrins'. He believes these problems are 'really part of chemical evolution and the origin of life' (Calvin 1961, 1969 ch. 10). But chloroplasts are enormously complex structures, only found in plant and eukaryotic algal cells. Besides the fact that they may have been originally intracellular bluegreen algal-like pro-karyotic symbionts (Ris and Plaut 1962; Edelman et al. 1967; Margulis 1970) there is no doubt that all membrane-bound plastids are relatively late in the evolutionary history of photosynthetic organisms and were preceded by anaerobic photosynthetic bacteria and bluegreen algae (Echlin and Morris 1965; Klein and Cronquist 1968; Echlin 1969). 'How the molecule chloro-phyll evolved' too, is not interpreted as a special case of 'chemical evolution' by biologists. Molecules do not evolve. 'Chemical evolution' like 'stellar' and 'cultural' evolution are helpful analogies to the very different process of biological or neodarwinian evolution. The former are a series of regular complex changes unidirectional in time. Biological evolution is a process only possible in populations of reproducing organisms. Differential survival, the production of permanently changed or mutated organisms again giving rise to their own kind are necessary elements of biological evolution. There is nothing on earth less complex than a population of living cells that is subject to this kind of evolution. There are microbial cells unable to make chlorophyll that share many other features, including specific protein sequences, in common with photosynthetic bacteria suggesting it is likely that biosynthesis of chlorophyll evolved in cells well after the type of cell ancestral to all life on earth evolved (Dayhoff 1969; Margulis 1969). This must also be true as well of many other classes of compounds mentioned by Calvin (1969) in connection with 'chemical evolution': membrane phospholipids, steroids and so forth. No matter how complex the organic chemical reactions of the primitive earth, there will never be a gradual increase in complexity of structure leading to a photosynthetic cell in the absence of some replication mechanism to insure the stability through time of basic components of that structure (Bernal 1969; Horowitz and Miller 1962).

2.2 *High level of mutagens in the environment are correlated with a high mutation rate*

The environment on the earth has never been deficient in either mutagens or lethal agents. In considering the origin of living systems the difficulty lies in explaining the origin of faithful reproduction from a more primitive presumably 'noisier' system (Crick 1968).

'It is a well-known fact, emphasized over and over again in discussions of genetics and evolution, that the vast majority of known mutations are inadaptive. They are almost always disadvantageous to the individual and ... to the population under the actually existing conditions. Just as often, it has been pointed out that this is precisely what would be expected. If adaptation is perfect, any change is disadvantageous. Even if adaptation is imperfect, mutations tending to improve adaptation and occurring frequently enough to be detected would in most cases already have been incorporated into the genotype of the population, i.e., would have become wild-type genes and not mutants with respect to the existing situation. This consideration applies to any population that has achieved a considerable degree of adaptive stability ...' (Simpson 1953). High mutation rates are no more directly correlated with rapid rates of evolution in well adapted extraterrestrial or Precambrian microbial than in other biological populations.

2.3. *Ultraviolet light was a significant selective agent in the late Precambrian; lack of oxygen limited the Precambrian development of life*

Statements such as these are accepted without argument in even the most recent biological literature: 'Berkner and Marshall (1965) pointed out that life could only have existed in the seas at depths below 10 m. Some life could also have existed in the soil in conditions where it was shielded from UV irradiation. Unavoidably, organisms must have been carried most closely to the surface by water currents and exposed from time to time to the potent mutagenic action of UV. Hence, we might expect a high mutation rate to have accelerated evolution at that time' (De Ley 1969, page 107). Berkner and Marshall (1965) and Fischer (1965) have suggested that lack of oxygen limited the development of life until the beginning of the Cambrian. Although it is true that in the absence of the atmospheric molecular oxygen no ozone layer would limit the ultraviolet light reaching the surface (Berkner and Marshall 1965), both the geological and biological evidence argue against central features of the Berkner–Marshall model (Schopf and Barghoorn 1969; Gutstadet and Schopf 1969). The transition to the oxidizing atmosphere probably occurred much before the Cambrian, most likely between 1 and 2 billion years ago (Cloud 1968); that the potential for oxygen-elimination was present at least 2.7 billion years ago is attested by algal stromatolites and 'oolites' (Ramsey 1963; Hoffman 1969) and microfossils (Schopf 1968; Schopf and Barghoorn 1969). Although an enormous range of sensitivity to oxygen is observable in microbial cells (see below), all cells, even the most primitive and smallest microbes, have nucleic acid repair mechanisms to UV damage (Hanawalt 1968), suggesting nucleic acid repair systems are extremely ancient. Distribution of oxygen sensitivities in microbes suggests analogous evolution in response to increasing levels of photosynthetically produced oxygen in many groups of bacteria and their relatives; there is no comparable evidence for selection pressure exerted by high UV flux except if it occurred in a population of universally ancestral microbes (before the transition to the oxidizing atmosphere). The universality of DNA repair mechanisms to excise pyrimidine dimers produced by UV suggests that if UV were ever a significant agent of selection it was in the most remote past. Ever-increasing amounts of data suggest a direct relationship between genetic recombination and at least one form of the repair system to ultraviolet induced damage in *Escherichia coli* (Witkin 1969). The possibility exists that UV nucleic acid repair mechanisms that originally evolved in bacteria in the anoxic terrestrial environment, were later modified and selected for in the origin of the various prokaryote recombination systems. This notion is consistent with its observation that certain mutations (*exr* and *lex*) showing reduced UV resistance coupled with reduced recombination ability do so by decreasing the efficiency of post-replication recombinational repair of DNA daughter strand gaps induced by pyrimidine dimers (Witkin 1969, page 493).

Since all nucleated, mitochondria-containing cells (e.g. eukaryotic cells of higher organisms) are aerobic and adapted to present conditions of atmospheric oxygen, there is no evidence that lack of oxygen prevented invasion of the land. It is possible that spore-forming anaerobic bacteria-like

assume that clearly recognizable fossil microbes had physiologies and lived in environments comparable to their modern counterparts. This assumption certainly is consistent with the comparison between algal stromatolitic remains in Northwest Canada and their modern tropical Australian counterparts (Hoffman 1968; McAlester 1968, page 14). Biological evidence is more consistent with the alternative view: the eukaryotic algae and green plants had an aerobic heterotrophic direct ancestor. These organisms only became photosynthetic after mitosis itself evolved and various protozoans symbiotically acquired blue-green algal-like structures that ultimately developed into chloroplasts (Ris and Plaut 1962; Sagan 1967; Margulis 1968). This alternative concept of the relation between blue-green algae and all other plants recognizes the depth of the evolutionary discontinuity between prokaryotes (blue-green algae and bacteria) and eukaryotes (nucleated algae, fungi, green plants, animals) (Stanier et al. 1963) (fig. 1).

All life has been dependent on 'green plant' photosynthesis since the UV (presumably the major source of energy for prebiotic organic syntheses) was

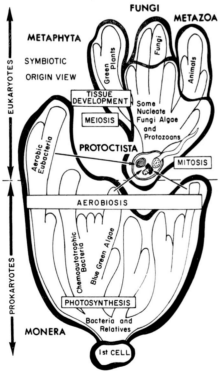

Fig. 1.

cut off and the atmosphere has become oxidizing. It is likely this event occurred at least as early as 1.2 billion years ago (Cloud 1965; Schopf and Barghoorn 1968) or perhaps even before 2 billion years ago (Davidson 1965). Yet there are no higher plant and animal fossils until about 0.6 billion years ago. The discrepancy between the evidence for ambient oxygen and the presence of many microfossil forms including blue-green algae close to 3 billion years ago, yet no green plants above algae until the Phanerozoic, in my opinion is entirely resolved by recognizing the depth of the evolutionary discontinuity between the prokaryotic blue-greens and the eukaryotic green algae. A substantial literature is available now relating the blue green algae and the photosynthetic bacteria (Echlin and Morris 1965). Consistent with both modern biology and Precambrian paleontology is the concept that the prokaryotic blue-greens produced the oxidizing atmosphere in response to which the numerous other prokaryotic aerobic microbes evolved. Not until the biological stage was set by the evolution of an aerobic (mitochondria-containing) nucleated eumitotic heterotroph 'higher' cell could the green algae and plants evolve (Sagan 1967; Margulis 1970a).

2.6. Phytane and pristane, isoprenoid hydrocarbons, are geochemical derivatives from chlorophyll. Chlorophyll comes from O_2 evolving photosynthetic organisms

Phytane and pristane, found to be present in sediments from 3 billion years (Schopf and Barghoorn 1967) are considered to be derived from the phytol chain on chlorophyll and therefore imply the early Precambrian potential for oxygen-eliminating photosynthesis (Echlin 1969). The isoprenoid alcohol derivative phytol on bacteriochlorophyll is indistinguishable from that on chlorophyll a (the chlorophyll of blue-green algae and plants). Thus phytane and pristane may be relics of a bacterial photosynthetic process. However, as Calvin himself notes, phytol itself may be found in living organisms (Calvin 1969, page 60) and phytyl-containing photosynthetic organisms (Calvin 1969, page 50) do not necessarily yield large relative quantities of phytane and pristane in mass spectrometric-gas chromatographic studies of their hydrocarbon fractions. Furthermore, both phytane and pristane have been detected in GC-MS studies of the hydrocarbons from non-photosynthetic (non phytyl-containing-microbes, e.g. *Micrococcus*: Calvin 1969, page 55).

Unusually high concentrations of pristane have also been found in the three copepods (*Calanus finmarchicus*, *C. glacialis*, and *C. hyperboreus*).

These crustaceans may be the primary source of pristane in liver oils of sharks and whales (Blumer et al. 1963). The recent observation that the cyclic isoprenoids farnesane and the isomers of phytane can be synthesized abiogenetically under plausible geochemical conditions (Munday et al. 1969) adds a new dimension of uncertainty to the whole question. It is therefore likely that the interpretation of the abundance of these branched hydrocarbons in ancient rocks and oil is far more complex than originally thought.

It seems to me that the use of any identified organic compound found in ancient sedimentary rocks as a marker (for example a hydrocarbon such as pristane, to identify fossil microbial populations, e.g., algae) depends on the establishment of at least the following criteria. Many recent studies (Eglinton 1970) have made an extremely important but as yet too uncritical contribution in this area.

(1) First, of course, it must be established that the compound is syngenetic with the rock. Since the migratory behavior of petroleum is well known, this is not trivial. The possibility of contamination from geologically younger rocks must be eliminated, as well as of course laboratory and other more modern contamination.

(2) The MS-GC profiles of hydrocarbons of living organisms under differing physiological conditions must be definitively established for comparison. Very little experimental work has been reported about the bulk of major naturally occurring hydrocarbons and other MS-GC detectable substances such as sterane (Burlingame et al. 1967), in most relevant microbial populations (Oro 1966). How do the distributions of, for example, alkanes change with death and conditions of sedimentation? To what extent is the persistence of detectable key compounds dependent on environmental variables such as radiation, temperature, a particular aquatic or terrestrial sedimentary environment, the amount of ambient oxygen and so forth? In a given ecological succession (including the microbial successions that must have characterized the dominant flora of the Precambrian) where does the 'ecological chain' end? Which species are most likely to be sedimented in large numbers to remain unchanged by decomposers? What happens to fatty acids, fats, carotenoids, carbohydrates, phospholipids and other extremely abundant, relatively small molecules that make up a significant percentage of the dry weight of most microbes? Amino acids are expected to be unstable over billions of years (Calvin 1969) therefore, the report that their presence is due to contemporaneous microfossils (2 billion years old) in ancient rocks can be challenged (Kvenvolden et al. 1969). Do major types of compounds react under mild conditions to form unsuspected derivatives?

Can we be sure Precambrian rocks were never heated to temperatures that destroy or alter these organics? All of these questions, and probably many others must be satisfactorily answered before the spectrographic identifications of certain detectable molecules can be interpreted to be molecular fossil keys to the histories of specific developed biological populations.

3. *Further alternative assumptions*

Although I am not as familiar with the natural groups (genera) of extant prokaryotes and their metabolic patterns as I someday hope to be, I believe that certain principles derived mainly from the advances of molecular biology can be used to order them phylogenetically. Admittedly, these depend on the recent published theoretical framework, the symbiotic theory of the origin of nucleated cells (eukaryotes) (Margulis 1970). The phylogeny presented here is based on the application of criteria presented in table 1 to known extant groups. Further assumptions include these (see table 1).

TABLE 1

Criteria for relating prokaryote microbes.*

Criterion	Techniques by which determined
Homology of DNA base pairs: both number and order	Direct DNA nucleotide sequence data (not yet available for complete genome of any organism) Agar-gel measurements of DNA–DNA and DNA–RNA homology (De Ley 1969) Genetic recombination (classical genetics, Gunsalus and Stanier 1964) (DNA base ratios on CsCl density gradients, DNA denaturation (melting point) data (Mandel 1969)
Homology of complete metabolic pathways and identity of the enzymes involved	Classical biochemistry (De Ley 1962)
Homology of individual cistrons	DNA–mRNA homologies DNA–DNA homologies (De Ley 1969)
Identity of genetic code	Same triplets (codons) determine amino acids, identity of transfer RNAs for each amino acid, identity of amino acid acetylating enzymes (Dayhoff 1969)
Ultrastructural morphology	Electron microscopy

* Roughly in order of importance (modified from Sagan 1967).

TABLE 1 *(continued)*

Criterion	Techniques by which determined
Morphology and life cycle	Light microscopy, classical cytology
Identity of molecular structure of single pigment or enzyme	Classical chemistry (Dayhoff 1969)
Individual identity of pigments, enzymes, etc.	Spectroscopy, classical biochemistry
Common phenotypic traits: ability to grow on same carbohydrate, production of same end-product, isolation from same type of enrichment medium, spore production, similar patterns of motility, etc.	Classical microbiology

(1) All extant cells are related to a common ancestor containing a replication mechanism based on reading a DNA nucleotide sequence by threes to form a colinear polypeptide (Dayhoff 1969). The most primitive ancestral cell to all extant life: (a) was heterotrophic, absorbing prebiotic organic matter from the primitive soup (Bernal 1968); (b) was fermentative; it used organic substances as electron acceptors in its metabolism (although this implies life in anoxic conditions, it is to be distinguished from 'anaerobic respiration' Stanier et al. (1970); (c) was minimally 0.1 μ radius. This roughly corresponds to a minimal number of independent biochemical functions of approximately 45. A rough estimate for the comparable 'number which measures functional complexity' in extant bacteria is closer to 1000 (*Haemophilus influenzae* and *Mycoplasma laidlawii*; Morowitz 1967); (d) contained those organic compounds known to be present in all extant free-living cells and in particular, the specific isomers of those compounds such as the L-series of amino acids and the D-series of pentoses, certain vitamins, etc. Compounds known to be ubiquitously distributed in cells (whether synthesized by them or ingested from the medium) are hypothesized to be necessary now and in the ancient past for replication itself (Margulis 1969).

(2) All biosynthetic pathways leading to major compounds or classes of compounds that are not universally distributed are considered secondary evolutionary advancement and their initial presence must be explicable in connection with the selective advantages conferred on a specific ancestral population of organisms in a specific environment at a specific point in time. Subsequent use of such compounds clearly can involve the same processes of radiative adaptation known for morphological innovations in higher organisms (Sagan 1967; Margulis 1970).

(3) There are no real 'missing links' in the sense of classes, phyla or higher taxa, no terrestrially unprecedented organisms must be hypothesized to have existed in order to explain evolutionary continuity between extant forms. In the absence of evidence to the contrary we shall assume a clearly distinguishable microbe, (such as the filamentous blue-green alga, *Archaeonema longicellularis*, Schopf and Barghoorn 1969, figs. 2–4), had the same basic pattern of metabolism a billion years ago as its modern descendants have now. Above we claimed that morphological conservatism does not imply lack of evolution. These are not contradictory. Mechanisms to preserve well-adapted organisms must themselves be subject to evolution. Certainly, mutations causing dramatic alterations in the measurable mutation rate are well known (Witkin 1969). It is reasonable to assume that blue-green algae, like the horse-shoe crab, was in the ancient past and is now optimally adapted to particular ecological conditions which have always prevailed, although their geographical positions have been shifted drastically (Hoffman 1968) (see McAlester 1968, page 14; Margulis 1970, ch. 5).

(4) Prokaryotes precede all eukaryotes and were well along into their diversification into genera when the production of significant quantities of free molecular oxygen required evolution to oxygen-tolerant and eventually oxygen-utilizing forms. This event, which accounts for the fact that many anaerobic prokaryotes have close relatives among aerobic forms (table 2), occurred before any eukaryotes (fungi, aminals, nucleated algae, green plants) evolved.

From the distribution of actual metabolic pathways in extant microbes my understanding of the literature, I have used the criteria in table 1 for placement on my phylogeny (fig. 1). I hope critics will be stimulated to correct my errors.

In principle, if we can develop a consistent phylogeny based on molecular biology and classical microbiology, we can make very explicit predictions for the nature of organic and inorganic depositions in sediment caused by microbes, at the time organisms died (Margulis 1969). A large input from comparative biochemistry and microbial ecology is required. The former field correlates the presence of specific organic compounds with closely related microbes (members of the same genus) and the latter provides us with the detailed knowledge of the types of deposition that occur in a given environment. The organic geochemists must be responsible for the determination of the effects of time and post-depositional geological events on the compounds originally sedimented by the organism. Scientists in these fields may soon cooperate toward the creation of a Precambrian paleontology

TABLE 2

Oxygen, intolerance, tolerance, utilization and indifference in natural groups of microbes
(prokaryotes).

Obligate anaerobes (after Prevot and Fredette 1966).

Genera	Comments
Eubacterial genera	
Neisseria (h)	
Veillonella (h)	
Methanococcus (h)	
Dialister (h)	
Ristella (h)	
Zuberella (h)	
Catenebacterium (h)	
Ramibacterium (h)	
Cillabacterium (h)	All species obligate anaerobes
Lachnospira (h)	
Paraplectrum (h)	
Desulfovibrio (h)	
Welchia (h)	
Terminosporus (h)	
Caduseus (h)	
Sporovibrio (h)	
Thiosarcina (p)	
Thiopedia (p)	
Chromatium (p)	
Chlorobium (p)	Some species obligate anaerobes
Rhodopseudomonas (p)	
Rhodospirillum (p)	
Mycobacterial genera	
Micromonospora (h)	
Corynebacteria (h)	
Actinobacteria (h)	
Bifidobacteria (h)	
Spirochaete genera	All species obligate anaerobes
Treponema (h)	
Borrelia (h)	
Microdentum (h)	

Genera containing both anaerobic and aerobic species

Blue-green algal genera	Anaerobic when photosynthesizing: many respire aerobically in the dark.
Eubacterial genera	
Diplococcus (h)	
Streptococcus (h)	

TABLE 2 *(continued)*

Genera	Comments
Gaffkya (Tetracoccus) (h)	
Staphlococcus (h)	
Sarcina (h)	
Micrococcus (h)	
Pasteaurellaceae (h)	
Actinobacillus (h)	
Plectridium (h)	
Acuformis (h)	
Escherichia (h)	
Aerobacter (h)	
Eubacterium (h)	
Vibrio (h)	
Endosporus (h)	
Clostridium (h)	
Chlorochromatium (p)	Oxidizes H_2S
Rhodospirillum (p)	
Rhodopseudomonas (p)	Oxidizes small organic molecules, facultative photo-
Rhodomicrobium (p)	autotrophs
Thiobacillus (c)	Oxidizes H_2S

	Obligate aerobes
Eubacterial genera	
Klebsiella (h)	
Bacteridium (h)	
Bacillus (h)	
Azotobacter (c)	Fixes N_2
Rhizobium (c)	Eubacteria fixes N_2, symbiotic with legumes
Mycobacterial genera	
Nocardia (h)	
Streptomyces (h)	
Mycobacterium (h)	
Sphaerotilus (c)	Iron hydroxide of organic sheaths, filamentous eubacteria
Others	
Clonothrix (c)	Organic sheath encrusted with iron or manganese
Leptothrix (c)	Sheaths encrusted with iron oxide, branching filaments
Crenothrix (c)	Visible sheath, thickly encrusted with iron at base
Caulobacter (c)	Asymmetric cells surrounded by ferric hydroxide gum
Spirillum (h)	Aerotactic but microaerophilic, large
Sapiospira (h)	Gyrating, marine, conspicuous carotenoids
Beggiatoa (c)	Filamentous, contain sulfur granules blue green alga analogue

TABLE 2 *(continued)*

Thiotrix (c)	Filaments attached by thin base surrounded by sheath
Thiogloea (c)	Blue-green alga analogue gelatinous colonies
Achromatium (c)	Ellipsoidal or cells with granule of $CaCO_3$
	Microaerophilic
	Oxidizes H_2S which it requires
Sporocytophaga (h)	Forms cysts from isolated vegetative cells
Cytophaga (h)	Breaks down polysaccharides
Some blue-green algae (p)	Use hydrogen from water to reduce CO_2, eliminating O_2
Myxobacterial genera	
Myxococcus (h)	Forms cysts and fruiting bodies
Chondromyces (h)	Forms cysts and fruiting bodies
Chondrococcus (h)	Forms fruiting bodies, pathogenic on fish

Key: h = heterotroph, p = photoautotroph or photoheterotroph, c = chemoautotroph.

concerned with the Age of Prokaryotes as extensive as our present Phanerozoic eukaryote paleontology.

The inclusion of fig. 2 is justified only on the principle that continual modifications of some phylogeny is better than none at all. The only other recent attempt (known to this author) to place a large number of prokaryote genera on a single phylogeny is that of De Ley (1968, fig. 12, page 147). Not drawn against time in its present form, De Ley's phylogeny is not useful for making predictions about Precambrian sediment.* De Ley's bacterial phylogeny is based upon total estimated numbers of cistrons (nucleotide sequences; genes as units of function) shared in common and reinforces the concept that the bacteria represent an extraordinarily diverse group of organisms that branched from common ancestors in the remote past.

For example, estimations of the prokaryote genome varies from less than 1000 (mycoplasms) to about 6000 cistrons per organism. Many bacteria share fewer than about 250 in common (De Ley, interpreted from fig. 8, page 142). All bacterial genera included in his review seem to differ in more than half their total cistrons (with the exception of *Salmonella*, *Shigella* and *Escherichia*, which differ from each other in about 2/5).

---

* Note added in proof: J. B. Hall (J. Theoret. Biol. 30, 429, 1971) has just published a scheme for prokaryote evolution based on the development of aerobic respiration remarkably consistent with this paper.

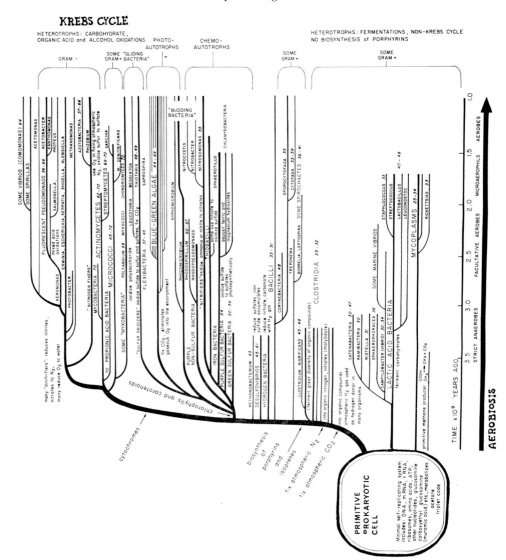

Fig. 2.

The criteria summarized in table 1 lead to tentative conclusions concerning the order of acquisition of metabolic virtuosities (table 4). These are used to order prokaryote organisms in time.

TABLE 3

Synthesis of isoprenoid derivatives microbes: showing relationship with aerobiosis.*

Organisms	Can synthesize CoQ (ubiquione)	Can cynthesize vitamin K	Response to oxygen
All obligate anaerobes investigated (e.g., *Clostridium*)	—	—	Can not tolerate
Rhodospirillum	+	—	
Pseudomonas	+	—	Facultative aerobes
Hydrogenomonas	+	—	
Escherichia coli	+	—	
Bacillus mesentericus	—	+	
Streptomycetes	—	+	Obligate aerobes
Mycobacterium	—	+	

* Compiled on the bases of data from (Miller 1961).

Several of these points have recently been discussed (Margulis 1969; Klein and Cronquist 1967); emphasis here will be on the outline of putative primitive-to-advanced multicistronically-determined traits in early cells and corrolaries based on these principles upon which there seems to be some agreement in the literature.

Even recently, several authors considering the origin of cells, and especially photosynthetic ones, have tried to explain simultaneously the origin of entire plant cells from complex solutions or coacervates of organic compounds. However, it seems more likely to biologists that the entire problem can be naturally analyzed into the following more limited problems at least, in approximate sequence from early to late:

(1) The origin of the ancestral population of anaerobic heterotrophic, triplet nucleic acid-protein coded prokaryotic cells from non-living precursors (Crick 1968; Orgel 1968; Dayhoff 1969).

(2) The evolution of mechanisms to repair nucleic acid damage induced by ultraviolet radiation.

(3) The evolution of pathways to reduce carbon dioxide (i.e. the dark reactions) and fix atmospheric nitrogen.

(4) The evolution in prokaryote microbes of the biosynthetic pathways

producing isoprenoid and tetrapyrrole derivatives. Development of diverse heterotrophic fermenting and anaerobic respiring organisms in the anoxigic environment.

(5) The evolution of anaerobic photosynthesis in prokaryotes.

(6) The evolution of 'green plant' photosynthesis, i.e., the use of water as hydrogen donor to reduce CO_2 in photosynthetic prokaryotes (ancestors to blue-green algae). This step resulted in the elimination of gaseous oxygen: $CO_2 + H_2O \rightarrow CH_2O + O_2\uparrow$.

(7) The evolution of aerobiosis in response to increasing partial pressures of oxygen in diverse and distantly related prokaryotes: heterotrophs, photoautotrophs, chemoautotrophs and so forth. (See footnote p. 357.)

(8) The evolution of metabolic reactions which require ambient molecular oxygen.

Since the much more complex eukaryote organisms probably appeared after the atmosphere was fully oxidizing (Sagan 1967; Margulis 1970), perhaps as long as two billion years ago (Licari and Cloud 1968), they will be omitted from any further consideration in this article. We can therefore proceed to discuss points 1–5 in connection with the anaerobic lower Precambrian, assuming 6 and 7 relate to the times at which there is evidence for a net accumulation of molecular oxygen in the environment.

From the recently developed principles of molecular biology (see Watson 1970 for introduction to the field) Morowitz has considered what, theoretically, must constitute the minimal self-replicating entity, the minimal cell. In a closely argued, highly recommended paper (Morowitz 1967), he concludes that the minimal radius predicted from general principles for the minimal cell is approximately 0.1 micron (1000 Å), that the minimal size of the cell genome (the entire set of genetic material required for all physiological processes leading to independent cell reproduction) is from 2.6 to 4.4 $\times 10^7$ daltons. These numbers are about a factor of ten less than the known parameters for the smallest living cells (*Dialister pneumocintes*, *Achromobacter parvalus*, several Veillonellas, some mycoplasms), probably because some functions required for replication were not accounted for in the calculation (Morowitz 1967).

As emphasized again by Morowitz, the proportion of the total known organic compounds actually found in living cells is astoundingly small. This fact may be useful in distinguishing primordial organic matter from that produced by living cells: the second class of substances presumably is far more restricted. It is reasonable to assume, as we have here, that all organic compounds found ubiquitously in all extant living cells are necessary for cellular replication and hence were present in the ancestral cell that gave rise to all terrestrial descendants (Margulis 1969, fig. 1). The fact that porphyrins, steroids, polyunsaturated fatty acids, polyterpenoids and flavonoids, as widespread as they are in nature, are not requirements for replication in all cells implies they are later evolutionary developments. The same argument applies to the various luciferan-luciferases, tetracyclines, macrolides, lichenic acids and other chemically related classes of compounds that are restricted in their distribution to certain limited groups of organisms.

Following the line of reasoning of Horowitz (1945) it has been assumed that those compounds such as ATP and amino acids required for reproduction were first available in the primitive environment, having been produced abiotically. Although it is not clear what organisms today are most unchanged from the primitive ancestral heterotroph it may be assumed that evolution proceeded from the use of such universal originally fermentable abiotically produced metabolites (lactate, malate, acetate, pyruvate, glycine etc.) toward the use of substances that could be converted into these. Because glycolysis proceeds by very different steps in unrelated bacteria (De Ley 1962) the progressive ability to utilize carbohydrates of increasing complexity for anaerobic fermentations must have evolved in several lines of microbes. Presumably the ability to give off H_2 and CO_2 as products of heterotrophic fermentations is very primitive. The universally distributed glyoxylate pathway (Klein and Cronquist 1967) requires acetyl coenzyme A (Mahler and Cordes 1966), clearly an extremely early product of cellular metabolism. CoA itself is synthesized from valine, aspartate and cysteine with pantathenic acid as an intermediate. These are all amino acids already required for functioning of the genetic code in cell reproduction. A geochemically stable derivative of acetyl CoA might provide a marker for acquisition of metabolic pathways in anaerobic fermenters.

Early in the evolution of anaerobic microbes must have come the carbon dark-cycle enzymes of the pathways to reduce carbon dioxide. This might have occurred in clostridial-like ancestral cells (Ljungdahl and Wood 1969) and may have been paralleled by the development of the biosynthesis of biotin. Closely related compounds to this universally required vitamin are known to be coenzymes in carboxylations. Precursors to biotin include only valine, aspartate, cysteine and require ATP and magnesium ions (Mahler and Cordes 1966).

Klein and Cronquist (1967) give excellent reasons for believing clostridia are more primitive than desulfovibrios. Sulfate reduction, carbon dioxide fixation and nitrogen fixation are processes known to be present with variations in clostridia; all three fixations are likely to have evolved in ancestors to this group. The clostridia make no hemes and contain no porphyrin derivatives. *C. nigrificans** does make the natural corrinoid related to vitamin B_{12} (Burnham 1969) and acetate synthesis from CO_2 involves cobalt methyl corrinoids (Ljundahl and Wood 1969). It is possible that this cobalt-chelated compound related to porphyrins in its synthesis functioned originally as it does now in methionine synthesis (Greenberg 1969). This is consistent with Crick's (1968) suggestion that 'certainly tryptophane and methionine look like later additions to the set of amino acids coded by triplet nucleotide sequences'. The non-heme iron protein, ferredoxin, always found in autotrophic organisms, probably evolved in the clostridia in connection with the evolution of pathways involving fixation of atmospheric gases: CO_2, H_2 and N_2. Iron chelated tetrapyrrole synthesis most likely evolved in a clostridia-like population ancestral to the desulfovibrios: these sulfate reducing, CO_2-fixing, N_2-fixing microbes are obligate anaerobes and contain only one heme: a c-type cytochrome (Klein and Cronquist 1967). The amino acid sequence in the heme-binding region on the protein (desulfovibrio cytochrome C_3) is related to this protein in other bacteria and higher organisms (Dayhoff 1969).

Desulfovibrios, essential today in the reduction of sulfates to sulfides in the sulfur cycle, may be a key group of anaerobic microbes that left fossil evidence. They need organic compounds such as sodium lactate and malate as hydrogen donors for the reduction of CO_2 even though they produce hydrogen sulfide. (Other anaerobes, such as photosynthetic bacteria, use H_2S to reduce CO_2.) Desulfovibrios are found in the sea, soil, sedimentary rocks and are ubiquitous in muds. They tolerate a pH range from 5.8–9.4 (Ferguson Wood 1958). Since sulfate reduction of H_2S leads to the formation of a variety of ferrous sulfides, called 'black FeS' by sedimentologists. This set of reactions is of extreme importance in the phosphate cycle today because much of the iron in sea water is bound as ferric phosphate in fecal pellets and particulate matter. Although ferric phosphate is insoluble, it is less so than the iron sulfides. When the iron phosphates are sulfidized because of the activity of desulfovibrios, phosphate ions are released from ferric phosphate and phosphoric acid goes into solution (Ferguson Wood 1958). It is possible that desulfovibrios have served to reduce sulfates and

---

* Some authors e.g. Campbell and Postgate (Bacteriological Reviews Sept. 1965) suggest removal of *C. nigrificans* from Clostridia and placing them with the spore-forming desulfovibrios.

release phosphate (but of course not from fecal pellets) since Precambrian times, perhaps (if post-depositional changes have not entirely obscured the picture) sulfides can be correlated with disappearing phosphates, (D. T. Rickard 1970 personal communication).

Isoprenoids, nearly ubiquitous components of cells, may have first been synthesized in response to selection pressures to protect replicative machinery from oxidation. They are synthesized metabolically from compounds that ultimately arise from acetate. Both carboxy and methyl labeled carbon atoms of acetyl coenzyme A are incorporated into an intermediate compound: mevalonic acid (Roberts and Caserio 1966). Both alcohol groups of mevalonic acid are phosphorylated, a decarboxylation follows and the universal 'biological isoprene unit': (isopentenyl pyrophosphate) the starting point for the many different isoprenoids is produced. Isoprene itself is never an intermediate in the biosynthesis of isoprene derivates, and was not thought to exist free in nature until recently when free isoprene was detected in gaseous emissions from leaves in strong light (News report *Bioscience*, vol. 19 page 928, October 1969). In anaerobic and other prokaryotes, the major end-products of isoprenoid biosyntheses are: coenzyme Q (ubiquinone), vitamin K, several types of saturated and mono-unsaturated fatty acids and other chemically related compounds. Are they made by any non-heme synthesizing primitive prokaryotes? They function now widely in oxidation-reduction reactions. The relation of their patterns of biosynthesis with aerobiosis (table 3) is consistent with the hypothesis that their primitive role was as antioxidants (Krinsky 1966). In photosynthetic prokaryotes, in addition to phytol isoprenoid on chlorophyll, are found a diverse assortment of orange, yellow and red derivatives: the carotenoids (Krinsky 1966). It seems reasonable that the evolution of the carotenoids paralleled the evolution of photoautotrophy in some population of CO_2-fixing, N_2-fixing ancestors to the anaerobic photosynthetic bacteria. It follows that one would expect evidence for microbial activity in Precambrian sediments antedating any evidence for cellular photosynthesis itself (Margulis 1969).

The experimental evidence that exogenous heme can protect against oxidation in bacteria unable even to produce it has been invoked in support of the idea that the original role for biological porphyrins was as antioxidants (Gaffron 1960).

Possible mutational steps from cytochrome (Fe^{++} chelated) synthesizing desulfovibrio-like organism to chlorophyll-synthesizing (Mg^{++} chelated) bacterial photosynthesizer have been reviewed and discussed lately (Klein and Cronquist 1967). Although the details are obscure, it is generally agreed

TABLE 4

Tentative order of acquisition of some major metabolic pathways in prokaryote cells.

Synthetic pathway	New synthetic capability (class of compounds: end result of metabolic pathway)	Probable environmental selection pressure	Possible ancestral population of microbes in which synthetic abilities evolved
1. Nucleic acid replication, etc. anaerobic	Ubiquitous compounds	Efficient reproduction	'Ancestral prokaryote'
2. Glycolysis to acetate	Enzymes of Embden–Meyerhof pathway	Depletion of available metabolites i.e., Horowitz hypothesis	Lactic acid bacteria
3. Other carbohydrate fermentations	Aldopentose breakdown, etc.	Depletion of available metabolites i.e., Horowitz hypothesis	Lactic acid bacteria
4. Fixation of atmospheric CO_2	'Dark cycle' enzyme's of CO_2 fixation	Depletion of available metabolites i.e., Horowitz hypothesis	Clostridia
5. Fixation of N_2	Molybdoferredoxin azoferredoxin	Depletion of available metabolites i.e., Horowitz hypothesis	Clostridia
6. Carotenoids	Isoprene derivatives	Protection of the genetic material from photo-oxidations	Clostridia
7. Porphyrin and corrinoid derivatives:	Enzymes involved in biosyntheses of porphyrins	Protection against photo-oxidation	Clostridia
8. Porphyrin-protein derivatives: cytochrome c and other cytochromes	Sulfate reduction to H_2S, anaerobic respiration, enzymes such as nitrate reductases	More efficient utilization of oxidizable substrates	Desulfovibrios
9. Porphyrin derivatives: chlorobium, bacterial chlorophylls, isoprenoid chain phytyl on chlorophyll	Chlorophyll synthesis, more isoprenoid derivatives	Photoproduction of ATP	Gram negative rods: purple and green sulfur bacteria, purple non-sulfur bacteria

TABLE 4 *(continued)*

Synthetic pathway	New synthetic capability (class of compounds: end result of metabolic pathway)	Probable environmental selection pressure	Possible ancestral population of microbes in which synthetic abilities evolved
10. Porphyrin derivatives: chlorophyll *a*	Enzymes of chlorophyll *a* synthetic pathway, utilization of H_2O as H donor in photosynthesis	More abundant and reliable source of H donor for photosynthesis	Bacterial photosynthesizers; blue-green algae
11. Porphyrin–protein derivatives: cytochrome oxidase	Carbohydrate metabolism via Entner–Douderoff pathway, enzymes of all Krebs cycle intermediates and enzymes of cytochrome oxidase synthetic pathway, complete electron transport chain, enzymes involved in direct carbohydrate oxidations	More efficient utilization of oxidizable substrates, complete oxidation of carbohydrates	Pseudomonads, micrococci, bacilli, mycobacteria, enterobacteria

that ancestors to photosynthetic bacteria and blue-green algae were from among a population of N_2 fixing anaerobic gram negative bacteria that evolved chlorophyll and carotenoid synthetic abilities (Echlin and Morris 1965). A critical set of mutational steps connecting such a photosynthetic bacterium to chlorophyll *a* containing blue-green algae has been suggested (Olson 1970). The devastating effect of the gradual elimination of gaseous oxygen into the atmosphere on prokaryote microbes has also been pointed out (De Ley 1969; Echlin 1969). Aerobic chemoautotrophs, nitrifying and denitrifying bacteria and many other microaerophils, facultative aerobes and obligate aerobes must be relatively recent (De Ley 1969).* Pathogenic, specific symbiotic prokaryote microbes presumably evolved after their animal and plant host themselves evolved (De Ley 1969). The phylogenetic tree (fig. 2) and table 4 are inserted here to illustrate and summarize these plausible

* The discovery of the absence of at least large quantities of atmospheric nitrogen in the atmospheres of Mars and Venus suggests that perhaps terrestrial N_2 is of microbial origin. Nitrogen is produced in the metabolism of many 'denitrifying' bacteria via anaerobic respiration. Nitrogen fixation is limited in the biosphere to a group of specialized microbes and plant-microbial symbioses. Is it possible that atmospheric N_2 has been accumulating steadily since the Precambrian origin of the denitrifiers from cytochrome-synthesizing ancestors?

but undoubtedly naive concepts. Their refinement awaits full scale cooperation between scientists, many of whom are still unaware that their disciplines
are related: bacterial taxonomists and ecologists, and Precambrian sedimentary geologists and geochemists.

References

AINSWORTH, G. C. and P. H. A. SNEATH, eds., 1963, Microbial Classification. (Cambridge
University Press, London).

BARGHOORN, E. S., 1970, In: Proceedings of the First Interdisciplinary Conference on the
Origins of Life (May, 1967), ed. L. Margulis. (Gordon and Breach, New York) (oral
communication).

BERKNER, L. V. and L. MARSHALL, 1965, J. Atm. Sci. 22, 225.

BERNAL, J. D., 1968, Origin of Life. (World Publishing Co., Cleveland).

BLUMER, M., M. M. MULLIN and W. THOMAS, 1963, Science 140, 974.

BURLINGAME, A. L., P. HAUG, T. BELSKY and M. CALVIN, 1965, Proc. Natl. Acad. Sci. 54,
1406.

BURNHAM, B. F., 1969, Metabolism of porphyrins and corrinoids in metabolic pathways;
In: Amino acids and Tetraphyrroles, Vol. 3, ed. D. M. Greenberg, 403.

CALVIN, M., 1969, Chemical Evolution. (Oxford University Press, New York and Oxford)
103.

CALVIN, M., 1962, Evolution of photosynthetic mechanisms. Perspectives in biology and
medicine, Vol. 5, 161.

CLOUD, P. E., Jr., 1968, Science 16, 729.

CLOUD, P. E., Jr., 1965, Science 148, 27-35.

CLOUD, P. E., Jr., Science 148, 27–35.

COX, E. C. and C. YANOFSKY, 1969, J. Bacteriol. 100, 390–397.

CRICK, F. H. C., 1968, J. Molecular Biology 38, 367–379.

DAVIDSON, C. F., 1965, Proc. Nat. Acad. Sci. 53, 1194.

DAYHOFF, M. O., 1969, Protein Sequence and Structure. National Biomedical Research
Foundation. (Silver Springs, Md.), Vol. 4, Chs. 11, 13.

DE LEY, J., 1968, Molecular biology and bacterial phylogeny; In: Evolutionary Biology,
eds. Dobzhansky, Hecht and Steere. (Appleton-Century-Crofts, New York), Vol. 2,
104–154.

ECHLIN, P., 1969, New Scientist, May 8, 286.

ECHLIN, P. and I. MORRIS, 1965, Biol. Reviews 40, 143.

EDELMAN, M., D. SWINTON, J. A. SCHIFF, H. T. EPSTEIN and B. ZELDIN, 1967, Bacteriological Reviews 31, 315.

EGLINTON, G. and M. T. J. MURPHY, eds., 1964, Organic Geochemistry. Methods and
Results (Springer, Berlin–Heidelberg–New York)

EHLERS, E. G., D. V. STILES and J. D. BIRLE, 1965, Science 148, 1719.

FERGUSON WOOD, E. J., 1967, Microbiology of Oceans and Estuaries. Elsevier Oceano-
graphy Series (Elsevier Publishing Co., Amsterdam, London, New York), Vol. 3, 261.

FERGUSON WOODS, E. J., 1958, Bacteriological Reviews 22, 1–19.

GAFFRON, H., 1960, The origin of life in evolution after Darwin. Evolution of Life (University of Chicago Press), Vol. I, 78.

GLAESSNER, M. F., 1968, Biological events and the Precambrian time scale. Canad. J. Earth Sciences 5, 585.

GREENBERG, D. M., 1969, Biosynthesis of amino acids and related compounds in metabolic pathways; In: Amino acids and Tetrapyrroles, ed., D. M. Greenberg, Vol. 3, 238.

GUNSALUS, I. and R. Y. STANIER, 1964, The Bacteria, Vol. 5. (Heredity Academic Press, New York).

GUTSTADET, A. M. and J. W. SCHOPF, 1969, Nature 223, 165.

HANWALT, P. C., D. E. PETTIJOHN. E. C. PAULING, C. F. BRUNK, D. W. SMITH, L. C. KANNER and J. L. COUCH, 1968, Repair replication of DNA *in vivo*. Cold Spring Harbor Symposium on Quantitative Biology, 1968. Vol. 3, 187.

HOFFMAN, P., 1968, Oral communication. Meeting on the Environment of the Primitive Earth cited in: R. Siever, 1968, Science 164, 711.

HOROWITZ, N., 1970, In: L. Margulis, ed., Origins of Life. Proceedings of the First Interdisciplinary Communication Program, May 1967. (Gordon and Breach, New York).

HOROWITZ, N. H. and S. L. MILLER 1962 Current Theories on the Origin of Life. Fortschr. Chem. Organ. Naturstoffe 20 423.

KLEIN R. M. and A. CONQUIST 1967 Quart. Rev. Biol. 42 105.

KRINSKY, N. I., 1966, The role of carotenoid pigments as protective agents against photosensitized oxidations in chloroplasts; In: Biochemistry of Chloroplasts, ed. T. W. Goodwin. (Academic Press, New York) Vol. 1, 423.

KVENVOLDEN. K. A. E. PETERSON and G. E. POLLOCK 1969 Nature 221, 141.

LA BERGE G. L., 1967, Geol. Soc. Am. Bull. 78, 331.

LJUNGDAHL L. G. and H. G. WOOD 1969, Ann. Rev. Microbiol. 23, 515.

MAHLER, H. R. and E. H. CORDES, Biological Chemistry 1966, Harper & Row, New York.

MANDEL, M., 1969, Ann. Rev. Microbiol. 23, 239.

MCALESTER, A. L., 1968, The History of Life. (Prentice-Hall, Englewood Cliffs, New Jersey.)

MARGULIS, L., 1970a, Origin of Eukaryotic Cells. (Yale University Press, New Haven, Conn.)

MARGULIS, L., ed., 1970b, Proceedings of the First Interdisciplinary Communications Program on the Origins of Life, New York Academy of Sciences. (Gordon and Breach, New York).

MARGULIS, L., 1969, J. Geol. 77, 606.

MARGULIS, L., 1968, Science 161, 1020.

MARGULIS, L., 1971, Evolution 25, 242.

MILLER, M. W., 1961, The Pfizer Handbook of Microbial Metabolites. (McGraw-Hill, New York) 722 pp.

MOROWITZ, H. J., 1967, Biological self-replication systems. Progress in Theoretical Biology. (Academic Press, New York) 1, 35.

MUNDAY, C., K. PERING and C. PONNAMPERUMA, 1969, Nature 223, 867.

OLSON, J. M., 1970, Science 168, 438.

OPARIN, A., 1969, Biogenesis and Early Development of Life. (Academic Press, New York).

ORGEL, L. E., 1968, J. Mol. Biol. 38, 381.

ORO, J., J. L. LASETER and D. WEBER, 1966, Science 154, 399.

ORO, J. and E. GELP, 1968, Science 161, 700.

PONNAMPERUMA, C. and N. G. GABEL, 1968, Space Life Sciences 1, 64.

RAMSEY, J. G., 1963, Trans. Geo. Soc. of South Africa. 66, 353.

RICH, A., 1962, On the problems of evolution and biochemical information transfer; In: Horizons in Biochemistry, eds. Kasha and Pullman, pp. 103–106.

RIS, H. and W. PLAUT, 1962, J. Cell. Biol. 13, 383.

ROBERTS, J. D. and M. C. CASERIO, 1965, Basic principles of organic chemistry. (W. A. Benjamin Co., New York and Amsterdam).

SAGAN, L. (Margulis), 1967, J. Theoret. Biol. 14, 225.

SCHOPF, J. W. and E. S. BARGHOORN, 1969, J. Paleontology 43, 11.

SCHOPF, J. W., 1968, J. Paleontology 42, 651.

SCHOPF, J. W., E. S. BARGHOORN, M. D. NASER and R. O. GORDON, 1965, Science 149, 1365.

SIMPSON, G. G., 1953, Major features of Evolution. (Columbia University Press, New York) p. 87.

STANIER, R. Y., 1964, Toward a definition of the bacteria; In: The Bacteria, eds. I. C. Gunsalus and R. Y. Stanier. (Academic Press, New York), Vol. 5, 445.

STANIER, R. Y., M. DOUDEROFF and F. ADELBERG, 1970, The Microbial World. (Prentice Hall, Englewood Cliffs, New Jersey). 3rd edition.

WATSON, J. D., 1970, Molecular Biology of the Gene, 2nd ed. W. A. Benjamin, Inc., New York.

WHITTAKER, R. H., 1969, Science 163, 150.

PART II

Life beyond the Earth

C. Ponnamperuma (ed.), Exobiology. © North-Holland Publishing Company

CHAPTER 11

Planetary atmospheres

S. I. RASOOL

1. Introduction

'We living things are a late outgrowth of the metabolism of our Galaxy. The carbon that enters so
importantly into our composition was cooked in the
remote past in a dying star. From it at lower temperatures nitrogen and oxygen were formed. These,
our indispensable elements, were spewed out into
space in the exhalations of red giants and such stellar
catastrophes as supernovae, there to be mixed with
hydrogen to form eventually the substance of the sun
and planets, and ourselves. The waters of ancient seas
set the pattern of ions in our blood. The ancient
atmospheres molded our metabolism.'

George Wald

The present composition of the earth's atmosphere is 78% nitrogen, 21%
oxygen, and 1% argon, with traces of carbon dioxide, water vapor, and
ozone. The atmospheres of Mars and Venus, on the other hand, are predominantly composed of carbon dioxide, while those of Jupiter and Saturn
contain mainly hydrogen, helium with small amounts of methane, and
ammonia. Such a wide variety in the composition of the atmospheres of the
planets appears most intriguing when one considers that all nine planets were
probably formed at the same time and out of the same chemically homogeneous mixture of gas and dust; that is, the primitive solar nebula.

The most likely explanation for this diversity in composition seems to be
that the planetary atmospheres have undergone important evolutionary
changes during their long history of about 4.5 billion years. If all planets

acquired a substantial amount of atmosphere at the time of their accretion, these atmospheres then went through profound transformations as a function of time because of the escape of lighter gas into space, continued replenishment of the atmosphere by outgassing from the interior and by chemical reactions of extreme complexity between the atmospheric gases and crustal rocks. The degree to which each of these processes has influenced the evolutionary history of a planet depends essentially on its size, mass, internal structure and the distance from the sun.

2. Early history of the earth's atmosphere

The first, and probably the most important, mechanism by which a planet acquires an atmosphere is by capturing gaseous atoms and molecules into its gravitational field at the time of its accretion. The gross composition and approximate amount of such an atmosphere can be estimated with reasonable confidence because the planet and its atmosphere should contain elements in the same relative abundances as they were present in the primitive solar nebula out of which the planets accumulated. The elemental composition of the primitive sun has been estimated by Cameron (fig. 1) with due consideration of the fact that the composition in the outer regions of a contracting proto star is different from that in its interior. From this table it is evident that hydrogen is the most abundant element followed by helium.

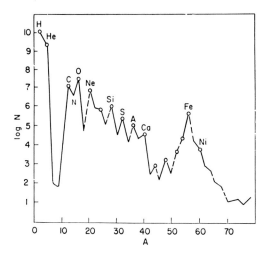

Fig. 1. Abundances of elements in the primitive solar nebula (after Cameron 1968).

Carbon, nitrogen, oxygen and neon, though each one of them about 1000 times less abundant than hydrogen, comprise the second most important group of elements in the solar nebula. Elements like Si, Mg and Fe, which today make up most of the solid earth, were present only in minute amounts ($<1\%$) in the primordial mixture from which the planets accumulated.

If the temperature in the primitive solar nebula, at planetary distances, were between 100–300°K, as suggested by Urey, a thermodynamic calculation would indicate the following initial composition of the gaseous envelope which the planet will acquire at the time of its formation (table 1).

TABLE 1

Gas	% by weight
H_2	63.5
He	34.9
H_2O	0.6
Ne	0.34
CH_4	0.26
NH_3	0.11
A^{36}	0.15

However, the total extent of this gaseous envelope will be extremely massive because from fig. 1 it is clear that H_2 is $\sim$ 400 times more abundant (by weight) than the group of elements Mg, Si and Fe which, along with oxygen, constitute the entire solid earth. In other words, by this process of acquisition of an atmosphere, the proto earth and also Mars and Venus would have to be 300–400 times more massive than their present values and essentially composed of gaseous H_2 and He. This is exactly the situation at the present time on Jupiter and Saturn. As deduced from spectroscopic observation, the composition is precisely the same as that of the primitive solar nebula deduced in table 2. On the other hand the composition of the earth is vastly different. Table 3 shows that, relative to the primitive solar nebula, earth is deficient by several orders of magnitude, not only in hydrogen and helium, which being the lightest gas can be assumed to have escaped from the gravitational field of the earth, but also in carbon, nitrogen and cosmically abundant rare gases (Ne, A^{36}, Kr and Xe) by as much as factor of 10^6! A comparison of the deficiency factors of these elements with those of non-volatiles, such as Na, Mg, Al, and Si, strongly suggests that the earth 'lost' only those elements which are 'volatile' at a few hundred degree temperature. It therefore appears that in the very early history earth became completely devoid of its gaseous

TABLE 2

(Atoms/10,000 atoms Si).

	Whole earth* (a)	Solar system** (b)	Deficiency factor log (b/a)
H	250	2.6×10^8	6.0
He	3.5×10^{-7}	2.1×10^7	13.8
C	14	135,000	4.0
N	0.21	24,000	5.1
O	35,000	236,000	0.8
Ne	1.2×10^{-6}	23,000	10.3
Na	460	632	~0
Mg	8,900	10,500	~0
Al	940	851	~0
Si	10,000	10,000	0
A^{36}	5.9×10^{-4}	2,280	6.6
Kr	6×10^{-8}	0.69	7.1
Xe	5×10^{-9}	0.07	7.1

* After Rubey, Ringwood and Mason.
** After Cameron.

envelope. How then did our planet acquire the present atmosphere and oceans?

Geochemists have presented convincing evidence that both the atmosphere and oceans of the earth developed slowly during the geological history through a number of well-defined processes (Mason 1966).

Additions during geological time

(1) Gases released in volcanic emanation and the crystallization of magmas (mainly H_2O and CO_2 with small amounts of N_2 and traces of HCl, HF, H_2S and SO_2).
(2) Oxygen produced by photochemical dissociation of water vapor.
(3) Oxygen produced by photosynthesis.
(4) Helium from the radioactive breakdown of U and Th.
(5) Argon from the radioactive breakdown of K^{40}.
(6) Additions by solar winds.

Losses during geological time

(1) Loss of H_2 and He by gravitational escape.
(2) Loss of CO_2 by the formation of coal, petroleum and organic burial.

(3) Loss of CO_2 by the formation of Ca and Mg carbonates.

(4) Loss of oxygen by oxidation of H_2 to H_2O and ferrous and ferric iron, sulphur to sulphates, etc.

(5) Loss of nitrogen by formation of oxides in air and nitrifying bacteria in the soil.

The most important process of input of gases into the atmosphere and ocean is the volcanic activity, which has probably been effective during the entire period of 4.5 billion years and has been the main source of water, carbon dioxide and nitrogen. Oxygen was produced later by photosynthesis and the traces of Ar which are today present in the atmosphere have been released slowly as a radioactive product during the geological history. Simultaneous to these inputs of the atmosphere there have been considerable losses of H_2 and He to the space by gravitational escape, of CO_2 to the crust to form carbonates and of O_2 to form oxides. Though the significance of these evolutionary processes have been fairly well established several important questions have yet to be answered. What has been the history of the volatiles now present at the surface of the earth? Has the C, N, O and H always been in the form of CO_2, N_2, and H_2O and H_2, or did carbon and nitrogen combine with hydrogen early in the earth's history to form CH_4 and NH_3? Under what atmospheric conditions did life originate on earth and how did the appearance of life change the atmosphere? These are some of the basic questions which must be answered in order to paint a coherent picture of the evolution of the earth's atmosphere.

Opinions on these questions are many and varied; sometimes they are almost diametrically opposed. The Oparin-Urey theory of the origin of life on the earth, supported by the laboratory experiments of Miller and more recently of Ponnamperuma and others, suggests a primitive atmosphere composed mainly of CH_4 with small amounts of NH_3, H_2 and H_2O vapor. On the other hand, the school of thought led by Abelson, and supported also by laboratory experiments on the synthesis of amino acids, holds that the early atmosphere of the earth was made of CO_2, CO, H_2, N_2 and H_2O vapor.

Geologists are also divided on the subject. Holland has presented a model for the evolution of the atmosphere in which, during the very early stage after the formation of the earth and at the commencement of outgassing, the major components of the volcanic emanation were CH_4 and H_2, rather than CO_2 and H_2O. This was so because oxygen was deficient in the volcanic 'melt', having been removed by free iron, which was more abundant in the crust at that time than now. Under these conditions the atmosphere of the earth will be largely composed of H_2 and CH_4 with small amounts of NH_3

and H_2O, *provided* free hydrogen did not 'escape' as rapidly as it does today. Rubey, on the other hand, believes that the early atmosphere was probably made of CO_2 and N_2 because not enough hydrogen was available to keep CH_4 from converting into CO_2. Holland's model is supported by the calculations of Rasool and McGovern, who have investigated the thermal properties of model primitive atmospheres of the earth. They find that in a 99 % CH_4-1 % H_2 atmosphere the average exospheric temperature may be as low as 650 °K (cf. present-day value 1500 °K), making the escape of hydrogen a relatively slow phenomenon (fig. 2). However, Abelson has argued that if methane were abundant in the primitive atmosphere, the earliest rocks should contain unusual amounts of organic matter, which apparently is not the case.

Despite the disagreement over the composition of the primitive atmosphere, it is almost certain that it was devoid of free oxygen. How and when did free oxygen become a major constituent of the atmosphere? There are two main sources of production of free oxygen, first the dissociation of H_2O

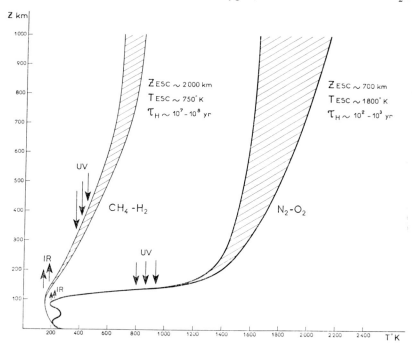

Fig. 2. Vertical distribution of temperature calculated for a methane-hydrogen primitive atmosphere. Also shown is the temperature profile in the present-day atmosphere of the earth.

vapor in the upper atmosphere and the subsequent escape of hydrogen and second, photosynthetic fixation of CO_2 and H_2O to produce carbohydrates and oxygen. During the prebiological history of the earth, however, the photodissociation of H_2O was the only source of free oxygen. Several calculations have been carried out to determine the amount of O_3 which could have been produced by this process. The results are discordant and estimates of the abundance of free oxygen in the primitive atmosphere range between 10^{-3} and 10^{-1} of the present amount. The principal reason for this uncertainty is the difficulty in obtaining a reliable estimate of the excape flux of hydrogen from such an atmosphere. At the same time the question of the abundance of O_2 in the primitive atmosphere is of considerable importance because an equilibrium amount in the atmosphere of as large as 10^{-1} of the present-day value will be amply sufficient to rapidly oxidize CH_4 and NH_3 into CO_2 and N_2. All theoretical attempts in calculating the abundance of O_2 in the earth's atmosphere have neglected the presence of NH_3. It is quite possible that even a small amount of NH_3 which is a strong absorber of the ultra-violet radiation will inhibit the dissociation of water vapor and the primitive atmosphere of earth may have been 'protected' against oxidation for the first several million years!

3. *Mars*

Viewed through a telescope, Mars presents a most fascinating and colorful picture. More than half the planet appears to be made of a reddish rocky material; about one-third is comprised of dark areas like the maria of the moon. The poles are covered with a bright whitish material whose extent waxes and wanes with the seasons. Some early optimistic observers have also noticed 'canals', which immediately led to speculation that Mars was a haven of life flourishing in those darker regions believed to be the remains of ancient oceans. Later observers seemed to support these ideas when the Greek and French astronomers reported observing seasonal changes in the intensity of the dark regions. A wave of darkening spread across the planet when the polar ice started to melt. It was believed that the additions of water every spring rejuvenated the Martian flora and fauna, resulting in the observed changes in color.

However, in the last decade, intense activity in space exploration and ground-based astronomy has provided us with new information that forces us to completely revise our long-held notions regarding the surface and the atmosphere of Mars. It now appears that the planet, instead of being a haven of life, is more like the moon, a rugged surface covered with craters; the dark areas may only be the sloping parts of the same desolate terrain; the 'canals' are nothing but linear hills or chains of craters, and even the polar caps which were believed to be the reservoir of water to feed the planetary 'canal' system are instead composed of frozen carbon dioxide at a temperature of 148 °K.

This extremely inhospitable nature of the planet, as revealed by spacecraft observations, is not confined to the surface alone. The atmosphere of Mars also appears to be entirely different than that of the earth. It was expected that, like earth, Mars has an abundance of nitrogen in the atmosphere, with small amounts of oxygen and carbon dioxide – a breathable atmosphere.

The latest results from the Mariner experiments suggest that not only the total amount of atmosphere on Mars is only 1/100th of that on earth, its composition is entirely different – it is almost pure CO_2 with little or no nitrogen.

The evidence for the unexpectedly different nature of the Martian environment has been obtained almost entirely by the exploration of the planet by Mariner spacecraft.

The first spacecraft (Mariner IV) which flew past Mars in 1964 at a distance of 20,000 km returned a large amount of new information. Twenty-two historic pictures taken from a distance of 20,000 to 40,000 km revealed the most unexpected feature of the Martian surface, namely, the existence of craters on that planet. The largest of the craters (300 km × 300 km) visible in the first photographs showed little erosion, indicating that primeval nature of surface, and suggesting that the planet has not had a heavy atmosphere of large oceans since early in its history.

As for the atmosphere, a team of scientists from NASA's Jet Propulsion Laboratory devised a brilliant experiment carried out on Mariner IV to determine the surface density and the atmospheric structure of Mars. This experiment, which is described in more detail in the section on Venus, revealed quite unexpectedly that the total atmospheric pressure may be only 6 mb (Earth = 1000 mb) and that the atmosphere may be largely composed of CO_2.

In July and August 1969, a much more detailed study of the Martian surface and atmosphere was accomplished by Mariners VI and VII, which flew past Mars as close as 2000 km. These spacecraft were equipped with high resolution TV cameras, capable of photographing large portions of the

Fig. 3. A mosaic of four pictures of Mars taken by Mariner 6 covering an area 4,000 km across and 700 km wide parallel to 15 °S latitude.

planet with a resolution of ~ 300 meters, and with a variety of other instruments designed to measure the surface and atmospheric temperatures, composition of the atmosphere, and structure of the ionosphere and the upper atmosphere. Both missions were successful.

Fig. 3 shows a typical Martian terrain photographed by Mariner VI, covered with craters of different sizes. About 200 photographs of resolution like those shown in the figure have now been carefully analyzed. They reveal that three principal types of terrain exist on Mars: (a) The cratered terrain, which resembles the surface of the moon; the absence of tectonic activity and extensive erosion suggest that this terrain may be as old as the planet itself. (b) The chaotic terrain, which consists of small ridges and depressions, especially in the region of 'meridiani sinus'. (c) The featureless terrain, which is devoid of any relief. The best example of this type of terrain is in the region of Hellas.

Existence of three different types of terrain strongly suggests that though the major part of the planetary surface may be primeval and has probably not undergone any substantial modification since its early history, there do exist regions on Mars where evidence for some surface activity is overwhelming. Mars may not be as 'dead' as the moon.

Mariner VII also extensively photographed the southern polar cap. Fig. 4 is an extraordinary view of the cap as seen from above the south pole. Craters are visible throughout the region, indicating that the material that covers the cap may be only a few meters thick. Also, an infrared radiometer carried by Mariner VII made precise measurements of the surface temperatures on Mars at different locations, including the south polar cap. Here the temperature was found to be $148 \pm 2\,°K$, which is exactly the condensation temperature of CO_2, implying that the material covering the two poles is largely solid CO_2.

The successful occultation experiments on both spacecrafts also provided new information regarding the temperature structure of the atmosphere. Temperature and pressure profiles are now available at four different points on the planet, two on the day side and two on the night side (table 3 and fig. 5). The results are close to the theoretical predictions of several authors who had suggested that the day-to-night and equator-to-pole variations in surface temperatures should be large because of the thinness of the atmosphere. In the atmosphere, the temperature decreases adiabatically with the altitude at the equator, while at the pole the atmosphere seems to be extremely cold and isothermal.

In summary, therefore, Mars appears to be a cold, dry planet; its surface,

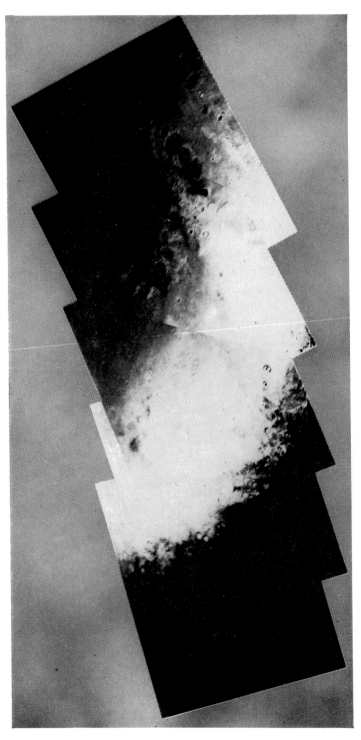

Fig. 4. The southern polar cap of Mars as photographed by Mariner 7.

TABLE 3

Summary of results.*

	Latitude	Longitude	Local time	Surface pressure (100% CO_2)	Surface temperature (100% CO_2)
Mariner 6 entry	3.7N	355.7E	15^{h}45^m	6.0 mb	250K
Mariner 6 exit	79.3N	87.1E	22^{h}10^m	7.6 mb	164K
Mariner 7 entry	58.2S	30.3E	14^{h}30^m	4.9 mb	224K
Mariner 7 exit	38.1N	211.7E	3^{h}10^m	7.5 mb	205K

* A preliminary error analysis indicates formal errors no greater than 0.5 mb in surface pressure and no greater than 5K in surface temperature.

a heavily cratered terrain dating back billions of years, and its atmosphere a thin envelope of almost pure CO_2.

The important question that emerges from this description of the surface and atmosphere of Mars is whether the new data on the planetary environment have increased or decreased the chances that life exists or has ever existed on Mars.

The answer is fairly straightforward. The new and more ample data from Mariner VI and VII have certainly *decreased* the possibility that Martian life

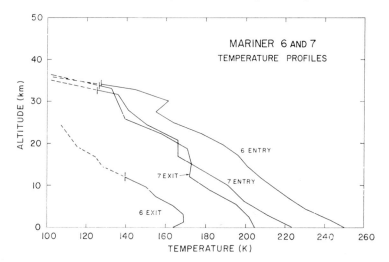

Fig. 5. Vertical temperature distributions in the atmosphere of Mars at the four occultation points of Mariners 6 and 7. Coordinates and local time of measurements are given in table 3. The temperature near the ground should be accurate within 5°K while near 30 km the estimated error is ±20°K.

ever reached the high state of evolution as on the earth. The TV pictures show that most of the surface of Mars is covered with old craters, and there is no evidence in these pictures that the terrain has been modified by 'large scale' tectonic activity or that liquid oceans once covered the planetary surface. The second result of the recent Mariner mission which is relevant to the problem of life on Mars is that polar caps are covered with CO_2 ice rather than H_2O ice. Both these observations force us to conclude that oceans of water have been absent on Mars from the very beginning. On the earth, oceans have played a vital role in the chemical and biological evolution of life. It is now generally accepted that chemical evolution, i.e. the synthesis of organic compounds like amino acids and nucleotides, which are the building blocks of all forms of life, essentially took place in the oceans. The random stirring and mixing of what is called the 'primordial soup' over millions of years is the only way the simple organic compounds could have joined together to form the complex chains of proteins, DNA and RNA. Without an enclosed fluid medium like the oceans, the probability that living cells would evolve out of simple inorganic compounds seems to be near zero. For this reason, the new evidence of the absence of oceans on Mars during the major part of its history strongly implies that life could not have gone through all the critical stages of evolution as it did on the earth. However, the possibility still remains that a *partial* evolution towards life *did* take place on Mars during the brief period in its early history when oceans might have existed. Perhaps the first few steps along the path of chemical evolution were taken, and then the process was stopped when, for some reason, the water disappeared. If this is confirmed by future analysis of the surface material on Mars, it will have the greatest impact on human philosophy. It will mean that given the proper conditions, evolution towards life *can* take place independently on a planet. Since there are certainly billions of planets in our galaxy alone, the probability of existence of 'intelligent' life elsewhere in our galaxy will suddenly increase a millionfold. On the other hand, a negative result on Mars will only mean that we must look elsewhere in the solar system for the existence of extraterrestrial life.

4. *Venus*

Venus, the closest planet to the Earth, is the third brightest object in the sky. It has been observed and studied for centuries and yields only to the Sun and the Moon in attracting attention. Yet, for all its popular appeal it had until recently remained an enigma to the astronomers $\frac{I}{N}$ a planet shrouded in mysteries.

The main reason for this lack of information has been that the planet Venus is permanently enveloped by a veil of clouds and consequently no surface features have ever been observed, even through the most powerful telescopes. On October 18, 1967, however, a Russian spacecraft, Venera 4, succeeded for the first time in the long history of planetary research, in penetrating the veil and unravelling many of the mysteries of the Venusian atmosphere. Only a day later, on October 19, an American probe, Mariner-5, flew past Venus at a distance of 4,100 km and explored the atmosphere of the planet by radio waves, obtaining precise values of the distribution of atmospheric density with height up to an altitude of several hundred kilometers. Combining the results of these two entirely different types of measurements, we now possess more information on the atmosphere of Venus than was available on the atmosphere of the Earth 25 years ago.

The two spacecraft measured the temperature in different ways, but their data led to the same conclusion. The surface of Venus is extremely hot, 700°K, and there is no reasonable chance of finding life on its surface.

How were these temperature measurements made and how sure are we of their accuracy? Venera 4 contained a simple thermometer, a barometer and several gas analyzers which made direct measurements of temperature, pressure and the composition at different points in the atmosphere as the capsule descended toward the surface. The first measurement was made at a point where the temperature in the atmosphere was 300°K. At this point, gas analyzers indicated that the atmosphere was mainly composed of CO_2 and contained only very small amounts of water vapor (0.1%). At approximately the same altitude, the parachute opened and the spacecraft began its slow descent toward the surface. Temperature was measured at frequent intervals throughout the ensuing parachute descent of the capsule and was found to be increasing at a rate of 10°K every kilometer. At the point where temperatures had attained a value of 550°K and the pressure about 20 times that of the earth's atmosphere, the signals from the spacecraft stopped abruptly. It was first believed that the signals ceased suddenly because the capsule had reached the surface of Venus; however, a critical analysis of this data, in comparison with that of Mariner 5, subsequently indicated that at the time when signals from Venera stopped, the capsule was still about 16 miles above the surface. The temperature and pressure values at the ground are therefore probably much higher than those indicated above.

Although Mariner 5 did not send a probe to the surface of Venus it ac-

quired information on conditions deep in the atmosphere by a radio propagation experiment. This experiment depended on the fact that the Mariner
5 trajectory carried the spacecraft behind Venus. As a result, signals from
the spacecraft to the earth were blocked for a period of about 20 minutes.
Prior to the passage of the spacecraft behind Venus, and again just after its
emergence on the other side of the planet, the signals from Mariner 5
traversed the Venus atmosphere en-route to the earth. As the beam penetrated the atmosphere, refraction caused the path of propagation to deviate
from a straight line, and the velocity of propagation to vary from the speed
of light in free space. In addition, because the density of the atmosphere
decreases with altitude, the power in the beam was spread over a greater
angular width, and caused the signal power received at Earth to decrease.
These effects were observed as changes in the frequency, phase, and signal
strength received during this period at the tracking stations on Earth. The
sign of the phase change differs for passage through the neutral atmosphere
and through the charged-particle ionosphere, respectively, hence it is possible
to construct separate density profiles for the atmosphere and the ionosphere.
From the rate of change of density with altitude, it is possible to deduce the
temperature and pressure profiles of the atmosphere if the composition is
known. As mentioned earlier, the Soviet probe provided this last data, in
that the Venus atmosphere is 90% CO_2: with this additional information,
temperature and pressure distribution in the atmosphere of Venus can be
obtained from the Mariner 5 data. This method should, in principle, determine atmospheric conditions down to the surface of the planet. However,
because of the high density of the Venus atmosphere, the radio signals from
the spacecraft were bent or refracted through so large an angle, as they
passed through the lowest layers of the atmosphere, that they never reached
the earth. At the lowest depth probed, the atmospheric temperature was
480°K and pressure about 7 atmospheres and both were still increasing.
Conditions at and near the surface were not determined by this experiment.

However more recently Veneras 5 and 6, identical spacecrafts to Venera 4,
probed the atmosphere of Venus to lower depths and determined the conditions down to a level where the temperature was 600°K and the pressure
~ 26 atmospheres. It has now been established that even this level in the
atmosphere is at least 10 km above the surface and the temperature and
pressure, if they continue to increase with depth at the same rate as above,
will reach to values as high as 720°K and 100 atmospheres at the surface.
As this article goes to press, the soviets have succeeded in measuring the
conditions at the surface of Venus by means of Venera 7. These are $T =$

same time out of similar materials, and situated at comparable distances from the sun, evolve along different paths? Why does one planet offer an excellent climate for life while the other offers conditions hostile to terrestrial organisms?

The answer to this problem requires the resolution of three important questions: Why is the surface of Venus so hot? Why does the atmosphere contain such large quantities of carbon dioxide instead of N_2 and O_2 as on the earth? And what happened to the oceans of water on Venus?

The measurement of the abundance of carbon dioxide on Venus is of great interest because it helps answer the first question: *Why is Venus so hot?* According to information radioed back to earth from the spacecraft, the atmosphere of Venus consists primarily of a heavy layer of carbon dioxide, about 70,000 times more than is in the atmosphere of the earth. The dense atmosphere of carbon dioxide acts as an insulating blanket which seals in the planet's heat and prevents it from escaping into space. The trapped heat raises the surface to a far higher temperature than it would have otherwise. This effect is caused by the absorbing properties of a planetary atmosphere and can best be explained by taking an example of the earth.

The solar radiation reaching the earth has a value of 2 cal/cm^2/min. Part of this energy (about 30%) is however directly reflected back to the space by the clouds present in the earth's atmosphere. Only 70% of the solar energy, therefore, penetrates through the atmosphere and reaches the surface. This energy is sufficient to heat the ground to a temperature of only 255°K. The earth's surface itself radiates, but as the temperature is not too high, the radiation is in the far infrared region of the electromagnetic spectrum — a dimly glowing object. The atmosphere of the earth, however, contains small quantities of water vapor, carbon dioxide and ozone. These gases have a property of absorbing the far infrared radiation with great efficiency. As a matter of fact, only 10% of the infrared actually gets through the earth's atmosphere. The atmosphere, having absorbed all this radiation, radiates itself in all directions, partly toward the surface and partly toward space. The radiation toward the surface increases the ground temperature by about 30°K, from 255°K to the observed value of 288°K. This phenomenon is called the greenhouse effect of the atmosphere, being an allusion to the greenhouse for plants where the glass cover acts like the atmosphere, transparent to the solar radiation but opaque to the radiation from the interior.

On Venus, the greenhouse effect of about 500°K is required to explain the observed temperature. Calculations based on the insulating properties of carbon dioxide show that the temperature of Venus could easily be raised by

500 degrees as a result of the greenhouse effect of a very heavy layer of CO_2 as reported to be present on Venus; 300,000 times more than in the earth's atmosphere.

The high temperature on the surface of Venus can, therefore, be understood by the large abundance of carbon dioxide in the atmosphere. The question which one then asks is why carbon dioxide is so abundant on both Venus and Mars and not on the earth. This problem is related to the evolutionary history of the three planets and in the following, we will attempt to discuss this question in more detail.

5.   *Early history of the atmospheres of Venus and Mars*

Table 4 summarizes the surface parameters and the abundances of major volatiles on Venus, Earth and Mars. Several interesting features are note-worthy. First, the amount of CO_2 in the atmosphere of Venus is approxi-mately equal to the amount buried in the crust of the Earth in the form of carbonates. Second, nitrogen, which is at the present time the major con-stituent of the Earth's atmosphere, is only 1–5% of the amount of CO_2, not only on Earth but also on Mars and Venus. Third, the large quantities of water which make up the oceans of the Earth are practically absent on Venus but may be present in a frozen state below or on the surface of Mars. Fourth, the amount of free oxygen in the atmosphere of the Earth is small in relation to the total amount of oxygen in the crust, but is even less abundant in the Venus and Mars atmospheres. Finally, when the abundances of the four gases in the atmosphere of Mars are compared to those on Venus and Earth,

TABLE 4

		Venus	Earth	Mars
Temperature		700 °K	300 °K	230 °K
Pressure		75 atm	1 atm	0.01 atm
CO_2	atm	90,000 g/cm²	0.3 g/cm²	~70 g/cm²
	crust	?	70,000	?
N_2	atm	<3000	800	<1
	crust	?	~2000?	?
H_2O	atm	~100	~1	~0.01
	oceans	0	300,000	?
O_2	atm	<10	200	~0.01
	crust	?	8 × 10⁶ total	?

Composition (row label spanning CO_2, N_2, H_2O, O_2)

they suggest that if the origin of these gases is the same (the outgassing from the interior), the volcanic activity on Mars has been ~1000 times less effective than that on either the Earth or Venus.

In order to investigate the possible evolutionary paths which the atmospheres of Venus, the Earth and Mars may have followed to arrive at their present diversity in composition, one can make three basic assumptions: (a) At some time in the early history of the solar system, the terrestrial planets (Mercury, Venus, the Earth and Mars) completely lost their primordial atmospheres. (b) The present atmospheres of these planets (table 4) have developed mostly from the degassing of the planetary interiors – free oxygen in the Earth's atmosphere being an exception related to the presence of life – and (c) the major constituents of the outgassing from the interior for all three planets are essentially the same; i.e., water vapor, and carbon dioxide with $H_2O/CO_2 = 4$, with nitrogen accounting for $<1\%$ of the volcanic emanations.

At the beginning of the outgassing, when the planets are more or less devoid of an atmosphere, their ground temperatures, T_G, are essentially determined by the amount of solar radiation absorbed by the surface, and are equal to the effective temperatures of the planet, T_e. For a rapidly rotating planet and for a given planetary albedo, A_p, we have

$$T_G = T_e = \frac{Sp}{4\sigma}(1 - A_p),$$

where Sp is the solar constant.

As the atmosphere accumulates, the ground temperature begins to exceed the effective temperature because of the additional heating of the surface by the atmospheric greenhouse effect. The magnitude of the greenhouse effect can be calculated by solving the equation of radiative transfer for a given atmosphere. However, the transfer problem becomes quite complicated when allowances are made for the strong frequency dependence of the infrared molecular absorption coefficient, especially for a gas such as water vapor. Therefore, for the first estimates of ground temperature, we assume a grey atmosphere and use the Eddington approximation for the solution of the radiative transfer equation.

We start the greenhouse calculations for Venus at time zero, when the outgassing has just commenced and the total amount of atmosphere is $<10^{-3}$ mb. At this point we assume that planetary albedo is determined only by the surface and, by analogy with the moon and Mercury, we assumed it to be equal to 7%. The average ground temperature at this time is equal to T_e and for an albedo of 7% it will be 330 °K for the fast rotation and 390 °K for slow rotation. As the atmosphere accumulates the greenhouse effect increases the ground temperature, provided the albedo remains constant. Because the initial surface temperature (T_e) is already quite high H_2O remains in the atmosphere as vapor and accelerates the greenhouse effect. The increase in ground temperature as a function of the build-up H_2O-CO_2 pressure in the atmosphere is shown in fig. 7. Each set of curves indicates the range of possible temperatures, considering the uncertainties in the greenhouse cal-culations and the effect of convection. However, in the case of Venus, by the time the atmosphere has accumulated to a total pressure of 10^{-1} atm the ground temperature has already risen to 430 °K for the case of $T_e = 330$ °K and is greater than 500 °K for $T_e = 390$ °K. One important implication of this result is that the temperature on the surface of Venus always remained *above* the boiling point of water at the pressures involved. This is illustrated by the position of phase diagram of water in fig. 7. The temperature curves for Venus miss the liquid phase of water by a wide margin. It may be noted here that the calculated Venus temperatures are so much higher than the boiling point of water that even if the albedo of the planet were 30–40% instead of 7%, water would not condense at the surface.

These initial high temperatures and the complete absence of liquid water from the surface had a significant bearing on the accumulation of CO_2 in the atmosphere of Venus. As the CO_2 is degassed from the interior into the atmosphere, its partial pressure should be buffered by reactions with the crust such as

$$Ca\ Mg\ Si_2\ O_6 + CO_2 \rightarrow Mg\ Si\ O_3 + Ca\ CO_3 + Si\ O_2.$$

These atmosphere-crust reactions are, however, temperature-dependent, and at high temperatures large quantities of CO_2 can accumulate in equilibrium with the silicates. Fig. 7 also shows the plots of CO_2 partial pressures as a function of temperature for reactions with $Ca\ Si\ O_3$, $Ca\ Mg\ Si_2\ O_6$ and $Mg\ Si\ O_3$. If at any time the amount of CO_2 in the atmosphere is higher than the equilibrium value at that temperature, then the situation is unstable and CO_2 should be removed from the atmosphere and deposited into the crust as carbonates. However, the atmosphere-crust reaction will proceed at a rapid rate *only* if liquid water is present at the surface to facilitate the contact. Both these conditions should be satisfied simultaneously in order to remove CO_2 from the atmosphere effectively. In the case of Venus the temperature was always high enough so that the CO_2 amount in the atmos-phere never substantially exceeded the equilibrium pressure at that tem-

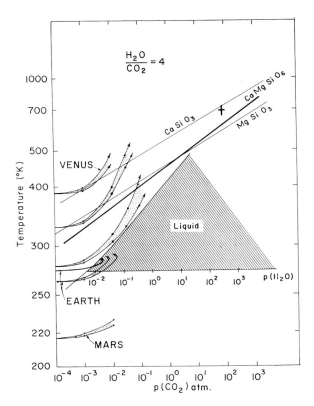

Fig. 7. The plot of increase in surface temperatures on Venus, Earth and Mars by the greenhouse effect of H_2O-CO_2 atmosphere during the evolution of the three planets. The initial temperatures on these planets equal the effective temperature for a planetary albedo of 7% and two different rates of rotation (in the case of Venus) and for two values of the planetary albedo (7% and 20%) for the Eearth. The phase diagram for water is shown and the region in which water can exist as liquid is represented by the hatched area. Also, plotted are the equilibrium values for the partial pressures of CO_2 as a function of temperature for three different silicate reactions (after Rasool and De Bergh 1970).

perature. The only exception is for the case of $T_e = 330°K$ and the reaction with $Ca Si O_3$. However, the absence of liquid water would have impeded the reaction from preceding rapidly and the temperature should have soon built up to a value at which the atmospheric CO_2 would be in equilibrium with the crust. In this way the CO_2 continued to accumulate in the atmosphere to the present value of ~ 75 atm at a temperature of $700°K$.

The important question which still remains is the absence of large quantities of water vapor in the present atmosphere of Venus. According to this model, the Venus atmosphere should contain ~ 300 atm of H_2O. However, only $\sim 10^{-1}$ atm of water appears to be present today (table 4). As mentioned earlier, most of the water could have escaped from the Venus atmosphere in the early stages when its exospheric temperature would have been quite high ($> 3000\,^\circ K$) and the escape of hydrogen and even oxygen very rapid.

If the present amount of H_2O in the Venus atmosphere ($\sim 10^{-1}$ atm) is the equilibrium value between the outgassing and loss of hydrogen to space, it would imply that the escape of water from Venus commenced at $P_{(H_2O)}$ $\sim 10^{-1}$ atm. At this point, however, according to fig. 7, the surface temperature is already $> 430\,^\circ K$ and CO_2 in the atmosphere is in equilibrium with the silicates. From this time to the present, if the partial pressure of water never exceeded 10^{-1} atm, the increase in the surface temperature to the present value of $700\,^\circ K$ would have been principally governed by the greenhouse effect of CO_2 alone. Consequently, the temperature would increase at a slower rate than shown by the arrows in fig. 7 and would probably follow the CO_2 equilibrium curve for $Ca\,Si\,O_3$.

When similar calculations are carried out for the Earth, a completely different evolutionary pattern emerges, explaining why the temperature on the surface of the Earth remains a comfortable $290\,^\circ K$ and why almost all of the CO_2 is in the crust and H_2O is in the oceans. For the Earth, the initial value of T_e for an albedo of 7% is $275\,^\circ K$. As the water vapor and CO_2 atmosphere begins to accumulate on the Earth, the ground temperature increases, as on Venus; but soon temperature and pressure conditions become such that liquid water can condense at the surface. (This is shown when the curve for Earth enters the hatched area for the liquid water phase in fig. 7.) From this point on, the Earth follows an evolutionary path completely different from that of Venus. With the volcanic steam condensing into liquid, the amount of water vapor remaining in the atmosphere is small and the increase in ground temperature is determined by the accumulation of CO_2 alone. However, the total amount of CO_2 soon becomes *greater* than the equilibrium value with the silicates. The reactions with crustal rocks proceed rapidly and are considerably accelerated by the presence of liquid water on the surface. Due to the erosion of rocks by liquid water, fresh silicates are brought into contact with the CO_2 in the atmosphere and in the oceans. The CO_2 therefore never accumulates in excess of the equilibrium value of $\sim 10^{-4}$ atm at $290\,^\circ K$. At the same time the volcanic steam continues to condense at the surface, slowly building up the oceans to their present depth of ~ 3 km. Nitrogen, the inert gas which constitutes only

$\sim 1\%$ of the volcanic gases, accumulates to make up the bulk of the present atmosphere.

This chain of events in the case of the Earth began only because the T_e was 275 °K and water vapor, the major constituent of volcanic emanations, was able to condense out of the atmosphere to mark the beginnings of the oceans. As is clearly evident from fig. 7, the initial temperature of the planet is an extremely critical parameter in determining which evolutionary path a planetary atmosphere will follow. In fact, a calculation for an initial temperature of 280 °K indicates that the surface temperature increase would have been rapid enough for the Earth to miss the liquid phase of water at its surface. A runaway greenhouse effect would have made the conditions on the Earth as hostile as on Venus. This situation could have occurred on the Earth if it were closer to the sun by only 6–10 million kilometers.

However, it is interesting to note that for a planet further away from the sun, like Mars, the changes are very small that a runaway greenhouse would ever take place. When the initial temperature of a planet is < 273 °K, the volcanic steam freezes at the surface and only CO_2 accumulates in the atmosphere (as is the case on Mars today). However, when large quantities of CO_2 have accumulated on the planet, the greenhouse effect due to CO_2 alone will raise the surface temperature above 273 °K melting the frozen water and thereby initiating the transfer of CO_2 from the atmosphere into the crust. When this evolutionary stage is reached on Mars, conditions on the surface may become very similar to those on the Earth today: water in a liquid state, CO_2 in the sediments, and an atmosphere consisting mainly of N_2.

This evolutionary path for Mars is completely different from the usual suggestions that Mars, in its early history, had oceans and a heavier atmosphere of which the present one is a remnant. According to the present model of the evolution, oceans have been absent on Mars from the very beginning but may eventually 'accumulate' by melting when the atmospheric pressure becomes large enough so that the greenhouse raises the temperature of the planet to > 273 °K.

6. *Jupiter*

The recent findings on Mars and Venus discussed earlier mark the beginning of the accumulation of basic data for the understanding of the history of the terrestrial planets. But to resolve the age-old problem of the origin and evolution of the solar system as a whole, it is exploration of Jupiter which will eventually provide information of prime significance. This is so, not only because Jupiter is the largest planet, several times more massive than the other eight planets combined, but because it presents such puzzling aspects of far reaching importance that their eventual solution will have direct bearing on our understanding of the primitive environments from which the planets were formed and the life on earth originated.

First, and perhaps the most interesting aspect of Jupiter, is that its present atmosphere seems to be composed of the same gases, hydrogen, methane and ammonia, out of which the first living organisms are believed to have been synthesized on the earth about four billion years ago. Is it possible that similar initial steps along the path of life are occurring now on Jupiter? To answer this question one needs to know the exact composition, the temperature and pressure conditions which exist at various levels in the atmosphere and at the surface. This leads us to the other puzzling aspect of Jupiter.

It has recently been found that Jupiter may have a source of heat in the interior, almost four times more intense than the sun at that distance. What is the source of this energy? Is it that Jupiter is still contracting towards its final size and thereby releasing gravitational energy? If so, what about the other giant planets? Are they all, 4.5 billion years after their birth, still in the process of accumulation and do not yet have a solid surface?

6.1. Composition

Jupiter and the other giant planets differ markedly from the terrestrial planets in regards to their composition, both in the interior and in the atmosphere. Though they are much larger and hundreds of times more massive than the earth they have suprisingly low density. In general, the density is about that of water. For Jupiter it is 1.33 g/cm³ and for Saturn only 0.71 g/cm³. This should be compared with the density of the earth which is 5.5 g/cm³. The low density of the giant planets is puzzling because the pressure of their great mass should compact them to a higher density than that of the earth. The explanation of this apparent paradox is connected with the composition of the giant planets; in contrast to the terrestrial planets which are made up of Fe, Ni and silicates, the major planet seems to be composed mainly of hydrogen and helium, the lightest of all the elements. In fact, the density of Jupiter is almost exactly the same as that of the sun, indicating that the ratio of hydrogen and helium to other heavier elements may be about the same on Jupiter as on the sun. This is what one would expect if Jupiter condensed out of a contracting solar nebula which had the same composition as the sun has today. Jupiter, being so massive did not lose any of the gases during its long history, and should, therefore, reflect the composition of the material out of which it was formed (table 1).

Spectroscopic observations of Jupiter have already discovered the presence of methane, ammonia and hydrogen above the clouds of Jupiter and it is believed that substantial amounts of water are present below. Methane and ammonia are easy to detect because of their strong absorption bands in the near infrared and careful analysis of the band structure can also give the concentration of these gases in the atmosphere. As early as 25 years ago, Kuiper successfully measured the amounts of methane and ammonia on Jupiter and their concentration appears to be roughly the same as mentioned above.

Hydrogen and helium, however, pose special problems. Helium being a rare gas is completely inert and under ordinary conditions cannot be detected by spectroscopic techniques employed in optical astronomy. It does, however, produce emission lines in the ultra-violet which can only be observed from the earth if the measurements are made from above the atmosphere. Hydrogen, however, under special conditions produces absorption lines in the visible part of the spectrum. Molecular hydrogen has a quadrupole moment and could produce a vibration-rotation spectrum which could be detected from the earth if sufficiently large amounts were present on Jupiter. Kiess, Corliess and Kiess were first to detect four such lines of hydrogen at around 8,200 Å. A comparison of their strengths with the theoretical values of their intensities can give an estimate of the amount of hydrogen on Jupiter above the reflecting level of 8,200 Å photons. Because of several inherent problems in this technique of measurement the estimate of hydrogen on Jupiter cannot be made with greater precision than a factor of three. A recent evaluation of this problem suggests that at a level in the atmosphere where hydrogen exerts a pressure of 1.8 atmospheres the 'total' pressure is between 2 and 2.8 atmospheres. If the 'other' gas is helium its abundance may be 6–30%, close to the computed solar composition (table 1).

6.2. Thermal structure

Situated at 5.2 A.U. from the sun, Jupiter receives 27 times less solar energy than the earth; only 12,500 ergs/cm²/sec distributed over the planetary globe. In addition, the albedo of Jupiter is 0.45 and, therefore, only 55% of this energy or 7,000 ergs/cm²/sec actually enters the planet, which should eventually be returned back to the space in the form of thermal radiation. The recent measurements of this energy, however, suggest that the flux emitted by Jupiter is of the order of 30,000 ergs/cm^2/sec, about four times higher than the expected value. How accurate are these measurements and what could be the source of this excess energy?

If it is assumed that Jupiter behaves as a blackbody then the effective radiating temperature for the amount of solar energy reaching the planet will be 103°K. At this low temperature the radiation will be mainly at wavelengths between 10 and 100 μ with a maximum at 30 μ. Measurement of this radiation is difficult to make from the ground because of the absorbing properties of the earth's atmosphere. The earth's atmosphere contains molecules of CO_2 and H_2O which absorb the far infrared with great efficiency. There are, however, 'windows' at $\lambda < 20 \mu$ where the atmospheric absorption is small, which allows ground based astronomers to measure energy from different celestial sources. Depending on the gases present in the atmosphere of Jupiter, such measurements made at different wavelengths will provide information on temperature at varying depths in the atmosphere. Fig. 8 shows a recent attempt in this direction. On the basis of these measurements one can derive a temperature profile of the atmosphere above the clouds and several such models are shown in fig. 9. The best fit to the data is the model (3) which corresponds to a distribution of gases in the atmosphere shown in fig. 10. Regarding the structure of the Jovian atmosphere *below* the visible clouds, the most comprehensive study to date is by Lewis who has carried out thermodynamical calculations for a NH_3-H_2O-H_2S system. The temperature is assumed to increase with depth adiabatically (~ 2°K/km). As the temperature increases, ammonia and water become important constituents. Several cloud layers are formed, the top most (at $T = 150$°K) being composed of NH_3 crystals while lower ones comprise of NH_4SH and aqueous NH_3 solution at temperature levels of 220°K and 310°K respectively (fig. 11).

6.3. Color and life

The orange and blue coloration of the cloud bands of Jupiter and the presence of a giant red spot has long fascinated the optical astronomers. A recent new interpretation of these features has made the problem all the more exciting. It has been proposed that the visible colorful 'surface' of Jupiter is the seat of intense pre-biological activity, where the first living organisms are being synthesized.

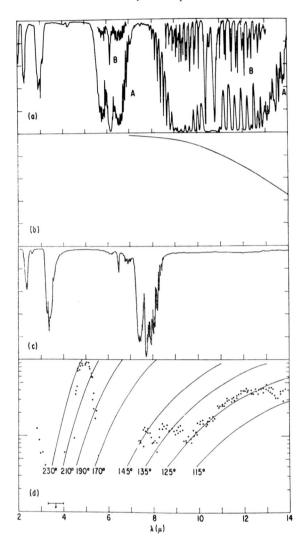

Fig. 8. (a) Absorption spectrum of ammonia from 2 to 14 microns for two different values
of pressure.

(b) Absorption spectrum of hydrogen from 8 to 14 microns for a pressure of 33 atmospheres and a temperature of 85°K.

(c) Room temperature absorption spectrum of CH_4 from 2 to 14 microns.

(d) 2.8 to 14 microns spectrum of Jupiter (after Gillet et al. 1969).

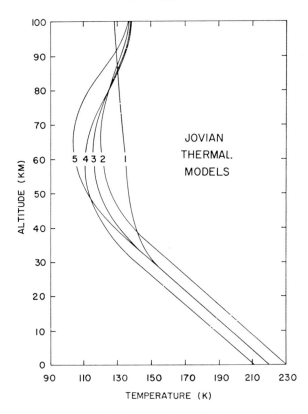

Fig. 9. Temperature distribution in the atmosphere of Jupiter above the cloud levels for five different models. Model 3 agrees best with the data shown in fig. 8 (after Hogan et al. 1969).

Three arguments have been advanced in favor of this hypothesis. First, the atmosphere of Jupiter is composed precisely of these gases, hydrogen, methane, ammonia and water vapor which have supposedly played a critical role in the events which have led to the development of life on the earth.

Second, Cyril Ponnamperuma has demonstrated that when a simulated Jovian atmosphere, at temperatures as low as 150°K, is exposed to ultra-violet radiation, it not only produces the complex organic molecules like amino acids and nucleotides but the color of the resulting products is very similar to the yellowish-orange red of the Jovian clouds and of the famous 'red spot'.

Third, an ultraviolet spectrum of Jupiter first taken by T. P. Stecher from a rocket, had indicated a significant absorption centered at $\lambda = 2600$ Å. None

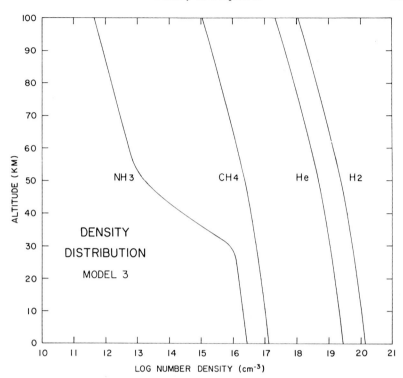

Fig. 10. Density distributions for H_2, He, CH_4 and NH_3, corresponding to thermal model 3 shown in fig. 9 (after Hogan et al. 1969).

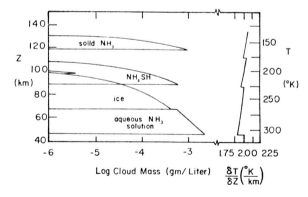

Fig. 11. Cloud masses and wet adiabatic lapse rate vs. altitude and temperature for NH_3-H_2O and NH_3-H_2S clouds in a solar composition model of the Jovian atmosphere. The prominance of the NH_4SH clouds and aqueous NH_3 clouds is noteworthy. Altitude scale in thjs fig. implies zero level at $\sim 400°K$ (after Lewis 1969).

of the known major atmospheric constituents of Jupiter can account for this feature. However, C. Sagan of Cornell University has pointed out that the absorption features observed on Jupiter match closely with those of adenine, which is a basic constituent of both RNA and DNA and, therefore, one of the most important chemicals in biological systems. In fact, laboratory experiments of Cyril Ponnamperuma have convincingly shown that in electron irradiation of methane, ammonia and water, the largest single non-volatile compound formed is adenine. In addition, the production of adenine is enhanced when hydrogen is deficient, as it seems to be the case at the cloudtops of Jupiter!

How can one test this extremely interesting hypothesis? First and foremost in this respect, is of course the ultraviolet spectroscopy of Jupiter from rockets and earth orbiting satellites. High resolution spectra of Jupiter in the 1800 to 3000 Å interval when matched with laboratory spectra of organic molecules, as suggested by C. Sagan, should provide highly significant information on this problem. Subsequent experimentations in the infra-red and a search for HCN by close-range fly-by missions should further clarify the question. The actual resolution of the problem will probably have to await in situ exploration of the Jovian atmosphere or return of the samples to the earth. It is difficult to predict the time-table involved for such experimentations. Because of the long travel to Jupiter and the crushing force of gravity of the planet, the in situ exploration and return of samples is perhaps another 20 years away. But numerous experiments of great value can certainly be performed from high altitude aircraft, balloons, rockets and earth-orbiting satellites starting immediately.

General bibliography

ABELSON, R. H., 1966, Proc. U.S. Nat. Acad. Sci. 55, 1365.

AVDUEVSKY, V. S., 1970, J. Atmospheric Sci. 27, July 1970.

BROWN, H., 1952, In: The atmospheres of the earth and planets, ed. G. P. Kuiper. (University of Chicage Press, Chicago) 258.

CAMERON, A. G. W., 1964, The origin of planetary atmospheres, Selections from the TRW Space Technology Laboratories, Lecture Series: Presented March 3rd, 1964.

HOLLAND, H. D., 1962, In: Petrologic studies: A volume in honor of A. F. Buddington, eds. A. E. Engel, H. James and B. F. Leonard. (Geological Society of America, New York) 447.

JASTROW, R. and S. I. RASOOL, eds., 1969, The Venus atmosphere. (Gordon and Breach, New York).

KLIORE, A., G. FJELDBO, B. SEIDEL and S. I. RASOOL, 1969, Science 166.

KUIPER, G. P., 1952, In: Atmospheres of the earth and planets, ed. G. P. Kuiper. (University of Chicago Press, Chicago).

LEIGHTON, R. B. 1970 Scientific American, May 1970.

LEIGHTON, R. B. et al., 1969, Science 166, 49.

MILLER, S. L., 1953, Science 117, 528.

MILLER, S. L. and H. C. UREY, 1959, Science 130, 245.

NEUGEBAUER, G. et al., 1969, In: Mariner-Mars, a preliminary report. NASA SP-225, Washington D.C., 105.

OPARIN, A. I., 1953, The origin of life. (Dover Publications, New York).

RASOOL, S. I. and C. DE BERGH, 1970, Nature 226.

RASOOL, S. I., J. HOGAN, R. STEWART and L. RUSSELL, 1970, J. Atmospheric Sci. August 1970.

RINGWOOD, A. E., 1964, In: Advances in earth science, ed. P. M. Hurley.

RUBEY, W. W., 1955, In: Crust of the Earth, ed. A. Poldervaart. (Geological Society of America, New York) 631.

SAGAN, C., 1967, International Dictionary of Geophysics, ed. K. Runcorn. (Pergamon Press, New York).

UREY, H. C., 1959, In: Handbuch der Physik 52, 363.

UREY, H. C., 1962, The planets, their origin and development. (University Press, New Haven).

C. Ponnamperuma (ed.), Exobiology. © North-Holland Publishing Company

CHAPTER 12

Distribution and significance of carbon compounds on the moon

SHERWOOD CHANG and KEITH A. KVENVOLDEN

1. Introduction

Carbon, the sixth member of the atomic table, is one of the most important of the elements. Its importance stems both from its cosmic abundance and from its chemical properties. In our solar system carbon ranks fourth. Only hydrogen, helium, and oxygen are present in greater concentration. Carbon and its compounds have been recognized in comets, interstellar gas clouds, stellar atmospheres, and the atmospheres of some planets. The chemical property which makes carbon so important is its ability to combine with itself and other elements by means of covalent bonds to form endless numbers of monomeric and polymeric compounds many of which are essential for all living systems. Because carbon-containing compounds are the basic building blocks of life, as we know it, carbon takes on an added significance.

2. Carbon in the Earth and meteorites

Before beginning to explore the available information concerning carbon on the moon, it seems reasonable to review what is known about carbon on the Earth and to consider briefly the results of studies of meteorites which have given us our first direct clues about extraterrestrial carbon.

To begin with, the average amount of carbon in the accessible regions of the Earth is small compared with the great abundance of carbon in the cosmos. Recently, Cameron (1968) has listed the relative abundance of the elements of the solar system. This new table was constructed based as much as possible on elemental measurements from C1 carbonaceous chondrites.

These meteorites are assumed to be the most representative source of abundance data for the solar system. For some elements abundances were determined from ordinary chondrites. The volatile elements are based on solar photospheric and cosmic ray abundance data. The six most abundant elements are listed in table 1. For comparison the six most abundant elements in the Orgueil C1 carbonaceous chondrite are listed in table 2, where carbon ranks fifth.

TABLE 1

The six most abundant elements of the solar system (normalized to $Si = 10^6$).

Rank	Atomic number	Element	Abundance
1	1	H	2.6×10^{10}
2	2	He	2.1×10^9
3	8	O	2.36×10^7
4	6	C	1.35×10^7
5	7	N	2.44×10^6
6	10	Ne	2.36×10^6

TABLE 2

The six most abundant elements in the Orgueil meteorite (Nagy 1966) (normalized to $Si = 10^6$).

Rank	Atomic number	Element	Abundance
1	8	O	7.5×10^6
2	1	H	4.7×10^6
3	12	Mg	1.1×10^6
4	14	Si	1×10^6
5	6	C	7.2×10^5
6	7	N	1.8×10^5

The average amount of carbon in the crustal rocks of the Earth has been estimated to be 200 parts per million (ppm) (Mason 1966), which equals an abundance of about 1.7×10^3 normalized to $Si = 10^6$. There are sixteen elements in the crust which are present in greater abundances than carbon. The present accessible Earth, therefore, has four orders of magnitude less carbon than would be expected from the abundance of carbon in the solar system and two and a half orders of magnitude less carbon than in carbonaceous chondrites.

Carbon occurs in several forms and states. Organic carbon generally is in the reduced state where carbon is bonded primarily to itself, and to hydrogen, nitrogen and oxygen. Inorganic carbon occurs in an oxidized state, an elemental state of several forms, and as carbides. Organic carbon is divided into that material which is extractable with ordinary organic solvents and the polymeric organic material which is not extractable. Oxidized forms of carbon include CO, CO_2 and the anions HCO_3^- and CO_3^{2-}. Elemental carbon can take the forms of graphite, diamond or amorphous carbon. Cohenite appears to be the most common carbide.

On the Earth carbon compounds are distributed between the atmosphere, hydrosphere, biosphere and lithosphere. The lithosphere is further divided into crust, mantle, and core. Only the crust and in rare cases the upper mantle are available for sampling. Table 3 summarizes the distribution of various forms of carbon on the Earth.

TABLE 3

Distribution of carbon on the Earth.

	Dominant form of carbon	Mass of C	Reference
Atmosphere	CO_2	6×10^{17} g	Borchert 1951
Hydrosphere	CO_2, CO_3^{2-}, HCO_3^-	4×10^{19} g	Borchert 1951
Biosphere	Organic carbon	3×10^{17} g	Borchert 1951
Lithosphere			
Crust		5×10^{21} g	Mason 1966
Sedimentary organic	Organic carbon	6×10^{20} g	Mason 1966
Sedimentary carbonate	CO_3^{2-}	2×10^{21} g	Mason 1966
Igneous and metamorphic rocks	see text	2×10^{21} g	Mason 1966
Mantle	unknown		
Core	unknown		

The estimates of carbon in the crust as given in table 3 are based on the average crustal abundance of carbon of 200 ppm and the distribution of carbon according to a synthesis of data by Mason (1966). The concentration of carbon in igneous and metamorphic rocks is about 100 ppm. On the other hand, work by Hoefs (1965) shows that the concentration of elemental carbon in igneous and metamorphic rocks averages about 200 ppm. The total concentration of carbon (carbonate carbon and elemental carbon) in these rocks would be even larger, but the difficulty in distinguishing primary and secondary carbonate carbon does not permit an estimate to be made of the amount of primary carbonate carbon in igneous and metamorphic rocks. Earlier, Rankama and Sahama (1950) calculated that the concentration of carbon in igneous rocks was 300 ppm. Most of this carbon was thought to be present as CO_3^{2-}. In volcanic emanations oxides of carbon are common. CO_2 has been found as a principal constituent while CO is of less quantitative importance. Elemental carbon in forms such as graphite and diamond are rare and quantitatively unimportant on Earth. Carbide carbon has not been found on the earth and, if present, must be in the mantle.

Sedimentary organic carbon is present either as extractable organic material or as a complex, highly polymeric, organic material, which is not extractable with ordinary solvents and which has been called kerogen. In sedimentary rocks usually less than 20 percent of the organic material is extractable. Analyses of about 1400 sedimentary rocks showed that the total organic carbon content of shales and limestones was about 10,000 and 2,000 ppm respectively, whereas the concentration of extractable organic material was less than 2000 ppm in the shales and less than 400 ppm in the limestones (Gehman 1962). Table 4 shows where the different forms of carbon can be found on Earth.

TABLE 4

Distribution of forms of carbon on Earth.

	Atmosphere	Hydrosphere	Biosphere	Lithosphere
Organic				
Extractable	+ (CH₄)	+	+	+
Polymeric	−	+	+	+
Elemental				
Graphite	−	−	−	+
Diamond	−	−	−	+
Amorphous	−	−	−	+
Forms of CO_2	+	+	+	+
Carbide	−	−	−	− (?mantle)

Meteorites contain carbon in concentrations ranging from less than 200 ppm to as much as 50,000 ppm. The concentration of carbon in unequilibrated chondrites seldom exceeds 2000 ppm and usually is less. Enstatite chondrites contain between 1000 and 5000 ppm carbon, while carbonaceous chondrites have carbon in concentrations as large as 50,000 ppm. In contrast to the Earth, where most carbon appears in igneous and metamorphic rocks in an oxidized state or perhaps in an elemental state, most carbon in carbonaceous chondrites is organic carbon (Hayes 1967). In C1 and C2 carbonaceous chondrites 30% of the organic material has been extracted with simple solvents. The remaining 70% is probably polymeric because there is no evidence for carbides or elemental carbon in forms of graphite, amorphous carbon or diamond. A small portion of the carbon, 0.21% and less, is present in the form of carbonates (CO_3^{2-}) (Smith and Kaplan 1970). Some C3 and C4 chondrites contain graphite and extractable organic carbon.

The carbon in ordinary chondrites is usually recorded as graphite (except for rare occurrences of diamonds), but much of the carbonaceous material does not give an X-ray pattern. This material is either amorphous carbon or polymeric organic material (Mason 1962). The principal form of carbon in enstatite chondrites is graphite with minor amounts of cohenite (Fe, Ni, Co)$_3$C (Mason 1966). Unequilibrated ordinary chondrites contain up to 1% carbon. The only reported forms are extractable organic carbon and graphite. The presence of these forms suggests that some polymeric organic carbon is also present. Much of the carbon, up to 50%, in ureilites is present as diamond and about 5% of the carbon is extractable organic material. Graphite, cohenite, and dissolved CO have been reported in iron meteorites. Table 5, modified from Hayes (1967), summarizes the distribution of forms of carbon in various chondrite classes.

TABLE 5

Distribution of carbon in meteorites.

	Carbonaceous chondrites	Enstatite chondrites	Ordinary chondrites	Unequilibrated chondrites	Ureilites
Organic					
Extractable	+	+	−	+	+
Polymeric	+		?	?	?
Elemental					
Graphite	− (+ for C$_3$ & C$_4$)	+	+	+	+
Diamond	−	−	−	−	+
Amorphous	−		?	?	?
Carbonate	+ (− for C$_3$)	−	−	−	−
Carbides	−	+	−	−	

Besides abundances and distributions of various forms of carbon, its isotopic composition can also be determined. This provides a measure of isotopic fractionation between C^{13} and C^{12}. Isotopic composition (δC^{13}) will be given in parts per thousand (‰) relative to the Peedee belemnite standard. Most carbon on earth has a range of isotopic compositions from about −36 to +10‰ (Craig 1953). Outside of this range falls carbonate carbon from some Teriary rocks of California having isotopic compositions as heavy as +20 (Murata et al. 1969). Methane gas can be isotopically very light with carbon isotopic compositions generally in the range −40 to −60‰.

The lightest isotopic composition yet measured on non-volatile organic material is $-89\permil$ for organic mats of southern Israel (Kaplan and Nissenbaum 1966). Carbon in meteorites ranges from -30 to $-4\permil$ and from $+40$ to $+70\permil$ (Smith and Kaplan 1970). In table 6 are ranges of carbon isotopic compositions that have been reported for various forms of carbon on the earth (summarized from Craig (1953) and Degens (1965)).

TABLE 6

Isotopic composition of carbon on the Earth.

	Dominant form of carbon	δC^{13} range (‰)
Atmosphere	CO_2	-10 to -6
Hydrosphere	HCO_3^-	-4 to -1
Biosphere	Organic carbon	-29 to -8
Lithosphere		
Crust		
Sedimentary organic	Organic carbon	-35 to -9
Sedimentary carbonate	CO_3^{2-}	-14 to $+9$
Igneous and metamorphic rocks	?	-26 to -19
Graphite	C	-36 to -3
Diamond	C	-5 to -2

3. Carbon on the moon

When the Apollo 11 astronauts returned to Earth from the first manned exploration of the moon, they brought with them the first extraterrestrial rock sample which had not been exposed for extended periods of time to the Earth's atmosphere and biosphere. Before this event, meteorites were our only examples of extraterrestrial material, and all meteorites had been exposed during their falls to the atmosphere of Earth and had been subjected to geological weathering processes at the Earth's surface. Samples from the moon were protected from the Earth's atmosphere by the capsule returning the astronauts and were never subjected to natural weathering processes. Carbon compounds in these samples should not contain the imprint of contamination by carbon compounds of the Earth.

During the last two decades many geochemists have determined the abundance and identity of organic compounds in modern and ancient sediments and in meteorites. Because of this experience, the analytical approach adopted by most groups interested in lunar carbon compounds was primarily that of the organic geochemist. In our laboratory, attention was directed toward determining the total carbon content, the isotopic composition of carbon, and the identity of indigenous carbon compounds in a sample of lunar fines (10086 Bulk A) from the Apollo 11 mission. Our analytical scheme is depicted in fig. 1 (Kvenvolden et al. 1970). Not included in the scheme is direct

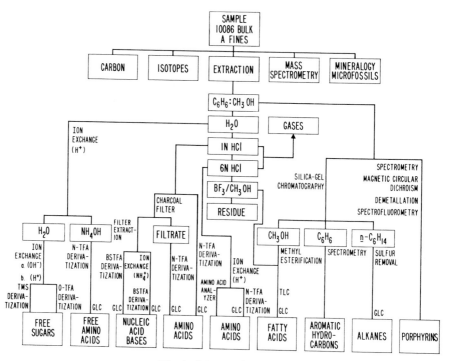

Fig. 1. Scheme of analysis.

pyrolysis of lunar fines and analysis of resulting volatile products, a procedure which provided much important data. The following discussion will center around results from our laboratory; however, we shall rely heavily on findings of others to provide a wider perspective of carbon on the moon. Unless cited otherwise, the experiments described were performed by the NASA-Ames Research Center Consortium. It must be remembered that our measurements and speculations concerning carbon on the moon are based at this time on analyses of samples from a single location in the Sea of Tranquillity.

carbon concentrations ranged from 102 ppm to 262 ppm, which is about the same range of values determined for the fines. Lowest values for total carbon were found in the basaltic rocks and amounted to about 70 ppm, except for a single low measurement of 16 ± 5 ppm (Kaplan et al. 1970). In a sample of sieved 10084 A lunar fines, particles from 60–140 mesh contained 92 ppm carbon; from 140–300 mesh, 183 ppm; and finer than 300 mesh, 261 ppm (Kaplan et al. 1970). The same trend was observed in 10086 A fines by Moore et al. (1970). Evidently, there is an anticorrelation between carbon content and particle size, carbon being more abundant in fine than coarse particles.

Measurements of δC^{13} in lunar fines cover a very narrow range from $+15$ to $+20\%_0$. The anomalously low value of $-17.6\%_0$ reported by Friedman et al. (1970) probably reflects serious contamination, possibly resulting from their sample being in contact with organic solvents prior to analysis. Contamination may also account for that group's negative δC^{13} data for breccia samples. Kaplan et al. (1970) and Epstein and Taylor (1970) report δC^{13} values ranging from $+2.2$ to $+10.8\%_0$ in breccia. To interpret these lighter δC^{13} figures as resulting from contamination does not appear reasonable because the breccias with the most carbon exhibit the highest δC^{13} values. Lowest δC^{13} measurements were obtained from the crystalline and fine-grained basaltic rocks. The fact that δC^{13} ranges from $-20\%_0$ to $-28\%_0$, coupled with the low carbon content in these rocks, raises the question of whether the carbon in these rocks is indigenous or primarily contamination.

Kaplan et al. (1970) examined the isotopic composition of carbon compounds evolved during stepwise pyrolysis of 10084 and 10086 A lunar fines over the range 150° to 1150 °C. From 150°–500 °C, δC^{13} values ranged from -41 to $-12\%_0$. Some of the evolved material is believed to be contamination. From 500°–750 °C, δC^{13} values increased to $+16$ to $+23\%_0$, and from 750° to 1150 °C, δC^{13} values decreased to the range $+9$ to $+13\%_0$. Although the δC^{13} values obtained at the mid-temperature range could be interpreted as the result of isotopic fractionation at lower temperatures involving preferential diffusion of compounds containing C^{12}, the observed decrease in the δC^{13} range at high temperatures argues against such a possibility and presents strong evidence for the presence in lunar fines of carbon with more than one range of isotopic values (Kaplan et al. 1970). A similar trend is evident in δC^{13} reported by Epstein and Taylor (1970).

Dissimilar experimental techniques and the occurrence of varying amounts of contamination between the times of sample collection and sample analysis probably account for some of the differences in total carbon and δC^{13} figures

within each rock group. There is evidence, however, that sample heterogeneity occurs which would ultimately be reflected in varying carbon contents and isotopic compositions. Much of the evidence for heterogeneity
consists of variations in the amount and composition of carbon compounds
isolated from lunar fines by separate investigating groups and will be discussed later. It should be kept in mind, though, that the identification of
some compounds could be made only if the investigators were actually
prepared to look for and find them. At times, the experimental conditions
required for analysis of a particular compound or class of compounds may
obscure or prevent identification of others, or the analytical schemes may
not have included any provisions for characterization of some compounds.
Consequently, some of the apparent discrepancies in the lunar analyses could
be attributed to experimental design; how many will be revealed when details
of the analyses are disclosed in the scientific journals.

The identity of lunar carbon compounds was established by a number of
chromatographic and spectrometric techniques, most important of which
were gas-liquid chromatography (GC) and mass spectroscopy (MS) (see
fig. 1). Release of volatile compounds from lunar materials was sought
through pyrolysis, often directly into the source of a mass spectrometer, and
by acid treatment. Extraction of non-volatile compounds was attempted
with benzene–methanol mixtures, water, and various acids.

5. Carbon compounds produced by pyrolysis

In a technique which presumably provides a measure of the total organic
carbon content, the lunar sample was pyrolysed at 800 °C in a stream of
hydrogen and helium (Johnson and Davis 1970). Resulting organic compounds were detected as a single peak with a flame ionization detector.
Under these conditions no response is obtained for CO, CO_2, graphite, ZrC,
or SiC. The amount of organic carbon in the 10086 A lunar fines was 40 ± 8
ppm. This figure is considerably higher than the preliminary value of 18 ppm
estimated by the same technique for a sample of 10002,73 fines. The latter
sample was removed from the same bulk container as 10086 A fines, but
subjected to much less handling (LSPET 1969). No correlation of organic
carbon content with rock type was reported. Taking contamination into
account, Johnson and Davis (1970) estimated that the total indigenous
organic carbon content in lunar material is less than 10 ppm.

Lunar fines were pyrolyzed in stepwise fashion in a closed, evacuated system. Gas chromatography of products obtained at $150°-250°C$, $250°-500°C$, and $500°-750°C$ revealed the presence of CO_2 amounting to about 50 ppm carbon, most appearing in the lower temperature ranges. Since large variations in the measurement occurred, this result was primarily of qualitative value. Carbon monoxide could not be determined under the gas chromatographic conditions used.

Abell et al. (1970), Burlingame et al. (1970), Epstein and Taylor (1970), and Oro et al. (1970) reported detection of CO during pyrolysis of lunar samples. In a later pyrolysis experiment we identified CO by high resolution MS only in the $500°-750°C$ range. The equivalent concentration of carbon was 121 ppm. Although no CO_2 was detected at each of three temperature stages in this experiment, a total of 24 ppm (3×8 ppm detection limit) carbon in the form of CO_2 could have gone undetected. The discrepancy in the CO_2 determinations does not appear so serious when the detection limits in the second experiment and the qualitative nature of the first are taken into consideration.

Release of CO_2 in lunar fines below $500°C$ has been reported by Abell et al. (1970), Lipsky et al. (1970), Murphy et al. (1970), Nagy et al. (1970) and Oro et al. (1970). The last group found most of the carbon in the form of oxides with CO_2 exceeding CO between $300°-500°C$, but CO exceeding CO_2 by a factor of 10 from $600°-750°C$. Epstein and Taylor (1970) reported similar results. In the range $150°-1150°C$, Burlingame et al. (1970) detected CO and CO_2 representing 156 ppm and 12 ppm carbon, respectively. Friedman et al. (1970) reported CO_2 from breccia between $300°-950°C$, but no CO. In another sample of breccia, Hintenberger et al. (1970) observed CO_2 and CO between $25°-1000°C$ in quantities corresponding to 78 ppm and 204 ppm carbon, respectively. Evidently CO and CO_2 determinations were reproducible between some laboratories, but not others.

In an experiment where CO_2 was detected by GC, no hydrocarbons were found at $150°-250°C$. Between $250°-500°C$, methane and an unidentified C_4 hydrocarbon were produced, and from $500°-750°C$, methane, ethene, propene, and an unidentified C_4 hydrocarbon were generated amounting to 1.6 ppm carbon. Similar results have been described by others. Pyrolysis of lunar fines to $400°C$ by Murphy et al. (1970) and above $400°C$ by Oro et al. (1970) afforded traces of methane. Other compounds were observed by the latter group at $600°C$ but the organic matter totaled less than 0.6 ppm. From $130°-700°C$, Abell et al. (1970) obtained small amounts (< 1 ppm) of methane and other unidentified hydrocarbons, and Burlingame et al. (1970), from $500°-1150°C$, detected traces of hydrocarbons. At $510°C$, Nagy et al. (1970) also reported traces, but at $700°C$ larger unstated amounts, predominantly methane, were found. In all instances except the last, only trace amounts (<1 ppm) of hydrocarbons were reported by others to have been produced by pyrolysis. The available data from pyrolysis experiments are summarized in table 8. Entries are listed in order of increasingly higher upper limits for pyrolysis temperature ranges.

TABLE 8

Results of pyrolysis experiments.

Pyrolysis temperature (°C)	ppm Carbon as			Reference
	CO	CO_2	Hydrocarbons	
50–250		+[a]		Lipsky et al. 1970
25–400		+	Trace[b]	Murphy et al. 1970
150–500	9	Trace	Trace	Burlingame et al. 1970
300–500	+	+		Oro et al. 1970
25–600			0.6	Oro et al. 1970
500–650	+	+		Epstein and Taylor 1970
500–700		+	+	Nagy et al. 1970
150–750		50	1.6	Chang et al. 1970
500–750	121			Chang et al. 1970
600–750	+ +[c]	+	Trace	Oro et al. 1960
130–900	+	+	Trace	Abell et al. 1970
25–1000	204	78		Hintenberger et al. 1970
150–1150	156	12	Trace	Burlingame et al. 1970
100–1200	+ +	+	Trace	Oro et al. 1970
550–1350		80–140		Friedman et al. 1970
650–1500	+ +	+		Epstein and Taylor 1970

[a] + indicates detection of unstated amounts.
[b] Trace corresponds to less than 1 ppm.
[c] + + corresponds to about 10 times +.

6. *Carbon compounds extracted with benzene-methanol*

Reference to fig. 1 shows that, if present, alkanes, aromatic hydrocarbons, fatty acids, and porphyrins are expected to be found in benzene-methanol extracts of lunar material. In our experiments spectrofluorometric evidence was obtained for porphyrin-like pigments (Hodgson et al. 1970) at an estimated concentration of 10^{-4} ppm of lunar fines. Rho et al. (1970), however, were not able to detect porphyrins using similar methods. None of the other classes of compounds were found at detection limits of less than

10^{-2} ppm. The absence of fatty acids, alkanes, aromatic hydrocarbons, as well as any other extractable non-polar compounds was corroborated by Burlingame et al. (1970), Meinschein et al. (1970), Murphy et al. (1970), and Oro et al. (1970). Furthermore, extractable polar compounds susceptible to trimethylsilylation were not detected by Murphy et al. (1970). According to detection limits established by some of these groups, the concentration in lunar fines of benzene–methanol soluble polar and non-polar organic compounds would be less than 1 ppm. On the other hand, Nagy et al. (1970) have observed alkyl ions $C_nH_{2n+1}^+$ (n as high as 40) in mass spectra obtained by direct probe analysis of evaporated benzene–methanol extracts. Saturated hydrocarbons were estimated to be present in concentrations ranging from 0.3 to 2.0 ppm.

7. Carbon compounds extracted with water

According to the scheme in fig. 1, free amino acids and sugars, if present in lunar fines, could be extracted with water. Sugars were not detected at a concentration of 2×10^{-3} ppm. Furthermore, using derivatization and gas chromatographic methods, neither we (Kvenvolden et al. 1970; Gehrke et al. 1970) nor Oro et al. (1970) were able to detect amino acids. Our minimum detectable concentration was 2×10^{-3} ppm. Using ion exchange chromatography, however, Fox et al. (1970) reported finding ninhydrin-reactive substances having chromatographic properties similar to glycine, alanine, threonine, serine, aspartic acid, and glutamic acid; and Nagy et al. (1970) reported glycine, alanine, urea, and ethanolamine. No positive identification was available, however, and the amounts of compounds in both cases amounted to less than 5×10^{-2} ppm organic carbon.

8. Carbon compounds freed by acid treatment

None of the expected organic compounds, amino acids and nucleic acid bases, were detected when lunar fines were treated with 1 N and 6 N HCl; minimum detectable concentrations of derivatives by means of GC were established at 2×10^{-3} ppm and 2×10^{-2} ppm, respectively (Kvenvolden et al. 1970). Furthermore, hexane extracts of HCl hydrolysates exhibited no peaks when examined by GC indicating that detectable amounts of non-polar compounds were not freed during acid hydrolysis. Similarly, Murphy et al. (1970) report that GC-MS analysis of methylene chloride extracts of HCl and HF hydrolysates revealed no significant peaks which could be attributed to material indigenous to the lunar fines. Clearly, ordinary polar and non-polar organic compounds were not freed during HCl and HF treatment.

Unusual derivatizable compounds were detected in 1 N and 6 N HCl hydrolysates, however (Gehrke et al. 1970). Although complete structural identification was not possible, analysis by GC-MS did indicate that the substances were organosiloxane derivatives. Based on hydrogen flame response to standard amino acid N-trifluoroacetyl *n*-butyl ester derivatives, the organosiloxanes were present at levels of 35 to 50 ppm. This corresponds to roughly 5 to 7 ppm carbon.

Treatment of lunar fines with HCl and HF led to evolution of CO_2 and C_1-C_4 hydrocarbons (including methane, ethane, ethene, acetylene, propane, propene, and unidentified C_3 and C_4 species). Only traces of CO_2 were collected amounting to less than 5 ppm carbon. The hydrocarbons represented substantial amounts of carbon, 5 to 20 ppm, in three experiments. Essentially the same hydrocarbons were obtained with a sample of 10084 fines and several samples of breccia and basaltic rock. The identification of hydrocarbons was obtained by GC and low resolution MS. Abell et al. (1970) and Burlingame et al. (1970) have examined volatile gases evolved during HF digestion. The former group reported 1 ppm methane, but no CO or CO_2. On the other hand, the latter group identified CO, amounting to 66 ppm carbon, but no CO_2 or hydrocarbons.

Carbides (Schmidt 1934) appear to be the only substances which would yield hydrocarbons during acid hydrolysis. Therefore, our results, combined with mineralogic observations made by Anderson et al. (1970), constitute convincing evidence for the existence of carbide carbon in lunar samples. That the hydrocarbons could have been freed from mineral enclosures by acid treatment is considered less plausible than their derivation from carbides. Data obtained from the pyrolysis experiments have some bearing on this question. During pyrolysis most of the total carbon was detected as oxides. Hydrocarbons should be as easily released from mineral enclosures as oxides of carbon; consequently, a major portion of the hydrocarbons should also have been released by pyrolysis. In fact the identifiable hydrocarbons amounted to less than 1 ppm carbon as compared to the much higher values of 5–20 ppm generated during HCl treatment. Furthermore, the composition of the hydrocarbon mixture produced by pyrolysis and hydrolysis were vastly different. Evidently, if hydrocarbons were trapped in mineral enclosures, and subsequently freed by HCl, they could have contributed only a small fraction of the hydrocarbons evolved.

9. Residual carbon

The residue remaining after pyrolysis of lunar fines to 750 °C contained 63 ppm carbon. When the corresponding pyrolysis residues from another sample of 10086 A and a sample of 10084 lunar fines were heated to 1050 °C and the resulting gases combusted in oxygen, measurement of the resulting CO_2 indicated that 100 % of the initial total carbon was accounted for. The residual carbon from the last two experiments had δC^{13} values of $+13$ and $+9.4\%_0$, respectively (Kaplan et al. 1970). Direct combustion of the remaining lunar material yielded no additional CO_2. Apparently, all the unpyrolyzed carbon was converted to volatile products at 1050 °C. Conversion of the carbon to carbon monoxide by way of the water-gas reaction ($C + H_2O \rightarrow CO + H_2$) at 750 °C, although thermodynamically favorable ($\Delta F = -2.13$ kcal), seems unlikely since most of the water evolved during pyrolysis of lunar material is expected to be driven off below 650 °C (Epstein and Taylor 1970). Some other possibilities at 1100 °C are shown below (all necessary thermochemical data obtained by interpolation from Dow Chemical Company, Thermal Research Laboratory 1965).

	ΔF (kcal)	P_{CO_2} (atmos)	P_{CO} (atmos)
$SiO_2 + 3C = SiC + 2CO$	28.5		5.4×10^{-3}
$TiO_2 + C = TiO + CO$	12.0		1.26×10^{-2}
$FeO + C = Fe + CO$	-11.8		74
$2FeO + C = 2Fe + CO_2$	-7.0	15	

In the first two equations the free energies are positive, but the equilibrium pressures of CO for the systems are appreciable. Confirmation that such processes were operating was provided in an experiment where 100 mg each of lunar fines and graphite were intimately mixed and pyrolyzed at 1100 °C. Oxidation of the resulting gases afforded an amount of CO_2 representing conversion of 3.1 mg of graphite to oxides of carbon, 200 times the quantity expected from pyrolysis of all the carbon in the lunar sample used in the experiment.

Material remaining after repeated treatment of lunar fines with HCl and HF contained 26 ± 6 ppm carbon with δC^{13} equal to $-6 \pm 2\%_0$. Since carbides would have been removed by this treatment, the residual carbon should be largely composed of elemental carbon and/or highly-condensed aromatic material resembling kerogen. The presence of 5–20 ppm carbon as carbides in the pyrolysis, but not in the hydrolysis, residues would account for part of the 27 ppm difference in residual carbon contents. Burlingame et al. (1970) reported that after digesting a sample of lunar fines in NF, pyrolysis of the residue to 1100 °C afforded 119 ppm carbon in the form of CO. Although, as they suggest, the CO could have been freed from the residual matrix by pyrolysis it seems highly probable that some was produced by oxidation of remaining non-extractable carbon by residual silica, metal oxides or metal salts as described above.

10. Discussion

Carbon and carbon compounds in lunar rocks and soils appear to be distributed heterogeneously. The total carbon measurements, roughly 50 to 250 ppm, cover approximately the same range estimated for the average concentration of carbon in igneous and metamorphic rock in the Earth's crust. The determinations fall in the low end of the wide range reported for carbon in meteorites.

The δC^{13} values for total carbon in lunar fines, $+15$ to $+20\%_0$ relative to the PDB standard, indicate anomalously heavy carbon when compared with the normal range $+2$ to $-30\%_0$ for terrestrial carbon and the range -4 to $-25\%_0$ for total meteorite carbon (Kaplan and Smith 1970). Only the carbon in carbonate phases of meteorites is more enriched in C^{13} with values in the range $+40$ to $+70\%_0$ (table 11). The varying δC^{13} values for products obtained at increasing pyrolysis temperatures and the different δC^{13} values for total carbon in fines, breccia, and rocks indicate the existence of carbon phases with more than one isotopic range, as is the case in meteorites and terrestrial samples. The anticorrelation between carbon content and lunar particle size could be a result of fine particles being more susceptible to contamination. Possible terrestrial contaminants are expected to exhibit δC^{13} values ranging from -20 to $-30\%_0$ so that their presence in lunar fines should bring δC^{13} values closer to that range. The δC^{13} measurements for carbon in the sieved particles, however, showed only slight variation from $+15$ to $+20\%_0$ in three size fractions, with a weighted mean value of $+17\%_0$. Clearly, the anticorrelation is real and terrestrial carbon contamination in that sample appears insignificant (Kaplan et al. 1970).

Of the variety of carbon-containing substances obtained from the lunar samples, gaseous oxidized carbon compounds, CO and CO_2, appear to be the most abundant. In most instances CO exceeded CO_2. Methane, the simplest reduced gas of carbon, was present only in trace amounts. Other significant inorganic forms of carbon include carbide and possible elemental carbon (graphite). With few exceptions, carbides (e.g., SiC, Cr_3C_2, CdC_2, Mg_2C_3) yield methane or acetylene or a mixture of both upon reaction with water or aqueous acids (Durrant and Durrant 1962). Notably, Fe_3C affords a mixture of hydrocarbons, as does cohenite, $(Fe, Ni, Co)_3C$ (Chang et al. 1970), which closely resembles the mixture obtained from lunar fines. In light of the 'complex texture of some iron' suggesting 'unmixed oxide, carbide, or phosphide, or all three' (Anderson et al. 1970), it is reasonable to conclude that iron carbides were present in lunar soil. Evidence for graphite was provided by Arrhenius et al. (1970) who reported finding a 1 mm^3 grain amidst lunar soil. The carbon remaining after exhaustive digestion of lunar fines with HF and HCl exhibited a δC^{13} value of $-6 \pm 2\%_0$. This δC^{13} value falls within the range determined for graphite in meteorites (table 11) and in some terrestrial ultramafic rocks (Vinogradov et al. 1967).

Except perhaps for small amounts (less than 0.6 ppm) of hydrocarbons produced during pyrolysis (Oro et al. 1970; Nagy et al. 1970), there appears to be little evidence for carbon in the form of organic polymers resembling kerogenic material. This is in direct contrast to the relative abundance of organic polymers in carbonaceous chondrites (Hayes 1967; Hayes and Biemann 1968). Murphy et al. (1970) reported finding none of the expected heteroaromatic pyrolysis products like pyridines, furanes and thiophenes at 400 °C, but Nagy et al. (1970) detected thiophenes at 700 °C in unstated amounts. If organic polymers were present in the lunar fines, they would have remained in sample residues after exhaustive treatment with HCl and HF or would have undergone transformation to graphitized or highly condensed aromatic systems after pyrolysis to 750 °C. δC^{13} values for residual carbon were $+11 \pm 2\%_0$ after pyrolysis and $-6 \pm 2\%_0$ after acid treatment (Kaplan et al. 1970). The ranges of δC^{13} for kerogen-like organic carbon in sedimentary rocks and meteorites are -9 to $-35\%_0$ (table 6) and -17 to $-15\%_0$ (table 11), respectively. Comparison of the isotope data indicate that if kerogen-like organic carbon is present in lunar fines, its isotopic composition differs dramatically from that of similarly constituted carbon in terrestrial rocks and meteorites.

Only a small fraction of the total lunar carbon is in the form of organic compounds. There is general agreement on the presence of methane and some other simple hydrocarbons. Organosiloxanes were detected, but their molecular structure is unknown. Compounds such as amino acids and porphyrins, for which there were no unambiguous identifications, were indicated in some instances, but not in others. No substantial evidence was found for any other classes of organic compounds, polar or non-polar.

If we assume that all organic compounds for which there is the slightest evidence were present, out of about 150 ppm total carbon in lunar fines, the organic carbon could amount of 11 ppm *at most*. The data which formed the basis for this estimate are presented in table 9. Entries were chosen from available data to maximize the possible organic content. The upper limit is, of course, highly tentative, and will remain so until confirmation of cited results and a more critical evaluation of contamination are provided in further investigations of lunar material. Interestingly, the average total organic carbon concentration in lunar material was estimated by Johnson and Davis (1970) to be less than 10 ppm. The seeming agreement between the two upper limits may be fortuitous, however.

TABLE 9

Maximum organic carbon content of Apollo 11 samples.

Organic compounds	Extraction method	Concentration ppm	Investigator
Hydrocarbons $< C_5$	Pyrolysis	1.6	Chang et al. 1970
Hydrocarbons $> C_8$	Benzene–methanol	0.3 to 2	Nagy et al. 1970
Organosiloxanes	HCl	5 to 7	Gehrke et al. 1970
Amino acids	H_2O	0.046	Nagy et al. 1970
Porphyrins	Benzene–methanol	0.0001	Hodgson et al. 1970
	Total	6.9 to 10.6	

The apparent discrepancy between the 40 ppm organic carbon found in 10086 A fines (Johnson and Davis 1970) and the maximum organic carbon as estimated above may be explained by the presence of carbides. Although SiC and ZrC were reported to give no response under the conditions for total organic carbon determination, Bahr and Bahr (1928 and 1930) reported that carbide carbon in Ni_3C was converted to methane and ethane when heated in the presence of hydrogen at 180°–350 °C. Quite possibly other carbides such as Fe_3C or cohenite behave in a similar fashion. In comparison to lunar material the concentration of organic carbon, determined by the method of Johnson and Davis (1970), in terrestrial soils ranges from 30–500 ppm in dry desert areas and 500–20000 ppm in agricultural areas. In Precambrian sediments, where much of the organic carbon has been incorporated in kerogen, concentrations as low as 50 ppm are obtained. Organic carbon in the recently fallen Pueblito de Allende meteorite amounted to 37 ppm.

In some cases identification of lunar carbon compounds was possible only after they were removed from the mineral matrix by rather drastic processes: hydrolysis with strong acid and pyrolysis. Therefore, in these cases, it is important to consider whether the isolated compounds were present originally or formed by degradation of or synthesis from other species during the experiments. Because the exact structure of the organosiloxanes obtained in HCl hydrolysates is unknown, nothing can be stated regarding their precursors or their possible molecular alterations during isolation with acid.

Methane, obtained during HF treatment of lunar fines, was thought to be released from enclosures by destruction of the mineral matrix (Abell et al. 1970); however, all of the methane could have been formed by hydrolysis of carbides. Wide differences in the compositions of hydrocarbons produced by pyrolysis and by hydrolysis indicate that hydrocarbons generated in pyrolysis experiments did not result from hydrolysis of carbides by water evolved during heating. Roedder and Weiblen (1970) observed gas-filled inclusions in olivine in lunar rocks and glass bubbles filled with non-condensable gases in lunar fines, and trapped gaseous hydrocarbons have been freed from meteorites by Studier et al. (1965) and by Belsky and Kaplan (1970). Therefore, it is possible that production of some hydrocarbons and other gases during pyrolysis may have been a result of liberation from mineral enclosures. On the other hand, it is well established that thermal degradation of many synthetic and natural organic monomers (Wolf 1966; Hurd 1929) and polymers (Stevens 1969; Frazer 1967; Society of Chemical Industry 1961) to hydrocarbons and other fragments occurs at temperatures as low as 300 °C. Thus, hydrocarbons evolved below 300 °C may have been originally trapped in the mineral matrix, but compounds identified at higher temperatures very probably resulted from thermal degradation of more complex organic species.

Another possible hydrocarbon origin was suggested by the work of Studier et al. (1968) who showed that in the presence of iron meteorite powder, hydrocarbons could be synthesized from CO and H_2 at temperatures around 150 °C and above. Lunar samples contain small amounts of native iron (LSPET 1969), and CO and H_2 (Epstein and Taylor 1970) were produced by pyrolysis up to 750 °C. Therefore, all the ingredients were available for the Fischer-Tropsch type synthesis. The free energy of the reaction $CO + 3H_2 = H_2O + CH_4$ is 7.8 kcal at 750 °C, -7.2 at 500 °C, and increasingly negative at lower temperatures. In the presence of appropriate catalysts, the Fischer-Tropsch reaction would be favored at the lower end of the range 150°–750 °C. Nevertheless, because most hydrocarbon production was observed at 750 °C in our experiments and at 700 °C in those of Nagy et al. (1970), we investigated the possibility of Fischer-Tropsch synthesis at 750 °C. When a previously pyrolyzed lunar sample was reheated to 750 °C in the presence of CO and H_2, only traces of methane, ethene and propene, which were absent in a control experiment, were detected (less than 0.65 ppm) by gas chromatography. Although the same gases were observed in our pyrolysis experiments, apparently, a Fischer-Tropsch type reaction could account for only a small fraction of the 1.6 ppm carbon in the hydrocarbons generated during pyrolysis at 750 °C. Some of the light hydrocarbons produced at lower temperatures, however, could have resulted from heating CO and H_2 together in cavities and pores in lunar material.

Pyrolysis and acid treatment of lunar samples also produced substantial quantities of oxides of carbon. Thermal degradation of organic compounds seems an unlikely source for the CO and CO_2 because the expected organic fragments, such as hydrocarbons, were always obtained in relatively minute amounts under circumstances when large amounts of CO and/or CO_2 were obtained. Formaldehyde, formic acid and methanol are reportedly susceptible to thermal degradation to oxides of carbon without concomitant formation of hydrocarbons (Wolf 1966), but no clear evidence exists for these compounds in lunar samples. Had they been present in the mineral matrix, they would have been released continuously from the low pyrolysis temperatures to temperatures just below their decomposition point. Formation of CO_2 by breakdown of carbonates is possible, but the fact that, at most, only traces of CO_2 were evolved when lunar material was digested with HCl and HF indicates that only traces of carbonate were present. This last observation also conflicts with the notion that CO_2 was originally trapped in mineral enclosures. Similarly, the fact that neither we nor Abell et al. (1970) detected CO during HF treatment of lunar fines contradicts the idea that CO is in mineral or glass bubbles. Conceivably, our experimental conditions were not sufficiently drastic to achieve adequate breakdown of the lunar mineral enclosures.

After HF treatment of lunar fines, Burlingame et al. (1970) reported finding CO amounting to 66 ppm carbon. They suggested that CO was freed from enclosures, but their estimate that glass bubbles in lunar fines had to contain 10^4 atmospheres partial pressure of CO to account for their results led them to suggest that some of the CO was in non-gaseous form, possibly metal carbonyls. Iron pentacarbonyl, $Fe(CO)_5$, like many other carbonyls decomposes to yield CO when heated at temperatures as low as 200 °C or when treated with aqueous acids. Thermal decomposition of carbonyls may indeed account for some of the CO produced in pyrolysis experiments. The existence of conditions on the lunar surface, now or in the past, compatible with those necessary for the synthesis of simple carbonyls (CO and metal at 200°–250 °C and 200–250 atmospheres, Durrant and Durrant 1962) is not unreasonable.

Although the CO and CO_2 determinations based on pyrolyses of lunar samples were not all consistent, there are some observable trends in the data (see table 8). In most instances, CO and CO_2 were evolved simultaneously. If both gases were not evolved, the single gas was always CO_2; and CO_2 was obtained generally at lower initial temperatures than CO. Most significantly, when the relative abundances of CO and CO_2 were observed at two temperatures (Oro et al. 1970; Epstein and Taylor 1970), the gas composition at the higher temperature reflected a serious change manifested by much more CO than CO_2, even though the gases were initially evolved in comparable quantities. If all CO and CO_2 were simply freed from enclosures in the mineral matrix, their relative abundance at any given pyrolysis temperature should not differ greatly from that at any other temperature. The data suggest, therefore, that in most cases, CO and CO_2 were not originally present in the lunar samples in the total amounts or relative abundances revealed by pyrolysis experiments. Some CO and CO_2 may have been trapped in mineral enclosures, but apparently, their release during pyrolysis is accompanied by chemical reactions in the lunar matrix which bring about changes in gas composition as the pyrolysis temperature is increased. Some possible reactions are shown in table 10. Included are the free energies at several temperatures and the equilibrium pressures of products where calculable (thermochemical data obtained by interpolation from Dow Chemical Company, Thermal Research Laboratory 1965).

At 500 °C reduction of CO_2 to CO by iron (reaction 2) and conversion of CO to methane by hydrogen (reaction 7) are thermodynamically feasible processes. Native iron is present in lunar material (LSPET 1969) and hydrogen is evolved up to 750 °C during pyrolysis (Epstein and Taylor 1970); therefore, it is possible for some entrapped CO_2 to be converted to CO and methane by reactions 2 and 7. At 750 °C reduction of CO_2 to CO by carbon (reaction 3) and the water–gas reaction (reaction 4) are possible. There is evidence for elemental carbon in lunar samples (see above); but water is expected to be expelled by 650 °C (Epstein and Taylor 1970), thereby reducing the probability of the water–gas reaction. At 1100 °C and above, reactions

(1–3, 5) involving conversion of elemental carbon or CO_2 to CO are very favorable. At these temperatures, carbides of chromium or iron, when heated with the oxides of these metals, produce CO (Chamberlin 1908).

TABLE 10

Thermodynamic data for possible reactions during lunar sample analysis.

Reaction	T (°C)	ΔF (kcal)	P_{CO} (atmos)	P_{CO_2} (atmos)	P_{CO}/P_{CO_2}
(1) $FeO + C = Fe + CO$	500	9.8	0.0013		
	750	0.9	0.64		
	1100	−11.8	74		
(2) $Fe + CO_2 = FeO + CO$	500	−1.3			1.6
	750	−2.9			2.8
	1100	−4.8			5.8
(3) $C + CO_2 = 2CO$	500	8.5			
	750	−2.1			
	1100	−16.6			
(4) $C + H_2O = H_2 + CO$	500	6.0			
	750	−2.6			
	1100	−14.6			
(5) $H_2 + CO_2 = H_2O + CO$	500	2.5			
	750	0.6			
	1100	2.0			
(6) $2FeO + C = 2Fe + CO_2$	500	11.1		0.0008	
	750	3.7		0.17	
	1100	−7.0		15	
(7) $CO + 3H_2 = H_2O + CH_4$	500	−7.2			
	750	7.8			
	1100	29.1			

Even though CO_2 may initially be in large abundance, and partial conversion of carbon to CO_2 by FeO may be occurring (reaction 6), the thermodynamic data indicate that as pyrolysis temperatures increase, CO will become the predominant gas through reduction of CO_2 by metals or oxidation of carbon by oxides. Because thermodynamic considerations reveal little about kinetics, we cannot be certain whether any of the indicated reactions proceed at meaningful rates during pyrolysis of lunar samples. On the other hand, results described in the section on residual carbon and those summarized in table 8, especially for temperatures around 1100°C or higher, support the view that some of the listed reactions, and perhaps similar ones involving other metals and metal oxides, played significant parts in determining CO and CO_2 compositions during pyrolysis experiments.

From the present data, we can only make suggestions regarding the generation of oxides of carbon and hydrocarbons during the lunar analyses. Most of the hydrocarbons evolved during acid treatment are probably derived from carbide hydrolysis, with possibly a small contribution from mineral enclosures. Hydrocarbons generated during pyrolysis could be derived from any or all of three processes: diffusion out of mineral enclosures, thermal degradation of organic compounds, Fischer-Tropsch synthesis from CO and H_2. The last process appears to be relatively unimportant at 750 °C, but no assessment of the overall importance of the three processes can be made. Much of the CO_2 produced during pyrolysis could have been present originally in the lunar samples. This may also have been true for CO; however, CO could have been produced by thermal or acid-induced breakdown of carbonyls. At high temperatures (750 °C and above) significant amounts of CO and probably CO_2 were likely to have been produced by chemical reactions during pyrolysis. The relative importance of the suggested sources for CO and CO_2 cannot be estimated at present.

An interesting comparison can be made between the lunar analyses and work completed more than 60 years ago. Chamberlin (1908) reported that heating a great many igneous rocks and meteorites under vacuum resulted in evolution of CO, CO_2 and methane (among other gases) in varying proportions. Generally, CO_2 was most abundant, except in iron meteorites where CO predominated. Although Chamberlin (1908) believed that CO_2 was produced by thermal degradation of carbonates, CO by reduction of CO_2 by iron and decomposition of carbonyls, and methane by hydrolysis of carbides and degradation of organic matter, his extensive experiments did not exclude the possibility that some of the gases were occluded in the rocks and meteorites. If more systematic and quantitative data were to become available for lunar samples, it might be possible to observe more clearly parallels or divergences between carbon-containing gases and their origins in lunar materials and such gases and their origins in terrestrial rocks and meteorites.

11.　*Origin of carbon and carbon compounds on the moon*

On the basis of present information, no conclusions can be drawn regarding the origin of the carbon in Apollo 11 lunar samples. However, it is interesting to explore the most likely possibilities: primordial carbon indigenous to the moon, carbon accumulated by means of meteorite infall, and carbon implanted by the solar wind. Almost certainly, all three sources contribute in some measure to the store of carbon on the lunar surface.

The difficulty with discussing primordial indigenous lunar carbon is that we have no firm idea of the forms or compounds in which it must originally have existed. If meteorites represent material condensed and accreted during cooling of the solar nebula (Larimer and Anders 1967; Wood 1966; Ringwood 1966; Anders 1964; Wood 1963), and if related condensation and accretion processes were responsible for the formation of the moon, then parallels may be drawn between carbon and carbon compounds found in meteorites (see table 5) and the state and constitution of carbon on the primitive moon. However, there is considerable petrological evidence (Wood et al. 1970; O'Hara et al. 1970; Ringwood and Essene 1970; Kushiro et al. 1970; Weil et al. 1970) that the area of the Apollo 11 landing underwent local melting and, possibly, crustal differentiation by melting. Under such circumstances, most, if not all, of the original volatile carbon compounds would have diffused or boiled out prior to or during melting and been lost from the moon's surface. Carbides, graphite, amorphous and interstitial carbon and highly condensed kerogen-like material would have been partially converted to CO and/or CO_2 by silica, silicates, metal oxides and metal salts, depending on the oxidation state, the temperature, and the degree of approach to thermodynamic equilibrium in the melts. Graphite and carbide isolated from silicate or oxide phases would have been temporarily preserved until remelting by meteorite impact or subcrustal magma flows promoted phase remixing and further oxidation to CO and CO_2. Wellman (1970) has predicted that upon cooling and crystallization of the melts, CO, CO_2, and methane would be entrapped in vesicles and vugs in relative abundances $CO > CO_2 \gg CH_4$. These three gases have been identified in lunar samples; however, since some proportion of them could also have been formed by synthesis from or degradation of other substances during the analyses, we cannot say whether their original abundances in lunar samples were consistent with Wellman's prediction.

In the present context, it is interesting to note that the amount of carbon in lunar rocks is roughly the same as that in the Earth's crust, which consists primarily of igneous and metamorphic rocks. Furthermore, olivine-bearing basalts from 72 locales on the Earth have been found to contain CO_2 inclusions (Roedder 1965). Although lunar materials appear to have a special geochemical history, they are more similar to basaltic achondrites and terrestrial basalts than to any other materials with which they were compared (Morrison et al. 1970).

Ample evidence for a meteoritic component in lunar fines has been reported (Keays et al. 1970; King et al. 1970; Mason et al. 1970; Arrhenius et al. 1970; Ramdohr and Goresey 1970; Quaide et al. 1970). Keays et al. (1970) estimate that a mixture of 1.5 to 2% carbonaceous chondritic material with lunar fines and breccias would account for the observed enrichment of a number of elements over amounts found in rock samples. Using 2% as a basis and recalling that carbonaceous chondrites have been found to contain 0.5 to 5% carbon, it can be calculated that a concentration of 150 to 1500 ppm carbon could have accumulated in the lunar fines from fragments of carbonaceous chondrites alone. Impact on the lunar surface, however, would result in partial or total melting and vaporization of the meteorite accompanied by loss of most, if not all, of the volatile carbon compounds, as well as some of the non-volatile carbon by conversion to CO and/or CO_2. The quantity of meteoritic carbonaceous material retained on the lunar surface after impact would depend on a number of factors, among which collision speed, original carbon content of meteorite, and cooling rates of melts and vapors are probably important.

Carbon in carbonaceous chondrites occurs in a variety of compounds including carbonate, extractable organic substances, and non-extractable heteroaromatic polymers (table 5; Hayes 1967; Hayes and Biemann 1968). Some evidence exists for these types of compounds in lunar fines, but their presence is not sufficient reason to assign them a meteoritic origin. The most significant evidence, as pointed out by Kaplan et al. (1970), lies in the δC^{13} data. In carbonaceous chondrites, δC^{13} values for total carbon lies in the range -7 to $-18\%_0$, and for non-extractable carbon, -15 to $-17\%_0$ (table 11). By comparison, δC^{13} for total carbon in lunar fines was $+15$ to

TABLE 11

δC^{13} in meteorites.

	δC^{13} range (%_0)
Total carbon	-18 to -7
Organic	
Extractable	-27 to -17 (-5.3 one value)
Polymeric?	-17 to -15
Elemental carbon	
Graphite	-8 to -4
Diamond	-5.7 (Vinogradov et al. 1967)
Amorphous	—
Carbonate	$+40$ to $+70$
Carbide	-8 to -4

$+20\%_0$, and the value of $-6 \pm 2\%_0$ was obtained for residual carbon remaining after exhaustive treatment of lunar fines with HCl and HF. Clearly, the isotope evidence is inconsistent with a major carbon contribution from carbonaceous chondrites. Moore et al. (1970) have indicated that a carbonaceous chondritic component could not account for nitrogen abundances in lunar material, and Kaplan et al. (1970) note that sulfur abundance and isotope data cannot be accomodated by the simple addition of meteoritic material to the lunar soil. On the other hand, the δC^{13} value for residual carbon does fall in the range for graphite from iron meteorites (table 11). Interestingly, iron meteorites reportedly contain extractable hydrocarbons in graphite nodules (Nooner and Oro 1967) and dissolved CO (Cohen 1894; Chamberlin 1908), substances that were detected in lunar samples. Ramdohr and Goresey (1970) have suggested that iron meteorites were most likely to survive impact on the lunar surface. If a small fraction of the carbon and carbon compounds found in lunar samples was actually derived from meteorites, it is possible that the contribution was largely from iron rather than carbonaceous meteorites.

The noble gases in lunar fines were thought to have been implanted by the solar wind (LSPET 1969). The anticorrelation between noble gas content and grain size of lunar fines was considered consistent with such an interpretation (Hintenberger et al. 1970; Heyman et al. 1970; Eberhardt et al. 1970; Kirsten et al. 1970). A similar anticorrelation between the carbon content and grain size was observed (Moore et al. 1970; Kaplan et al. 1970), suggesting a solar wind origin for the carbon. Kaplan et al. (1970) observed close similarities in δC^{13} values for different grain sizes and suggested the possibility of a common source of carbon for the lunar fines. Abell et al. (1970) and Moore et al. (1970) have estimated that 1 to 1000 ppm carbon could have been deposited on the moon's surface by the solar wind.

The kinetic energies of solar wind ions were estimated at 470–2900 eV/ nucleon by Lal et al. (1969). Interaction of comparably energetic protons with various surfaces during laboratory tests showed that: metals, metal oxides and a variety of rocks, including basalt, suffer sputtering losses (Wehner et al. 1963; Kenknight and Wehner 1964); metal oxides and other compounds undergo metal atom enrichment, probably through reduction by hydrogen (Wehner et al. 1963); hydrogen and helium are implanted in minerals (Lord 1968) and metals (Lal et al. 1969; Grant and Carter 1965); and protons induce hydroxyl formation in glasses (Zeller et al. 1966). In the absence of knowledge concerning the products to be expected from reactions between solar wind carbon ions with minerals and other materials on the lunar surface, little can be inferred about a solar wind origin from the identity of lunar carbon compounds. If one may draw analogies from the experimental work with hydrogen ions, one might speculate that organosiloxanes could conceivably result from reaction between solar wind hydrogen and carbon ions and silicate minerals, or that the carbon ions would be neutralized by the lunar surface and remain imbedded in the soil as elemental carbon.

Carbon isotope data may contain valuable information, but Oro et al. (1970) have pointed out that, although the lunar isotope ratios are consistent with some solar carbon isotope ratios (Greenstein et al. 1950; Herzberg et al. 1967), the latter are not known with sufficient accuracy to warrant drawing conclusions from them. Kaplan et al. (1970), on the other hand, point out that the notion of the solar wind as a major source of carbon in the lunar fines cannot account for the observation that the carbon content in breccia, from which lunar fines were presumably derived, is as high as in the fines, whereas the δC^{13} values are lower.

Kaplan and Smith (1970) have made the interesting suggestion that penetration into and interaction with the carbon or carbon compounds in the lunar fines by protons in the solar wind may have led to the heavy isotope enrichment of the carbon in the fines over that in breccia and basaltic rocks. A 'hydrogen-stripping' process was proposed which involved conversion of carbon, CO, and CO_2 into methane by solar wind hydrogen followed by preferential diffusion loss of isotopically light methane. Kaplan et al. (1970) have calculated that kinetic isotope effects in 'hydrogen-stripping' could account for the anomalously heavy carbon in the lunar fines. Such a process is expected to be more efficient for fine particles than breccia or rocks because the fines have a relatively high surface to volume ratio and can be easily overturned by micrometeorite impact to constantly expose fresh surfaces. Solar wind ions undoubtedly interact with the lunar surface; however, there are little data to indicate their actual contribution to and effect on carbon on the moon.

12.   Abiotic organic synthesis on the moon

Evidence supporting the hypothesis of abiotic primordial synthesis of organic matter on the Earth has accumulated since Miller (1953) first demonstrated the synthesis of organic compounds from methane, ammonia and water (Fox 1965; Ponnamperuma and Gabel 1968). The occurrence of extraterrestrial synthesis is indicated by carbonaceous material isolated from meteorites

(Hayes 1967), and the possibility of cosmochemical organic synthesis has been raised by the identification of simple compounds containing carbon, hydrogen, nitrogen, and oxygen in comets (Swings and Haser 1956), interstellar dust clouds (Cheung et al. 1968; Snyder et al. 1969) and in stellar atmospheres (Vardya 1966).

It was hoped that analysis of the organic substances in lunar material could confirm or shed new light on the mechanisms for synthesis of carbon compounds in the solar system (Sagan 1961; Dayhoff et al. 1964; Eck et al. 1966; Studier et al. 1968). Since high temperatures were apparently involved in formation of soils and rocks in the Sea of Tranquillity, products of primordial organic synthesis are not likely to have survived. This is consistent with the low concentration of organic compounds found in lunar samples (see above).

Pathways appear to be available for more recent organic synthesis on the moon, however. Carbon monoxide, metals, metal oxides, and hydrogen were found indigenous to lunar material; therefore, under appropriate conditions Fischer-Tropsch reactions may lead to formation of hydrocarbons. Water is evolved from lunar soil by pyrolysis so that hydrolysis of carbides to hydrocarbons is also conceivable. Similarly, hydrolysis of nitrides would afford ammonia, which in combination with CO, hydrogen, and metal catalysts could lead to formation of nitrogenous organic compounds under favorable conditions (Hayatsu 1968). The thermal energy required for these processes could have been provided by impact of meteorites and comets and volcanism. Furthermore, carbon compounds originally in the impacting bodies would be converted, at least partially, to CO and hydrogen. Upon cooling, Fischer-Tropsch type reactions could take place in molten meteorite fragments. Entrapment of products derived from these processes in the cooling mineral matrices (meteoritic and lunar) could preserve some of the organic compounds from loss by volatilization or destruction by solar and cosmic radiation. Despite all these possibilities, the lunar analyses thus far indicate that if organic synthesis had occurred in the area of the Apollo 11 landing, very little evidence remains.

References

ABELL, P. I., G. H. DRAFFAN, G. EGLINTON, J. M. HAYES, J. R. MAXWELL and C. T. PILLINGER, 1970, Science 167, 757.

ANDERS, E., 1964, Space Sci. Rev. 3, 583.

ANDERSON, A. T., Jr., A. V. CREWE, J. R. GOLDSMITH, P. B. MOORE, J. C. NEWTON, E. J. OLSEN, J. V. SMITH and P. J. WYLLIE, 1970, Science 167, 587.

ARRHENIUS, G., S. ASUNMAA, J. I. DREVER, J. EVERSON, R. W. FITZGERALD, J. Z. FRAZER, H. FUJITA, J. S. HANOR, D. LAL, S. S. LIANG, D. MACDOUGALL, A. M. REID, J. SINKANKAS and J. WILKENING, 1970, Science 167, 659.

BAHR, H. A. and T. BAHR, 1928. Ber. 61, 2177.

BAHR, H. A. and T. BAHR, 1930, Ber. 6313, 99.

BELSKY, T. and I. R. KAPLAN, 1970, Geochim. Cosmochim. Acta 34, 257.

BORCHERT, H., 1951. Geochim. Cosmochim. Acta 2, 62.

BURLINGAME, A. L., M. CALVIN, J. HAN, W. HENDERSON, W. REED and B. R. SIMONEIT, 1970, Science 167, 751.

CAMERON, A. G. W., 1968, In: L. H. Ahrens, ed., Origin and distribution of the elements. (Pergamon Press, New York) 125.

CHAMBERLIN, R. T., 1908, The gases in rocks. (Carnegie Institute of Washington, publication No. 106, Washington, D.C.).

CHANG, S., J. W. SMITH, I. KAPLAN, J. LAWLESS, K. A. KVENVOLDEN, and C. PONNAM-PERUMA, 1970, Geochim. Cosmochim. Acta, Supplement 1, 1857.

CHEUNG, A. C., D. M. RANK, C. H. TOWNES, D. D. THOMSON and W. J. WELCH, 1968, Phys. Rev. Letters 21, 1701.

COHEN, E., 1894, 1903, 1905, Meteoritenkunde, Vols. I-III. (E. Schweizerbart, Stuttgart).

CRAIG, H., 1953, Geochim. Cosmochim. Acta 3, 53.

DAYHOFF, M. O., E. R. LIPPINCOTT and R. V. ECK, 1964, Science 146, 1461.

DEGENS, E., 1965, Geochemistry of sediments. (Prentice-Hall, Inc., New Jersey) 342.

DOW CHEMICAL COMPANY, THERMAL RESEARCH LABORATORY, 1965, JANAF Thermochemical tables, National Bureau of Standards, Institute for Applied Technology.

DURRANT, P. J. and B. DURRANT, 1962, Introduction to advanced inorganic chemistry. (Longmans, Green and Co. Ltd., London).

EBERHARDT, P., J. GEISS, H. GRAF, N. GROGLER, U. KRAHENBUHL, H. SCHWALLER, J. SWARZMULLER and A. STETTLER, 1970, Science 167, 558.

ECK, R. V., E. R. LIPPINCOTT, M. O. DAYHOFF and Y. T. PRATT, 1966, Science 153, 628.

EPSTEIN, S. and H. P. TAYLOR Jr., 1970, Science 167, 533.

FOX, S. W., ed., 1965, The origins of prebiological systems and of their molecular matrices; Proceedings of a conference held at Wakulla Springs, Oct. 1963. (Academic Press, N.Y.).

FOX, S. W., K. HARADA, P. E. HARE, G. HINSCH and G. MUELLER, 1970, Science 167, 767.

FRAZER, A. H., ed., 1967, High temperature resistant fibers. (Interscience Publishers, New York).

FRIEDMAN, I., J. R. O'NEIL, L. H. ADAMI, J. D. GLEASON and K. HARDCASTLE, 1970, Science 167, 538.

GEHMAN, H. M., Jr., 1962, Geochim. Cosmochim. Acta 26, 885.

GEHRKE, C. W., R. W. ZUMWALT, W. A. AUE, D. L. STALLING, A. DUFFIELD, K. A. KVEN-VOLDEN and C. PONNAMPERUMA, 1970, Geochim. Cosmochim. Acta, Supplement 1, 1845.

GRANT, W. A. and G. CARTER, 1965, Vacuum 15, 477.

GREENSTEIN, J. L., R. S. RICHARDSON and M. SCHWARZCHILD, 1950, Publ. Astron. Soc. Pacific 62, 15.

HAYATSU, R., M. H. STUDIER, A. ODA, K. FUSE and E. ANDERS, 1968, Geochim. Cosmochim. Acta 32, 175.

HAYES, J. M., 1967, Geochim. Cosmochim. Acta 31, 1395.

HAYES, J. M. and K. BIEMANN, 1968, Geochim. Cosmochim. Acta 32, 239.

HERZBERG, L., L. DELBOUILLE and G. ROLAND, 1967, Astrophys. J. 147, 697.

HEYMANN, D., A. YANIV, J. A. S. ADAMS and G. E. FRYER, 1970, Science 167, 555.

HINTENBERGER, H., H. W. WEBER, H. VOSHAGE, H. WANKE, F. BEGEMANN, E. VILSECK and F. WLOTZKA, 1970, Science 167, 543.

HODGSON, G. W., E. BUNNENBERG, B. HALPERN, E. PETERSON, K. A. KVENVOLDEN and C. PONNAMPERUMA, 1970, Geochim. Cosmochim. Acta, Supplement 1, 1829.

HOEFS, J., 1965, Geochim. Cosmochim. Acta 29, 399.

HURD, C. D., 1929, ACS Monograph No. 50 (The Chemical Catalogue Co., Inc., New York).

JOHNSON, R. D. and C. C. DAVIS, 1970, Geochim. Cosmochim. Acta, Supplement 1, 1805.

KAPLAN, I. R. and A. NISSENBAUM, 1966, Science 153, 744.

KAPLAN, I. R. and J. W. SMITH, 1970, Science 167, 541.

KAPLAN, I. R., J. W. SMITH and E. RUTH, 1970, Geochim. Cosmochim. Acta, Supplement 1, 1317.

KEAYS, R. R., R. GANAPATHY, J. C. LAUL, E. ANDERS, G. F. HERZOG and P. M. JEFFREY, 1970, Science 167, 490.

KENKNIGHT, C. E. and G. K. WEHNER, 1964, J. Appl. Phys. 35, 322.

KING, E. A., Jr., M. F. CARMAN and J. C. BUTLER, 1970, Science 167, 650.

KIRSTEN, T., F. STEINBRUNN and J. ZAHRINGER, 1970, Science 167, 571.

KUSHIRO, I., Y. NAKAMURA, H. HARAMURA and S. AKIMOTO, 1970, Science 167, 610.

KVENVOLDEN, K. A., S. CHANG, J. W. SMITH, J. FLORES, K. PERING, C. SAXINGER, F. WOELLER, K. KEIL, I. BREGER and C. PONNAMPERUMA, 1970, Geochim. Cosmochim. Acta, Supplement 1, 1813.

LAL, D., W. F. LIBBY, G. WETHERILL and J. LEVENTHAL, 1969, J. Appl. Phys. 40, 3257.

LARIMER, J. W. and E. ANDERS, 1967, Geochim. Cosmochim. Acta 31, 1239.

LIPSKY, S. R., R. J. CUSHLEY, C. G. HORVATH and W. J. MCMURRAY, 1970, Science 167, 778.

LORD, H. C., 1968, J. Geophys. Res. 73, 5271.

LSPET (Lunar Sample Preliminary Examination Team), 1969, Science 165, 1211.

MASON, B., 1962, Meteorites. (John Wiley and Sons, New York) 274.

MASON, B., 1966, Geochim. Cosmochim. Acta 30, 23.

MASON, B., 1966, Principles of Geochemistry. 3rd ed. (John Wiley and Sons, New York) 329.

MASON, B., K. FREDRIKSSON, E. P. HENDERSON, E. JAROSEWICH, W. G. MELSON, K. M. TOWE and J. S. WHITE, Jr., 1970, Science 167, 656.

MEINSCHEIN, W., E. CORDES and V. J. SHINER, Jr., 1970, Science 167, 753.

MILLER, S. L., 1953, Science 117, 528.

MOORE, C. B., C. F. LEWIS, E. K. GIBSON and W. NICHIPORUK, 1970, Science 167, 495.

MORRISON, G. H., J. T. GERARD, T. A. KASHUBA, E. V. GANGADHARA, A. M. ROTHEN-BERG, N. M. POTTER and G. B. MILLER, 1970, Science 167, 505.

MURATA, K. J., I. FRIEDMAN and B. M. MADSEN, 1969, U.S.G.S. Prof. Paper 614-B, 1.

MURPHY, R. C., G. PRETI, M. NAFISSI and K. BIEMANN, 1970, Science 167, 755.

NAGY, B., 1966, Geol. Foren. Stockholm Forh. 88, 235.

NAGY, B., C. M. DREW, P. B. HAMILTON, V. E. MODZELESKI, M. E. MURPHY, W. M. SCOTT, H. C. UREY and M. YOUNG, 1970, Science 167, 770.

NOONER, P. W. and J. ORO, 1967, Geochim. Cosmochim. Acta 31, 1359.

O'HARA, M. J., G. M. BIGGAR and S. W. RICHARDSON 1970, Science 167, 605.

ORO, J., W. J. UPDEGROVE J. GILBERT, J. MCREYNOLDS, E. GIL-AV, J. IBANEZ and A. ZLATKIS, 1970, Science 167, 765.

PONNAMPERUMA, C. and N. GABEL, 1968, Space Life Sci. 1, 64.

QUAIDE, W., T. BUNCH and R. WRIGLEY, 1970, Science 167, 671.

RAMDOHR, P. and A. E. GORESEY, 1970, Science 167, 615.

RANKAMA, K. and TH. G. SAHAMA, 1950, Geochemistry. (The Univ. of Chicago Press) 912.

RHO, H. J., A. J. BAUMAN and T. F. YEN, 1970, Science 167, 754.

RINGWOOD, A. E., 1966, Rev. Geophys. 4, 113.

RINGWOOD, A. E. and E. ESSENE, 1970, Science 167, 607.

ROEDDER, E., 1965, Am. Minerologist 50, 1746.

ROEDDER, E. and P. W. WEIBLEN, 1970, Science 167, 641.

SAGAN, C., 1961, Nat. Acad. Sci., Nat. Res. Council Publ. 757, 49.

SCHMIDT, J., 1934, Z. Elecktrochem. 40, 170.

SMITH, J. W. and I. R. KAPLAN, 1970, Science 167. 1367.

SNYDER, L. E., D. BUHL, B. ZUCKERMAN and P. PALMER, 1969, Phys. Rev. Letters 22, 679.

SOCIETY OF CHEMICAL INDUSTRY, 1961, S.C. 1. Monograph no. 13. (MacMillan Company, New York).

STEVENS, M. P., 1969, Characterization and analysis of polymers by gas chromatography. (M. Dekker, New York).

STUDIER, M. H., R. HAYATSU and E. ANDERS, 1965, Science 149, 1455.

STUDIER, M. H., R. HAYATSU and E. ANDERS, 1968, Geochim. Cosmochim. Acta 32, 151.

SWINGS, P. and L. HASER, 1956, Atlas of representative cometary spectra. (University of Liège Astrophysical Institute, Louvain).

VARDYA, M. S., 1966, Monthly Notices. Roy. Astron. Soc. 134, 347.

VINOGRADOV, A. P., O. I. KROPOTOVA, G. P. VDOVYKIN and V. A. GRINENKO, 1967, Geokhimiya 3, 267.

WEHNER, G. K., C. KENKNIGHT and D. ROSENBERG, 1963, Planet. Space Sci. 11, 1257.

WEIL, D. F., I. S. MCCALLUM, Y. BOTTINGA, M. J. DRAKE and G. A. MCKAY, 1970, Science 167, 635.

WELLMAN, T. R., 1970, Nature 225, 716.

WOLF, T., 1966, Ph.D. Thesis, University of Rhode Island.

WOOD, A. J., 1963, Icarus 2, 152.

WOOD, A. J., 1966, Icarus 6, 1.

WOOD, A. J., J. S. DICKEY, Jr., U. B. MARVIN and B. N. POWELL, 1970, Science 167, 602.

ZELLER, E. J., L. B. RONCA and P. W. LEVY, 1966, J. Geophys. Res. 71, 4855.

CHAPTER 13

Organic molecules in space

BERTRAM DONN

1. Introduction

The detection of complex organic compounds in meteorites and in ancient rocks plus the synthesis of a variety of organic molecules including amino acids from simple gases has been described in previous chapters. These observations have resulted in the intensive and so far fruitful efforts to fill the gap between the simple gases, water, ammonia and methane, and the simplest living cell. In this chapter, I indicate the possibility of a somewhat different approach to the question of chemical evolution. Observational evidence for the occurrence of some organic compounds in space is presented and the likely existence of more complex molecules as yet undetected or unidentified is pointed out. This implies that massive solid objects, namely planets and asteroids were accumulated from material which already contained a variety of organic compounds. The extent to which such molecules were preserved throughout the accumulation process is a major question. However, only a brief discussion will be given, as it takes us into new areas of cosmogony of an essentially non-chemical nature.

The objects or regions of the galaxy that will concern us are: (1) comets, (2) interstellar space (3) pre-stellar nebulae and (4) cool stellar atmospheres. The primordial solar nebula in which the planets accumulated is a particular case of a pre-stellar nebulae but is by no means unique. During the lifetime of the galaxy similar objects must have appeared billions of times.

It is appropriate to conclude this introduction with an account of a cosmogonic hypothesis which relates all of the possible sources of molecules.

The brightest, hot stars in the Milky Way cannot have been shining for more than a few hundred thousand years with their available energy supply.

This is an insignificant fraction of the age of the Galaxy which is about 10 billion years. Other stellar age indicators require continuous star formation throughout the galaxy, Spitzer (1968a). Star formation takes place by the development of interstellar clouds of gas and solid grains and their subsequent collapse into stars. Several lines of evidence demonstrate the ejection of matter from stars into space either continuously or in bursts as with novae and supernovae. Thus, there is a sequence of stars forming from interstellar matter and then ejecting material back into space. This material later takes part in subsequent star formation.

At the end of the collapse stage of the prestellar cloud a fairly dense nebula with a temperature in the neighborhood of $1000\,°K$ must develop. An approximate molecular equilibrium will occur with the composition depending upon density and temperature. When the collapse ceases the temperature will drop to a low value determined by the energy input from the newly formed star. At some point the temperature becomes sufficiently low that a composition becomes 'frozen in'. Planets and meteorites could accumulate throughout the entire later stages of the nebula but comets must form during the cold stage because of their volatile nature. Because of their small mass and low temperatures comets would be expected to preserve with minimum transformation the composition of the material out of which they and planets formed. Cometary molecules thus represent some mixture from all sources of molecules in the galaxy.

The material which went into the formation of the planets also represents matter from many sources of molecules. The extent to which complex molecules could withstand temperatures and excitation conditions which were experienced from their time of formation to the time they became part of the earth's crust determined the organic composition of the primordial crust before terrestrial chemical evolution started.

2. Cool stellar atmospheres

Stars show an extremely wide range of atmospheric temperatures, Aller (1963). For the coolest stars, some of which are variable in brightness, the lowest temperatures found are about $1500\,°K$. These are stars of spectral class M, which have carbon/oxygen ratios less than unity and whose spectrum is dominated by absorption bands of TiO. For our purpose, the important class of stars are the so-called carbon stars. Their spectra show a marked difference in appearance from the stars of type M. The spectrum is dominated by the C_2 Swan bands and an analysis indicates a C/O ratio greater than 1.5. Because of the excess of carbon over oxygen, carbon is not nearly completely combined as CO and a large number of carbon compounds are possible.

A major factor in molecular equilibrium calculations is the temperature of the atmosphere. Effective temperatures of carbon stars, based on the total radiation from a star, are about 2500 °K for the coolest stars. Vibrational temperatures are several hundred degrees less and may be 2000 °K or somewhat lower.

No detailed models of carbon star atmospheres have been computed but an estimate of their characteristics can be obtained from the M type giant stars with comparable temperatures and luminosities. For these stars the pressure region of interest is $10-10^4$ d/cm^2 or 10^{-2} to 10 Torr. The importance of carbon compounds in the carbon star sequence is emphasized by the identification of C_3 and SiC_2 in addition to the major role of C_2 in their spectrum.

The usual procedure for computing chemical equilibrium in stellar atmospheres is to write an equation for each element considered in the analysis. For hydrogen this has the form

$$P(\text{H}) = p(\text{H})\left[1 + \frac{2p(\text{H})}{K(\text{H}_2)} + \frac{p(\text{O})}{K(\text{OH})} + \frac{p(\text{N})}{K(\text{NH})} + \frac{p(\text{C})}{K(\text{CH})} + \frac{p(\text{O})p(\text{H})}{K(\text{H}_2\text{O})} + \right.$$

$$\left. + \frac{p(\text{C})p(\text{H})}{K(\text{CH}_2)} + \frac{p(\text{H})p(\text{C})p(\text{O})}{K(\text{H}_2\text{CO})} + \cdots \right],$$

where $P(\text{H})$ is the fictitious pressure of the element which would occur if the element existed only as the neutral atom and $p(\text{H})$ is the actual partial pressure at equilibrium. The K's are the equilibrium constants defined by

$$K_{\text{ABC}} = \frac{p(\text{A})p(\text{B})p(\text{C}).}{p(\text{ABC})}$$

Similar equations occur for each element. There is a term in the analyses for each molecule in which the element occurs. Equilibrium constants are taken from experiments when possible, or are calculated from the molecular constants. Detailed treatments of this procedure are given by Tsuji (1964) and by Morris and Wyller (1967) for carbon stars. A different approach to determination of a complex equilibrium has been formulated by White et al. (1958) in which the free energy function of the system is minimized by varying the concentration of each species. This method has been employed by Dayhoff et al. (1967) for studying prebiological atmospheres.

Table 1 has been prepared from unpublished calculations by Tsuji for the composition $H:C:N:O = 10^4:18:1:10$ for pressures of 10^2 and 10^4 d/cm^2 and a variety of temperatures. In these calculations only H, C, N and O were considered. No metals, silicon or sulfur were introduced in the atmosphere. These are the cosmically abundant elements which would affect the chemistry. Because relative abundances in carbon stars are not the same as in the sun, for example the alkaline earths and rare earths appear enriched, some of these elements may also be important. However, the trend of these calculations should give a proper indication of carbon compounds to be expected. Morris and Wyller included silicon and used atomic abundances based on later and improved estimates of carbon star composition. However, only a few triatomic molecules were included. Dolan (1965) calculated molecular equilibrium in stars including one carbon star composition. His results were essentially limited to H, C, N and O and included diatomics plus HCN and CH_2. In addition to these calculations, an extrapolation of the calculations of Duff and Bauer (1962) suggests that benzene may occur to the extent of 1 part in about 10^8.

All the above assumed thermodynamic equilibrium which is probably a reasonable approximation. A more serious restriction may be the use of constant temperature and pressure instead of a model atmosphere in which both physical parameters increase from the boundary of the atmosphere inward. This refinement has been carried out for M stars by Vardya (1966).

Several analyses of carbon condensation in carbon stars have been carried out following the initial work of Hoyle and Wickramasinghe (1962). Later papers, Donn et al. (1968), Kamijo (1966), (1969), and Friedemann and Schmidt (1967) have improved the calculations and taken the nucleation process into account. It was shown in these investigations that graphite grains of a few hundred angstroms could well be expected to form in the cooler stars of the carbon sequence. Friedemann and Schmidt and Wickramasinghe et al. (1966) showed that such grains could be ejected from the atmosphere by radiation pressure. The latter paper also pointed out that the grains could carry a small but significant part of the atmosphere with them. The process of mass ejection in M giant stars is well established, Deutsch (1960) Weymann (1963). A similar process in the carbon stars would supply a copious source of carbonaceous material to interstellar space as was pointed out by Tsuji (1964).

Closely related to the phenomena of molecular equilibrium and particle

TABLE 1

Abundance of carbon compounds in carbon stars $H:C:N:O = 10^4:18:1:10$ (log of partial pressures d/cm^2).

Com-pounds	$P = 10^4$ d/cm² $T(°K)$					$P = 10^2$ d/cm² $T(°K)$				e
	2520	2290	1940	1440	1008	2520	2290	1940	1440	1008
H_2O	0.51 −6	0.51 −6	0.58 −6	0.91 −6	0.66 −3	0.18 −10	0.49 −10	0.62 −10	0.86 −10	0.73 −9
CO	0.00 1	0.00 1	0.00 1	0.00 1	0.00 1	0.00 −1	0.00 −1	0.00 −1	0.00 −1	0.00 −1
H_2CO	0.42 −8	0.49 −8	0.55 −8	0.66 −8	0.87 −8	0.74 −13	0.23 −12	0.51 −12	0.66 −12	0.87 −12
C	0.81 −2	0.66 −3	0.19 −5	0.43 −11	0.50 −23	0.46 −2	0.43 −3	0.12 −5	0.48 −11	0.56 −21
CH	0.69 −3	0.83 −4	0.90 −6	0.33 −10	0.47 −20	0.00 −4	0.47 −5	0.82 −7	0.38 −11	0.39 −19
CH_2	0.46 −2	0.06 −2	0.00 −3	0.38 −6	0.79 −13	0.42 −5	0.56 −5	0.91 −6	0.43 −8	0.71 −13
CH_4	0.82 −4	0.45 −3	0.37 −2	0.15 −0	0.90 0	0.11 −9	0.70 −8	0.25 −6	0.20 −4	0.83 −2
C_2	0.55 −3	0.50 −4	0.04 −6	0.12 −12	0.80 −27	0.85 −4	0.03 −4	0.91 −7	0.21 −12	0.65 −23
C_2H	0.29 0	0.98 −1	0.96 −2	0.26 −4	0.33 −13	0.25 −2	0.38 −2	0.82 −3	0.35 −5	0.18 −10
C_2H_4	0.69 −5	0.17 −4	0.92 −4	0.23 −2	0.40 −0	0.94 −12	0.70 −10	0.39 −8	0.16 −6	0.98 −5
C_3	0.84 −3	0.22 −3	0.45 −5	0.85 −10	0.57 −24	0.79 −4	0.51 −4	0.26 −5	0.99 −10	0.34 −18
C_3H_3	0.48 −5	0.91 −5	0.14 −4	0.04 −3	0.23 −11	0.40 −10	0.81 −9	0.91 −8	0.16 −7	0.01 −8
C_4H	0.25 −2	0.27 −2	0.59 −3	0.26 −5	0.57 −18	0.50 −5	0.20 −4	0.32 −4	0.45 −6	0.28 −11
CN	0.04 −1	0.54 −2	0.45 −3	0.88 −6	0.36 −12	0.75 −3	0.41 −3	0.42 −4	0.91 −7	0.25 −11
HCN	0.81 −1	0.89 −1	0.89 −1	0.79 −1	0.40 −3	0.18 −3	0.63 −3	0.85 −3	0.82 −3	0.27 −3

formation in cool stellar atmospheres is the role of similar phenomena in pre-stellar and circumstellar clouds. Infrared observations have detected objects with strong intensity at wavelengths of several microns but with weak or no visible radiation. Brief reviews have been given by Johnson (1967) and Feldmann et al. (1966). There appear to be three classes of infrared objects. These are: very cool M giant variables; extremely dense interstellar clouds, and cool circumstellar clouds that may be planetary systems in an early stage of formation. There is much evidence for solid particles and molecules associated with these clouds as well as evidence for the ejection of material from these clouds into interstellar space, Herbig (1969). These objects presumably are associated with solar type stars and primordial nebula with a carbon to oxygen ratio less than unity. However, the discussion of section 3 shows that complex organic compounds formed early in the history of the solar system, probably prior to or during the accumulation of comets and meteoritic objects. Consequently, organic molecules could also form in the infrared objects associated with star and planet formation.

3.   Comets and the primordial solar nebula

The comet family (Richter 1963) consists of astronomically small ($\sim$ 40 km diameter, 10^{15}–10^{19} g) objects which readily evolve large quantities of gas and dust when they approach the sun. About 600 individual comets have been identified when they came within 4 A.U. of the sun. The total number must be enormously larger. There must be a large storehouse of comets at larger distances from which they gradually feed into the inner solar system where they gradually disintegrate. Various statistical analyses suggested from 10^7 to 10^{11}.

No complete theory of the origin, structure and composition of comets exists. It is generally believed that they were formed in the early stages of the solar system although an alternative hypothesis of formation in interstellar clouds and subsequent capture by the sun has been proposed. If they are indeed associated with the early solar system, comets tell us much about the composition of the primordial solar nebula. Because the solar nebula is believed to have formed from an interstellar cloud the alternative explanation still tells something about the early solar system, although the interpretation is not as direct.

The characteristic observational phenomenon indicating the presence of a comet is the head or coma, an extremely diffuse cloud of gas and dust. A tail, which is the most prominent feature of bright comets, is generally missing from the fainter, short period comets. Embedded in the coma is the nucleus. This is the permanent part of a comet which revolves around the sun and is the source of coma and tail material. Our only knowledge of the composition of the nucleus comes from spectroscopic observations of coma and tail.

Table 2 shows the molecules identified in comets, listed in the coma in order of appearance as the comet approaches the sun and in the tail the order is generally that of decreasing intensity. The ultraviolet Lyman alpha line of atomic hydrogen has been detected in great strength in two recent bright comets by orbiting spacecraft. The dust continuum arises from scattered sunlight. A very close correspondence of the intensity distribution within molecular emission bands with that of the solar continuum indicates that the molecular emissions result from resonance fluorescence.

TABLE 2

Atoms and molecules identified in comets.

Coma	Tail
Dust continuum	Dust continuum
H, CN	CO^+
C_3, NH_2	N_2^+
C_2	CO_2^+
OH, NH, CH	CH^+
Na	OH^+
Fe, Ni, Ca, K	

The composition and structure of the nucleus of comets must be such as to yield the observed radicals, account for the gas and dust production, and be able to explain the dynamical properties of comets. Whipple's (1950, 1963) icy nucleus and its further development, Donn (1963, 1968b), provides a model that appears to meet these requirements and also to be consistent with an accumulation process in the primordial solar nebula. Because the nucleus is a volatile object the temperature had to be low at all times during and since its formation, Huebner (1965). Further, because of its small mass, pressures were also generally low, hence no substantial chemical changes in comets have taken place since their origin.

The composition of icy cometary nuclei can be estimated from molecular equilibrium calculations of the primordial nebula. In 'The Planets' Urey

(1952) considered the equilibrium of the solar nebula under several conditions and cites earlier work. Since then more extensive investigations have been carried out for arrays of about one hundred molecular species. The most useful for the present purpose are those of Tsuji (1964) and of Lord (1965). Table 3 summarizes these results and is a modified form of Urey's calculations.

TABLE 3

Compounds in the primordial solar nebula.

	$P = 10^{-3}$ atm	
Element	298 °K	1200 °K
H	H_2, CH_4, NH_3, H_2O	H_2, H_2O, H_2S
He	He	He
C	CH_4	CO
N	NH_3, NH_4^+	N_2
O	H_2O	H_2O
Ne	Ne	Ne
Si	SiO_2	SiO_2, SiC
S	FeS	H_2S
Fe	FeS, Fe_3O_4	Fe, Fe^3C
	Fe_2SiO_4	

Two factors may have seriously modified the composition of the solar nebula from the equilibrium values, Fowler et al. (1962), Donn (1968). Radioactive nuclei were much more abundant at the time of formation of the solar system about 5 billion years ago than they are at present. The half life of U^{238} is 7×10^8 years and therefore $N_5 = 1100\ N_0$. In addition, there is evidence for short lived radioactive elements with half lives $t \sim 10^5$ years that were present in the early solar system.

Of more significance for producing the non-equilibrium effect may have been the energetic corpuscular radiation from the primitive sun. Qualitative considerations indicate that significant chemical effects should have occurred. During this period the energy flux of energetic protons has been estimated at 10^{45} ergs. If all the energy were absorbed by the nebula, it would correspond to 10^{57} eV. The effects of radiation is measured in terms of the G value for a reaction, the number of molecules destroyed or formed per 100 eV of radiation absorbed. Although G values show a considerable spread and depend upon the medium and its detailed composition, a reasonable value is $G \sim 1$. This leads to a conversion of the thermodynamically stable compounds to the extent of 10^{55} molecules. With a mean molecular weight of 20, this is equivalent to 3.5×10^{32} g or about two solar masses. If only a fraction of the radiation were absorbed, a substantial proportional of the material in the solar nebula would have been transformed.

Some idea of the ultimate composition of comets may be obtained from experiments on chemical composition of irradiated, condensed gases and the related experiments on the condensation of dissociated gases, Bass and Broida (1960). In the latter case, a small concentration of radicals are obtained in addition to more complex and more reactive molecules. Warm-up of the condensed films yielded an additional array of molecules. It is not clear in most instances whether these existed as such in the condensed matrix or were recombination products produced as the solid warmed and vaporized. Condensation of discharged methane–nitrogen mixtures has yielded the following molecules, Glasel (1961): acetylene, ethylene, ethane, butadiene, propylene, propane, and butene.

Dissociated water vapor, when condensed in a cold trap, produces a high percentage of hydrogen peroxide, H_2O_2, plus the HO_2 radical. The formation of the super-peroxide H_2O_4 has been claimed but recent work does not support such a molecule. The production of such unstable species as HO_2, HCO, HONO, NH, NH_2 and almost certainly similar as yet unidentified compounds by photolysis of simple mixtures in the laboratory again suggests the variety of molecules to be expected in the primordial condensed gases.

All these possible contributions to cometary material are combined in table 4 to yield a suggested composition of the cometary nucleus. The species

TABLE 4

A suggested composition of the cometary nucleus.

Type of material	Examples
Volatile, inorganic	N_2, Ne, Ar, H_2O, NH_3, CO, H_2O_2, N_2H_4, H_2S
Volatile, organic	CH_4, C_2H_2, C_2H_4, HCN, CH_3OH, (C_6H_6)
Non-volatile, organic	Higher aromatics, complex organic molecules, polymerized material
Non-volatile, inorganic	Silicates, metallic oxides (metals, carbon grains)
Radicals	OH, NH, C_2, C_3, CH_2, HO_2

listed in table 2 demonstrate the occurrence of significant amounts of carbon compounds in comets. The restriction of observations to the visible spectrum and the absence of laboratory spectra for carbon polymers larger than C_3, suggest that parent molecules containing more than three carbon atoms may also occur in comets. It must be emphasized that no single model can explain all comets. For Comet Morehouse 1908 III, and Comet Humason 1962 VIII, carbon monoxide appears to have been the dominant constituent. The organic compounds in meteorites described in part II, chapter 3, is strong evidence for their occurrence in the primordial nebula. There is therefore good reason to expect complex organic molecules in comets as well.

4. Interstellar space

The sun is about two-thirds of the distance from the center of our galaxy to the outer edge and lies almost in its central plane. Although the overall star distribution is spherical, the most luminous and hottest stars, and the diffuse matter between the stars are highly concentrated to the galactic plane. Interstellar chemistry is concerned with this median region or galactic disc as it is generally known.

Several modes of analyses, Spitzer, (1968b) yield total densities of interstellar matter as about 5×10^{-24} g/cm^3. Wide angle photographs of the Milky Way show the very irregular distribution of obscuring clouds and of luminous regions in the galaxy. In addition, high resolution spectra show that the absorption lines of interstellar molecules may contain several Doppler displaced components. This is evidence for a discrete velocity distribution of the gas and very likely a discrete spatial distribution also. It is convenient, although oversimplified, to describe the distribution of interstellar matter by a random cloud model. According to this picture a kiloparsec (3×10^{21} cm) line of sight intercepts five to ten clouds with mean diameters of 10 parsecs each. Within the cloud there are ten hydrogen atoms per cm^3 and between clouds about 0.1 atom per cm^3. The radio line of the hydroxyl radical, OH, has recently been detected in obscuring or dust clouds with radii as small as 0.5 pc and densities probably near 1000 H atoms/cm^3.

Most of the information of the composition of the gas is based on spectra obtained in the visible region. With the development of radio astronomy and more recently, vacuum ultraviolet spectroscopy from rockets, additional atoms and molecules have been observed. The list of all elements and compounds observed in interstellar space is presented in tables 5, 6, and 7. Tables 5 and 6 list the species observed in the visible, in absorption and in emission respectively. Table 7 shows the constituents detected more recently by radio or ultraviolet observations. Abundances in interstellar space conform generally to a so-called cosmic abundance distribution, Suess and Urey (1956), Cameron (1959) derived from a variety of celestial sources, including

TABLE 5

Interstellar atoms and molecules: visible spectrum in absorption.

Atoms		Molecules	
Na I	3302–3 Å	CH	4300 Å
	5690–6		3890–86–77
K I	7665–99		3146–43–38
Ca I	4227	CN	3874–5–6
Ca II	3933–68	CH$^+$	4233, 4
Ti II	3073		3957
	3229–42		3745
	3283		
Fe I	3720		
	3859		

TABLE 6

Interstellar atoms: visible spectrum in emission (H II regions, emission nebula and planetary nebulae).

Hydrogen	Silicon
Helium	Sulfur
Carbon	Chlorine
Nitrogen	Argon
Oxygen	Pottasium
Fluorine	Calcium
Neon	Iron
Sodium	

TABLE 7*

Interstellar atoms and molecules: ultraviolet and radio spectrum.*

U.V.		Radio	
Species	wavelength	Species	wavelength
O I	1302 Å	H	21 cm
C II	1335	OH	18 cm, 5 cm
Si II	1526	NH_3	1.25 cm
Al II	1671	H_2O	1.35 cm
H_2	1000–1100	H_2CO	6.2 cm , 2.07 cm
CO	1280–1550	CO	2.6 mm

* See note added in proof.

the earth. It is reasonable to apply this result to obtain an approximate complete composition for the interstellar medium. The more abundant elements and their approximate average interstellar number density appear in table 8.

TABLE 8

Cosmically abundant elements (Suess and Urey 1956).

Z	Element	Interstellar density (atoms/cm^3)	Rel abundance $Si = 10^6$
1	Hydrogen	1	4×10^{10}
2	Helium	1×10^{-1}	4×10^9
8	Oxygen	5×10^{-4}	2×10^7
10	Neon	2×10^{-4}	9×10^6
7	Nitrogen	2×10^{-4}	7×10^6
6	Carbon	1×10^{-4}	4×10^6
14	Silicon	3×10^{-5}	1×10^6
12	Magnesium	3×10^{-5}	9×10^5
26	Iron	2×10^{-5}	6×10^5
16	Sulfur	1×10^{-5}	4×10^5
18	Argon	5×10^{-6}	2×10^5
13	Aluminium	3×10^{-6}	9×10^4
20	Calcium	1×10^{-6}	5×10^4
11	Sodium	1×10^{-6}	4×10^4
28	Nickel	5×10^{-7}	3×10^4
15	Phosphorus	2×10^{-7}	1×10^4
17	Chlorine	2×10^{-7}	9×10^3
18	Potassium	1×10^{-8}	3×10^3

The significant presence of carbon compounds appears in interstellar space also. Not only are two of the three observed diatomic radicals the compounds CH and CN, but the most complex molecule so far found in interstellar space is the organic compound, formaldehyde.

In addition to the sharp lines, optical spectra reveal some 25 broader absorption features. None have yet been identified but several possibilities have been considered a few of which, Duley (1969a, b), Herzberg (1969), Johnson (1965), are related to the subject of this chapter.

The third constituent of the interstellar medium is the material producing the obscuration and reddening of starlight. Again, many suggestions have been made as to the nature of this material, generally referred to as interstellar grains. The proposals include (a) mixtures of condensed gases, mainly water, ammonia and methane, (b) silicates, (c) graphite (d) metals (e) condensed gas mixture on graphite core. Detailed discussions of such particles are given by Greenberg (1968) and Wickramasinghe (1967).

The prevailing theory considers classical scattering and extinction by small particles and requires particles with dimensions from about 300 Å to 3000 Å depending upon composition. A recent proposal by Platt (1956) based upon a quantum mechanical interaction indicated that disordered molecular aggregates of about 20 Å diameter could produce the extinction. This suggestion has been further developed by Donn and his associates (1968, 1969, 1970). It was shown that arrays of polycyclic hydrocarbons containing about 5 or more rings have continuous extinction through the visible region of the spectrum very similar to the interstellar extinction as shown in fig. 1 by Donn and Krishna Swamy (1969). Infrared and ultraviolet absorption coefficients for such molecules are consistent with the extinction ratio for these spectral regions observed in interstellar space. From the discussion in section 2 on molecules in carbon stars, such stars may be the source of these compounds. The analysis of graphite formation reviewed there also makes the occurrence of polycyclic aromatics likely.

Clar (1964) has called attention to the sequence of isomers with maximum chemical and thermal stability (fig. 2): benzene, diphenyl, triphenylene, dibenzopyrene, tribenzoperylene, tetrabenzoanthanthrene and hexabenzo-

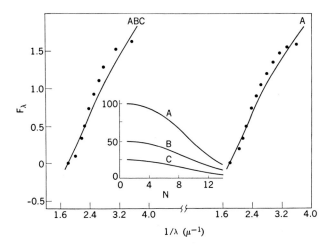

Fig. 1. Comparison of interstellar extinction curve and extinction by array of polycyclic hydrocarbons (Donn and Swamy 1969). The ordinate is the extinction in magnitudes compared to that at 5600 Å. The insert shows the size distribution of molecules comprising the array. The curves A, B and C are for successively less stable groups. The abscissa N is the number of rings in the molecule and the ordinate is the relative number of molecules.

GRAPHITE PRECURSORS

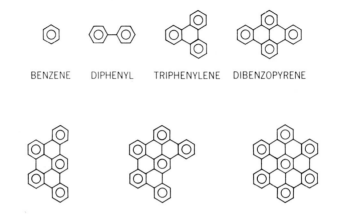

Fig. 2. Sequence of most stable polycyclic isomers (graphite precursors).

coronene. The largest member has been described by Clar in the following way. 'Hexabenzocoronene which does not melt even above 700 °C, is insoluble in all conventional solvents and can be formulated with sextets only, can therefore be considered as a particle of graphite.' In a crystal nucleation and growth process for the formation of graphite grains in stellar atmospheres as described in section 2, the carbon skeleton of these or similar polycyclics or heterocyclic aromatics could well be among the growing clusters. Conditions for the formation of such molecules are less stringent than for the growth of graphite grains of a few hundred angstroms. Whatever process would eject graphite into space also works for these species. It appears possible to explain all features of the wavelength dependence of interstellar extinction, as well as interstellar polarization by these molecules. They have the advantage of being stable up to about 600 °K for the smaller molecules and to 1000 °K for the largest. Further experiment on the optical, thermal and photochemical properties of these types of molecules are under way at Goddard. Thus, in addition to small organic molecules and radicals now known to be present in interstellar space, there is a possibility of very much larger molecules also. An array of aromatics containing at least about five rings but with a much larger upper limit to the number of rings could be present. These would not be pure hydrocarbons but should include other atoms both in the ring structure and in the side chains replacing hydrogen.

Irradiation of the Van de Hulst ice model of the insterstellar grain is another, somewhat more hypothetical, source of organic material in interstellar space. The suggested composition given by Van de Hulst (1948), Greenberg, (1968) was 100 parts H_2O, 30 part H_2, 20 parts CH_4, 10 parts NH_3. 5 parts MgH and other metallic compounds. With a radius under 0.5 μ, the particle dimensions are the order of the skin depth and the ultra flux will penetrate the entire grain. In addition, violet low energy cosmic rays would contribute to chemical changes in the grain. Measurements from spacecraft indicate a cosmic ray flux near one Mev that may be four orders of magnitude greater than previously considered. Both types of radiation, acting over grain lifetimes of 10^7 to 10^8 years would produce radicals like OH, NH, CH_3, etc. These would recombine in a random fashion within the grain to gradually polymerize the particle and build up more complex organic molecules. Experiments by Oro (1963) on the irradiation of a water, ammonia, methane film by 5 Mev electron yielded about 20% non-volatile organic matter including simple amino acids. Further experiments to determine the type of change and the rate are being started at Goddard. Such chemical changes would also effect the thermal stability and optical properties of the grain and have significant astrophysical consequences.

5. Summary

Spectroscopic studies of comets and interstellar matter demonstrate the presence of simple organic molecules in interplanetary and interstellar space. Difficulties in detecting or identifying larger carbon compounds in either case suggest that presently known molecules are not the most complex that occur.

Several proposals attempting to explain the extinction of starlight in interstellar space suggest some type of carbonaceous material may be present. The form of the carbon runs from pure graphite flakes through polymerized water, ammonia, methane, etc. grains to polycycle aromatic molecules.

Laboratory analyses of meteorites described in part II, chapter 3, indicate the large variety of quite complex organic material that is found. Although the origin is unknown, their presence in meteorites is indicative of formation in the solar nebula or earlier stage of galactic evolution. On the basis of their laboratory experiments Studier et al. (1968) have proposed hydrocarbon formation via a Fisher-Tropsch mechanism on particulate matter which condensed in the primordial nebula.

As indicated in the Introduction, the presence of organic matter from a variety of cosmic sources may all be closely related. Our knowledge of galactic evolution and space chemistry is too incomplete at present to draw definite conclusions. But whether organic matter throughout the galaxy has a common origin or formed independently in several places, we know it does occur. The question of how much and in what form these compounds survived during the process of earth accumulation and crustal formation is the most significant issue remaining. However, the thermal stability of the more complex molecules and the likelihood of some polymerization suggests that a primordial organic residue remained on the surface of the earth after the crust and atmosphere formed. Some consideration of this problem appears in the paper by Oro (1965) and the discussion following his paper. This raises the final questions of this discussion of organic molecules in space: What was the probable nature of the residual organic material and to what extent could a scheme of terrestrial biochemical evolution proceed from such initial circumstances?

Note added in proof. The discovery of interstellar molecules has been going on at an accelerated rate. In 1970 the following were detected: H_2, HCN, $CN–C \equiv C–H$, CH_3OH, CHOOH, and one unidentified radio line. In 1971, through November the following additional molecules were discovered: CS, NH_2CO, SiO, OCS, CH_3CN, HNCO, $CH_3–C \equiv C–H$, CH_3CHO, H_2CS and one tentatively identified as HNC. In order to keep up with the rapid proliferation of interstellar molecules the reader must resort to the appropriate journals, particularly the Astrophysical Journal, Part 2; Letters, Nature and Science.

References

ALLER, L. H., 1963, Astrophysics, the atmospheres of the sun and the stars. 2nd ed. (Ronald Press, N.Y.), Ch. 3.

BASS, A. and H. BROIDA, 1960, Production and trapping of free radicals. (Academic Press, New York).

CAMERON, A. G. W., 1959, Astrophys. J. 129, 676.

CLAR, E., 1964, Polycyclic hydrocarbons. (Academic Press, N.Y., London), Vol. 1, Ch. 6.

DAYHOFF, M. O., E. R. LIPPENCOTT, R. V. ECK and G. NAGARAJAN, 1967, Thermodynamic equilibrium in prebiological atmospheres of C, H, O, N, P, S and Cl, NASA SP-3040. (National Aeronautics and Space Administration, Washington, D.C.).

DEUTSCH, A. J., 1960, The loss of mass from red giant stars; In: J. L. Greenstein, ed., Stellar atmospheres. (Univ. of Chicago Press, Chicago) 543.

DOLAN, J. F., 1965, Astrophys. J. 142, 1621.

DONN, B., 1963, Icarus 2, 396.

DONN, B., 1968a, Astrophys. J. 152, L 129.

DONN, B., 1968b, Cosmic chemistry; In: G. Mead and W. N. Hess, eds., Introduction to space science. (Gordon and Breach, N.Y.), 501.

DONN, B. and K. S. KRISHNA SWAMY, 1969, Physica 41, 144.

DONN, B., L. J. STIEF and W. A. PAYNE, 1970, Spectroscopy, pyrolysis and photolysis of polycyclic aromatic molecules (to be published).

DONN, B., N. C. WICKRAMASINGHE, J. P. HUDSON and T. P. STECHER, 1968, Astrophys. J. 153, 451.

DUFAY, J., 1957, Galactic nebulae and interstellar matter, transl. A. J. Pomerans. (Philosophical Library, N.Y.).

DUFF, R. E. and S. H. Bauer, 1962, J. Chem. Phys. 36, 1754.

DULEY, W. W., 1969a, Physica 41, 135.

DULEY, W. W., 1969b, Nature 224, 785.

FELDMAN, P. A., M. T. REES and M. W. WERNER, Nature 224, 752.

FOWLER, W., J. L. GREENSTEIN and F. HOYLE, 1962, Geophys. J. 148.

FRIEDEMANN, CHR. and K. H. SCHMIDT, 1967, Astron. Nachs. 289, 223.

GLASEL, T. 1961, Proc. Nat. Acad. Sci. 47, 174.

GREENBERG, J. M., 1968, Interstellar grains; In: B. M. Middlehurst and L. H. Aller, nebulae and interstellar matter. (University of Chicago, Chicago, Ill.) 221.

HERZBERG, G., 1969, Symposium on laboratory astrophysics, Lunteren. (Unpublished).

HOYLE, F. and N. C. WICKRAMASINGHE, 1962, Roy. Astron. Soc. Monthly Notices 124, 417.

HERBIG, G. H., 1969, In: Liege Symposium, Premain-Sequence stellar evolution, Liege, 1969, to be published.

HUEBNER, W. F., 1965, Zt. Astroph. 63, 22.

JOHNSON, F., 1965, In: J. M. Greenberg and T. P. Roark, eds., Interstellar grains, NASA SP-140. (NASA, Washington, D.C.) 229.

JOHNSON, H. L., 1967, Infrared stars; Science 157, 635.

KAMIJO, F., 1966, Supersaturation of carbon vapour in the carbon stars; In: M. Hack, ed., Colloquium on late type stars. (Obs. Astronomica di Trieste, Trieste) 252.

KAMIJO, F., 1969, Physica 41, 163.

KIMURA, H., 1962, Proc. Astron. Soc. Jap. 14, 374.

LORD, H. C., 1965, Icarus 4, 279.

MORRIS, S. and A. A. WYLLER, 1967, Astrophys. J. 150, 877.

ORO, J., 1963, Nature 197, 971.

ORO, J., 1965, In: S. W. Fox, ed., The origins of prebiological systems. (Academic Press, New York) 137.

PLATT, J. R., 1956, Astrophysics J. 123, 486.

RICHTER, N. B., 1963, The comets, transl. by A. Beer. (Dover Publications, N.Y., Methuen and Co., Ltd., London).

SPITZER, L., 1968a, Dynamics of interstellar matter and the formation of stars; In: B. M. Middlehurst and L. H. Aller, Nebulae and interstellar matter. (Univ. of Chicago Press, Chicago) 1.

SPITZER, L., 1968b, Diffuse matter in space (Interscience, N.Y.).

STUDIER, M. H., R. HAYATSU and E. ANDERS, 1968, Geochim. Cosmochim. Acta 32, 151.

SUESS, H. E. and H. C. UREY, 1956, Rev. Mod. Phys. 28, 53.

Tsuji, T., 1964, Molecular abundances in stellar atmospheres; Ann. Tokyo Astron. Obs. 2nd Ser. 9, no. 1.

Urey, H. C., 1952, The planets. (Yale University Press, New Haven, Conn.), Ch. 4.

Van de Hulst, H. C., 1949, Rech. Astron. Observ. Utrecht 11, pt. 2.

Vardya, M. S., 1966, Monthly Notices Roy. Astron. Soc. 134, 347.

Weymann, R., 1963, Mass loss from stars; In: Ann. Rev. Ast. Astrophys. (Annual Reviews, Inc., Palo Alto, Calif.) 97.

Whipple, F. L., 1950, Astroph. J. 111, 375.

Whipple, F. L., 1963, In: B. Middlehurst and G. P. Kuiper, eds., The moon, meteorites and comets. (Univ. of Chicago Press, Chicago, Ill.) 639.

Wickramasinghe, N. C., B. Donn and T. P. Stecher, 1966, Astrophys. J. 146, 590.

Wickramasinghe, N. C., 1967, Insterstellar grains. (Chapman and Hall, London).

White, W. B., S. M. Johnson and G. B. Dantzig, 1958 J. Chem. Phys. 28, 751.

C. Ponnamperuma (ed.), Exobiology. © North-Holland Publishing Company

CHAPTER 14

Potential targets in the search for extraterrestrial life

HAROLD P. KLEIN

1. Introduction

Speculations and experimental studies on the origin of terrestrial life are interesting greater numbers of scientists today than ever before. At the same time, the development of spacecraft technology has also been expanding and becoming more sophisticated. At present, these two fields of endeavor are converging towards a common goal in the search for extraterrestrial life.

From the point of view of the origins of terrestrial life, the possibilities of examining other planets hold great potential for increasing our understanding. If we sampled the surface of another planet and found living organisms, we would, of course, want to know how they got there. In this case, there are only two possibilities: transport from another inhabited planet or independent origin. The first of these – the so-called 'panspermia' theory – was proposed by Arrhenius (1908) as a mechanism for seeding the planets of our solar system. According to this hypothesis, spores or other resistant structures of microorganisms escape the gravitational field of an inhabited planet, and are transported to other planets over long time periods, under the pressure of solar winds. Sagan (1966) has analyzed this proposed mechanism for the dispersion of organisms and has concluded that the hazards of interplanetary travel for microorganisms are too formidable to allow acceptance of this idea. The rigors of space – the near perfect vacuum, the intense ultraviolet flux and the presence of other types of destructive solar radiation – would appear to make it a virtual certainty that any unprotected organism would be killed soon after leaving the protection of a planetary atmosphere. (Of course, 'panspermia' becomes much more feasible if organisms are transported in a manner to protect them from the harsh environment of interplanetary space, e.g., within a spacecraft!)

An independent origin for life on another planet would be difficult to prove. In trying to assess this probability, much would depend on the fundamental chemistry and morphology of these hypothetical extraterrestrial organisms. Radically different types of constituent chemicals – for example, a silicon-based 'bio'-chemistry (Pimentel et al. 1966) – would be regarded as strong evidence of such independent origin. Similarly, the presence of truly 'exotic' structural attributes, or the absence of 'standard' terrestrial structures could be persuasive. For example, all terrestrial cells contain ribosomes and a variety of membranous elements. The total absence of these would be most intriguing and strongly suggestive of an origin separate from that of terrestrial organisms. Sets of 'building block' amino acids, or purines or pyrimidines, different from those found on the earth, would be interpreted similarly, though with less assurance. Finding d- rather than l-amino acids in functional proteins, or L- rather than D-sugars also would tend to support the contention of independent origin.

Whether or not an independent origin for an extraterrestrial biota could be deduced, it is precisely considerations such as these that make the search for extraterrestrial life so important to the general biologist. *The fundamental chemical and structural attributes of terrestrial organisms are so remarkably uniform that any living forms outside the terrestrial blueprint would almost certainly be regarded as an alien organism.* The extent to which an alien organism diverged from the terrestrial model would be regarded, by biologists, as of the highest consequence. There is no doubt that the prospect of sampling other planets could significantly extend current concepts of the nature of living systems.

The present level of our knowledge does not allow the a priori conclusion that independent origins of life on two different planets would proceed via significantly different chemical and biochemical processes. It is by no means clear to what extent chemical evolution is a random process. Pullman and Pullman (1962), for example, have shown that, on quantum mechanical grounds, adenine is the most likely purine to arise from precursor molecules. Future studies may extend these contentions, and it may well turn out that independent origins on separate planets would be expected to proceed along similar general lines, given the same kinds of energy sources and precursors.

It is also a distinct possibility that, in attempting to obtain evidence for living organisms on another planet, none will be found, but that organic compounds will be present. We have only rudimentary knowledge concerning the stability of organic molecules over long periods of time (Abelson 1959), but it is plausible that on other planets we may find evidence supporting the hypothesis of chemical evolution even in the absence of a biota. This could occur because chemical evolution had not yet reached the stage of molecular self-replication at the time of sampling, or because planetary conditions became too severe to sustain a living system, following a brief period of chemical evolution.

Perhaps even more intriguing is the possibility that on some planets chemical evolution has been followed by prolonged periods of biological evolution, and that the organisms on these planets include some that are much further advanced than those on the earth (Drake 1962; Oliver 1966).

2. *Possible targets*

The probable number of sites in the universe where chemical evolution leading to living systems may have taken place has been the subject of treatises by many astronomers. For example, Brown (1964) came to the conclusion that each star in our galaxy should, on the average, have two planets at distances that would receive sufficient radiation to support life. Huang (1959), and Shklovskii and Sagan (1966), also considered the probable occurrence of planets capable of supporting life outside of this solar system. The latter authors concluded that at least a billion stars with potentially habitable planets exist within our own galaxy. With numbers such as these, it is certainly possible to envision a broad range of possibilities for the evolutionary processes discussed above – from some places where the emergence of living organisms may not yet have occurred, to others where this process has led to highly advanced technical civilizations (Shklovskii and Sagan 1966).

In view of these considerations, and the fact that space technology is at hand, it might seem reasonable to conclude that man is at the threshold of an age of intensive interstellar exploration, and that biology is soon to experience an era in which our knowledge and the basic concepts of biology will be greatly extended. However, the vastness of space may well limit these explorations to the confines of our own solar system. The enormous distances involved, together with the corresponding energy requirements, to reach even the nearest stars in our own galaxy, are staggering (Purcell 1960; Von Hoerner 1962). Indeed, the bleak prospects for travelling to points

outside this solar system have been cited as overriding arguments for programs designed to contact distant civilizations by the techniques of radioastronomy (Oliver 1966).

3.   *Assessing the probabilities*

As we narrow our sights to targets within this solar system, two kinds of
questions must be looked into in attempting to determine the most attractive
targets in the search for living organisms. First, what are the physical and
chemical limits within which living organisms can be expected to perpetuate
themselves? Secondly, what is known about the current environmental
conditions on the other objects in the solar system? Clearly, without precise
answers to both of these questions, we cannot adequately gauge the probabilities of finding non-terrestrial life somewhere in the solar system.

At best, we have more information about the first of these questions than
we have about the second. But it should be emphasized that our information
on the former is still far from complete. For example, the temperature limits
for terrestrial organisms are not precisely determined. Until relatively
recently, algae and bacteria, capable of growing at about 70 °C, were thought
to represent the upper limit for thermoresistance in terrestrial organisms.
However, Brock (1967) showed that extremely thermoresistant bacteria were
growing in the superheated pools of Yellowstone National Park at the boiling
point of water, approximately 96 °C. As a result of his work, Brock has
written, '... there seems to be no reason why bacteria could not live in
nature at any temperature where there is liquid water'. Similarly, the minimum temperature for growth of terrestrial organisms has not been clearly
established. Larkin and Stokes (1968) have grown various microorganisms
at temperatures as low as about − 10 °C in the presence of antifreeze agents,
such as glycerol. But since these agents also exhibited some toxicity towards
the test organisms, these authors felt that further research might uncover
more suitable antifreeze compounds for the cultivation of extremely psychro-
philic organisms at even lower temperatures.

Since most terrestrial organisms are confined to the surface and near-
surface of the earth, at one time it was thought that this distribution of
organisms was, in part, the result of a predilection for sea-level pressures.
However, it is now known that a great variety of organisms are present, and
in large numbers, even at such great depths as the Marianas trench in the
Pacific Ocean – over 10,000 meters deep, with a hydrostatic pressure of over

1,000 atmospheres (Marshall 1954; Zobell and Morita 1957). Furthermore, it is known that a large variety of animals, normally inhabiting sea level atmosphere habitats, can tolerate several hundred atmospheres of pressure (Cattell 1936). While few studies have been made of the growth of organisms at pressures lower than ambient terrestrial pressures, numerous workers have found that microorganisms can survive for long periods of time at considerably reduced pressures. In one such study (Brueschke et al. 1961), it was found that spores of several microorganisms, including fungi and a species of Bacillus, survived 10 days exposure to pressures reaching as low as 8×10^{-8} mm Hg. (At pressures lower than this, however, the organisms tested failed to survive.) In any case, it is evident that the range of pressures – atmospheric, hydrostatic, or osmotic (see below) – that can be tolerated by 'life', as represented by different terrestrial organisms, is enormous.

Any reasonable model for an effective biological infection of a planet requires as an ultimate energy source a continuing source of radiation to supply energy. On the earth, only the visible wave lengths of the solar spectrum are utilized in this way, the more energetic ultraviolet light being largely absorbed in the earth's atmosphere. On the other hand, it has been known for many years that ultraviolet radiation below 3,000 Å is highly injurious to all types of terrestrial organisms. Even though many kinds of cells, including mammalian cells (Rasmussen and Painter 1964), are known to have repair mechanisms, sensitivity to large doses of ultraviolet light is characteristic of terrestrial organisms which are essentially protected from this radiation by the atmosphere. But it may well be that on another planet where such protection is absent or reduced, organisms will have developed greater resistance to the injurious effects of these radiations, perhaps by utilizing ultraviolet absorbing pigments on their surface. Recognizing the adaptability of living organisms, this 'trick' does not seem far-fetched. Indeed, it has been suggested that organisms on a 'cold' planet, with a limited atmosphere like Mars, might absorb some of the energy of ultraviolet radiation, through a pigment system, in order to maintain an elevated internal temperature.

In considering the susceptibility of terrestrial organisms to ionizing radiation such as gamma rays and X-rays, and to solar particulate radiation, all of these radiations are known to be deleterious to terrestrial organisms, affecting different organisms to varying degrees. For example, it is has been reported (Alexander 1957) that, compared with mammalian cells, cells of *Escherichia coli* are 200 times more resistant, and cells of Paramecium are 6,000 times more resistant, to given levels of ionizing radiation. It should be remembered that these radiations are largely prevented from reaching the surface of the earth because of its atmosphere and magnetic field. As with ultraviolet radiation, terrestrial organisms have evolved under conditions where they were essentially protected from these kinds of radiation, and did not have to evolve major mechanisms of adaptation. Mechanisms affording some increased resistance are conceivable, but more important as a consideration is that even as thin an atmosphere as exists on Mars would attenuate most of these radiations reaching the surface.

The cellular effects of other terrestrial physical parameters, such as magnetic fields, tidal forces, etc., are incompletely understood at present. That gravity could play an important critical role in biological systems has been challenged (Pollard 1965).

While the presence of an atmosphere is clearly necessary, it is known that considerable variation in the composition and density of the atmosphere can be tolerated in the growth of terrestrial organisms. For example, there is no a priori need for free oxygen or nitrogen. Most terrestrial organisms do not metabolize free nitrogen and many do not utilize atmospheric oxygen. On the other hand, among the 'unusual' gases that are efficiently metabolized by some terrestrial species are: hydrogen, hydrogen sulfide, ammonia, methane, and carbon monoxide – found in traces in our own atmosphere, but generally regarded as rather toxic materials for most organisms. Various oxides of sulphur and of nitrogen are also metabolized by certain terrestrial microorganisms. It would therefore appear that wide divergences from the average terrestrial atmosphere are compatible with the sustenance of some living systems. In the absence of complete information on the atmospheres of other planets, it is reasonable to conclude that, within wide limits, atmospheres significantly different in composition from that of the earth could still sustain living organisms.

Water accounts for about 80% of the cellular materials of terrestrial organisms. Numerous treatises have been written (e.g., Henderson 1927) concerning the properties of water that make it essential for the activities of terrestrial organisms. Indeed, the availability of water on earth often is the limiting factor that determines whether organisms will grow in a given environment. It is common to find that microorganisms, including bacteria, molds, and protozoa, go into resistant stages when they are deprived of water. Higher organisms, such as lichens, rotifers and nematodes, resist desiccation for years and are revived when placed into water. Despite this obvious necessity for available water, however, there is considerable diversification among terrestrial organisms in the manner in which they take up water.

Thus, many microorganisms can grow in extremely dilute solutions, as for example in ordinary laboratory distilled water, while other organisms – e.g., the osmophilic fungi – require solutions of high osmotic pressure for growth. Similarly the obligate halophiles require approximately 30% salt solutions for optimal growth, and will die when exposed to ordinary bacteriological media (i.e., to higher water activity). As far as is known, however, all terrestrial organisms require at least the intermittent presence of liquid water for active metabolic purposes. By terrestrial analogy, it would appear that water, in liquid form, would be a requirement at least at certain stages in the life of organisms on other planets. This constraint dictates that nonterrestrial organisms would have worked out mechanisms to obtain water from their environment. In the view of many biologists, the complete absence of water would almost certainly preclude the presence of organisms on another planet. Some, however, have held this position to be too restrictive (Pimentel et al. 1966), arguing that other solvents might take the place of water on planets having organisms with 'exotic 'biochemical mechanisms.

4. *Possibilities for life in nonterrestrial environments*

The moon became the first extraterrestrial body to be subjected to extensive investigations for the presence of living organisms, when samples of lunar material were returned to earth in July, 1969. Investigations, both at the Manned Spacecraft Center in Houston and at the Ames Research Center (Oyama et al. 1970) failed to reveal any indigenous organisms. But these findings came as no suprise to biologists, since the present physical environment on the moon is almost certainly outside the limits of tolerance for living organisms. The most significant factor in this regard is the absence of an atmosphere on the moon, thus affording no opportunity for establishing an equilibrium between organisms and their gaseous products, eliminating any possibility of liquid water on the surface, permitting large temperature fluctuations between daytime and nighttime, and also permitting the complete spectrum of solar radiation to reach the surface.

Since the moon would be expected to collect meteorites and other interplanetary debris, without significant alteration, it has been proposed that the moon may be a gravitational trap for organic matter (Lederberg and Cowie 1958). Oyama et al. (1970) suggested also that extensive samplings from the moon might be useful in testing the 'panspermia' hypothesis. Chemical analyses of the material returned on the Apollo 11 mission revealed very low levels of organic matter (Ponnamperuma et al. 1970; Oro et al. 1970). Whether these observations have a direct bearing on the general concepts of chemical evolution is uncertain, however. As pointed out above, our knowledge about the stability of organic compounds over long periods of time is incomplete, and this uncertainty is compounded by ignorance of the environmental (e.g., thermal) history of the lunar surface. Secondly, since these samples were obtained from the uppermost regions of the lunar surface, exposed to highly energetic ultraviolet and ionizing radiations from the sun, it is reasonable to assume that, had any organic compounds been deposited in this region some billions of years ago, they would have been destroyed by this time. In this connection, Sagan (1961) suggested that there may still remain organic compounds of primordial origin at depths of the order of 50 meters below the surface.

Of the other planets in the solar system, the inner planets, Mercury and Venus, do not, on the basis of current data, appear to have conditions favorable to sustain life on their surfaces. Although relatively little is known about Mercury, it is expected to sustain extremely high temperatures on the side facing the sun, both because of proximity to the sun and because of the fact that it has virtually no atmosphere (Belton et al. 1967).

On the basis of earth-based observations, Venus is known to have a very dense atmosphere and extremely high surface temperatures (Sagan 1967). In the spring of 1969, the Soviet Venera 5 and 6 spacecraft entered the Venusian atmosphere and made a series of measurements of pertinence to this discussion (TASS 1969). Surface temperatures of 400–530 °C were reported, with concomitant atmospheric pressures of 60–140 atmospheres. CO_2 was present at high concentrations (93–97%); oxygen was less than 0.4%; water in the atmosphere was reported as 4–11 mg per liter. Nitrogen and other inert gases accounted for 2–5% of the atmosphere. In addition, large variations in surface elevations, of the order of tens of kilometers, were deemed likely. These results are in essential agreement with earlier Venera 4 and Mariner 5 data (Jastrow 1968). Despite the presence of carbon dioxide, oxygen and water in the atmosphere, the high surface temperatures on Venus would appear to make life improbable there. On the other hand, Morowitz and Sagan (1967) have speculated that moderate temperatures and atmospheric pressures are likely in the lower zones of the Venusian atmosphere, and suggested that some types of organisms could float in the atmosphere under these conditions. Such organisms could use the available water, carbon dioxide and sunlight for photosynthesis, and carry out a complete metabolism, provided that small amounts of inorganic material were stirred up occasionally into these cloud layers from the surface. It is of interest to note the statement of these authors that '... the conditions in the lower clouds of Venus resemble those on earth more than any other extraterrestrial environment now known'.

Beyond the planet Mars, the remaining planets are thought to resemble one another in their general physical characteristics (Sagan 1966). Jupiter is known to have a very extensive and turbulent atmosphere, in which abundant quantities of ammonia, methane, hydrogen and, probably, water vapor are present (Michaux 1967). Sagan (1964) brought attention to the fact that the current atmosphere of Jupiter is analogous to the 'primitive atmosphere' postulated to contain the precursors of terrestrial life, and that the Jovian atmosphere might well be a reservoir of organic material resulting from abiogenic condensation processes. This contention has received some experimental support from the studies of Woeller and Ponnamperuma (1969), who showed that a number of amino acid anhydrides and other organic compounds were produced under simulated Jovian conditions (i.e., at the temperature of liquid nitrogen). It has also been suggested (Michaux 1967) that in some regions of the atmosphere of that planet, zones occur having temperatures of 0° to 80 °C, leaving open the possibility of 'floating' organisms similar to that described above for Venus. That the individual components of the Jovian atmosphere are not necessarily toxic to living organisms has been known for a long time. Indeed, the combination of these gases appears to be utilized, as documented by the studies of Siegel and Giumarro (1965), who described the isolation, from terrestrial sources, of organisms that were grown in the presence of high concentrations of ammonia, together with methane, hydrogen, and water. Experiments such as these, as well as speculations about the possibilities for life on Jupiter, only serve to stimulate the imagination. Our information about Jupiter is too fragmentary at present to allow for more precise assessments about the likelihood of finding life on that planet.

Like Jupiter, the atmosphere of Saturn is known to contain methane, ammonia, and hydrogen and it, too, presumably also contains water (Sagan 1966). We know so little about Saturn, and even less about the more distant planets of the solar system, that very little can be said – beyond mere speculation – about the chances of finding organisms on these planets. Nor will consideration be given here to the satellites of the planets, other than the earth, for this reason.

5. Mars

Mars has been the subject of extensive astronomical observation coupled, in recent years, with observations from spacecraft. As a result of both ground-based and remote observations, much information has been obtained that bears critically on the environmental characteristics, and the probabilities of life, on that planet. The gross physical characteristics of Mars have been known for some time (Glasstone 1968). Mars orbits the sun at a mean distance of approximately 140 million miles; has a diameter approximately one-half that of the earth; and one-tenth the mass of the earth. This planet is inclined at an angle of approximately 24° perpendicular to its orbit, compared with an angle of 23.5° for the earth. Since the Martian period of rotation is also very similar to that of the earth (24 hours and 37 minutes), both seasonal and day–night cycles, analogous to those experienced on the earth, must be present there.

The temperature of Mars, integrated over large areas, has been determined by various techniques, both from the earth and from spacecraft, and these measurements agree on an average surface temperature, across the planet, that is about 50° colder than the average for the earth. In the equatorial regions, daytime temperatures of approximately 300°K at noon have been obtained by a variety of methods of determination. Nighttime temperatures in this same region as low as 200°K have been measured by the Mariner 6 spacecraft (Neugebauer et al. 1969).

The surface of Mars appears, under telescopic scrutiny, to be composed of three general areas: the polar caps, and the so-called 'bright' and 'dark' areas. The white polar regions, previously thought to be composed of a thin layer of ice, are now generally regarded to be composed primarily of solid carbon dioxide, based on measurements made on Mariners 6 and 7 by infrared spectrometry and infrared radiometry (Herr and Pimentel 1969; Neugebauer et al. 1969). Water, which is known to be present in the Martian atmosphere (Glasstone 1968), almost certainly is also present in these pole caps where temperatures of the order of 150°K have been recorded.

The 'dark' areas, covering about a quarter of the surface of the planet – generally arranged along the equatorial zone – were once thought to be bodies of water. Later, it was thought that these areas were covered with vegetation. At present, all that can be accepted is that these regions are darker by contrast with the remaining, 'bright' areas.

Earlier investigators reported that, concomitant with the waning of the polar cap during the summer season in a particular hemisphere, there was a 'wave of darkening' in that hemisphere – proceeding from the pole toward the equator (Glasstone 1968). This does not seem to be uniformly accepted at present by most astronomers. That there are seasonal variations in the contrast between 'bright' and 'dark' areas of Mars is still held to be true, but there is no concurrence on a possible mechanism for these observed changes.

The atmosphere of Mars has been probed spectroscopically from earth-based telescopes, from balloons, by radio occultation methods (Kliore et al. 1965), and directly by ultraviolet and infrared spectroscopes on spacecraft. The total atmospheric pressure is currently thought to be approximately 6.5 mb. Since topographical differences of 12 to 15 km have been deduced from radar measurements, the atmospheric pressure at any particular location could thus vary over a large range, low areas having a total atmospheric pressure as high as 15 mb, while high areas could have very low pressures. Indeed, during the Mariner 7 mission, as the occultation experiments were performed over the region of Hellespontus, the calculated total atmospheric pressure was only about 3.8 mb, suggesting that this region is considerably elevated over the average terrain (Kliore et al. 1969).

From ground-based spectroscopic observations (Glasstone, 1968) carbon dioxide, carbon monoxide, and water have been detected in the atmosphere. The major component, carbon dioxide, appears to comprise almost all the atmosphere (Belton et al. 1968). Data obtained from the orbiting astronomical observatory have yielded for other gases the following upper limits: H_2S (2 ppm), SO_2 (2 ppm), NH_3 (2 ppm), N_2O_4 (5 ppm), NO_2 (10 ppm), NO (10 ppm), O_3 (0.025 ppm) (T. Owen, personal communication). Ultraviolet spectroscopy experiments on the Mariners 6 and 7 (Barth et al. 1969) also showed carbon dioxide to be the major constituent of the lower atmosphere, with CO_2^+ as the major ionized gas in the upper atmosphere. At these higher levels, small amounts of carbon monoxide and traces of oxygen were found in addition. These ultraviolet experiments were also designed to detect nitrogen in the Martian atmosphere, using equipment whose sensitivty was such that nitrogen could be detected in amounts as low as a few percent (1 to 4%) of the atmosphere. But none was detected. Of interest also is that analysis of the data from these experiments of Mariners 6 and 7 indicated that considerable high energy ultraviolet radiation – down to wavelengths of approximately 1,900 Å – penetrated the thin Martian atmosphere and reached the surface of the planet. In this connection, it should be pointed out that solar protons and X-rays would be largely prevented from reaching the Martian surface, even by the thin atmosphere of that planet.

Water is known to be present in the Martian atmosphere (Glasstone 1968)

in extremely small amounts. If all the water in the atmosphere were precipitated over the entire surface of Mars, a layer 15–50 μ thick would result
(Owen and Mason 1969). For comparison, the atmosphere of the earth
would yield, on the average, about 2000 μ of precipitable water. It is also
noteworthy that there is great variability in the amount of water seen in the
atmosphere of Mars from one time to the next, in any one particular area.

From pictures obtained by all three Mariner spacecraft, it is clear that the
surface of Mars is heavily cratered. While most of the surface area examined
was covered by this type of terrain, two other types of terrain were seen:
some areas, covering thousands of square kilometers were featureless; others
exhibited a very irregular, jumbled topography, characterized as 'chaotic
terrain' (Leighton et al. 1969).

An analysis of the crater frequencies on Mars relative to that found on the
moon, combined with information recently obtained (Albee et al. 1970)
showing that Apollo 11 lunar samples contained surface material that had
remained on the surface, undisturbed, for periods of the order of 4.6×10^9
years, led to the conclusion that the present surface of Mars is essentially
primordial (Leighton et al. 1969). If these contentions are correct, it would
appear that large bodies of water have been absent from the surface of Mars
going back as far as the very earliest history of that planet. In this connection, Chapman et al. (1969) claimed that cratering statistics cannot be relied
upon to give information regarding standing bodies of water early in the
history of Mars.

That the data from the Mariner spacecraft did not detect life on Mars came
as no surprise to biologists. Even before these missions, it was recognized
that surface and atmospheric conditions on that planet were likely to be
austere by terrestrial standards. The space flight experiments have corroborated and extended many of our estimates of the physical and chemical
properties of that planet. Nevertheless, in many cases our information is
still not precise. For example, we are not certain at this time whether
free nitrogen exists in the atmosphere of Mars; the complete array of compounds in the Martian atmosphere is still not determined; we have virtually
no knowledge of the composition of the surface materials other than on the
surface in the polar regions; our topographical information (now covering
approximately 20% of the surface area) is known only to rather gross spatial
resolution; and we have very little information about the seasonal variations
in any of these important parameters.

Despite the uncertainties, it is obvious that the present physical environment on Mars is very bleak from the point of view of most terrestrial organ-

isms. There is little or no oxygen; there are large daily temperature extremes; the average temperature of the planet is low; the atmosphere is, on the average, thin and contains extremely small quantities of water; there is probably no liquid water on the surface; and there would appear to be a high flux of ultraviolet radiation at the surface.

6. *Biological adaptation*

But if the biologist is impressed with any general property of life, it is its ability to adapt to different environments. The observations of Barghoorn and his collaborators (1966; Schopf and Barghoorn 1967) and of Engel (1968), that fossilized microorganisms resembling bacilli and algae are present in terrestrial samples 3.0 to 3.4 billion years old, would indicate that processes leading from the precursor molecules of the primitive atmosphere to living organisms, as highly evolved as microorganisms, took place rapidly after the earth was formed. From the first surviving types of organisms, living systems have subsequently adapted to an enormous range of environmental niches on this planet. As indicated above, organisms are found to grow under such widely divergent conditions as in the boiling hot springs of Yellowstone National Park and at 7,000 to 10,000 meters below the surface of the Pacific Ocean.

Some astronomers (Shklovskii and Sagan 1966; Urey 1959) have argued that Mars may, at one time, have been warmer and may have contained a heavier atmosphere than it does at present. Thus, the early stages of chemical evolution on Mars may have proceeded under conditions much like that on the earth. Once started, early biological evolution on Mars could have proceeded under much more favorable conditions than exist on that planet today. Given a foothold on the planet, living organisms might well have adapted, through evolutionary processes, to lower temperatures, to increasing ultraviolet fluxes, to decreasing availability of water, etc., as Mars began to lose its atmosphere over relatively long periods of time.

7. *Concluding comments*

It seems almost unbelievable that, in only a half dozen years, the United States and the Soviet Union have progressed so rapidly in the direct exploration of our solar system. Three Mariner spacecraft have visited the vicinity of Mars and sent back enormous amounts of new information about that planet; another Mariner and three Venera spacecraft have very substantially increased our knowledge about Venus. The moon has been sampled and directly studied by hundreds of investigators. Coupled with intensified activity on the part of ground-based astronomers, our information about our neighbors in the solar system has increased manifold in this period. The next half dozen years must surely be as productive. As we look to the future, the United States is expected to send two Mariner spacecraft to orbit Mars in 1971 in order to map thoroughly the planet topographically, and to obtain further information on the temporal and spatial variations in its atmosphere. It can be anticipated that the Soviet Union will continue its studies of the planet, Venus. Also in this time frame, we may see spacecraft fly past Mercury and others land on Mars. In addition, two Pioneer spacecraft are scheduled to fly past Jupiter in 1974 and 1975 for a look at that interesting and, to date, mysterious planet.

As we look toward the future, from the vantage point of our relative ignorance about the micro-environments that are possible on the other planets of our solar system and, recognizing that the limits even for terrestrial life have not been sharply defined, it would be a tragic mistake not to take advantage of all reasonable opportunities to search for life on other planets.

References

ABELSON, P. H., 1959, In: P. H. Abelson, ed., Researches in geochemistry. (John Wiley and Sons, Inc., New York) pp. 79.

ALBEE, A. L., D. S. BURNETT, A. A. CHODOS, O. J. EUGSTER, J. C. HUNEKE, D. A. PAPANASTASSIOU, F. A. PODOSEK, G. P. RUSS, II, H. G. SANZ, F. TERA and G. J. WASSERBURG, 1970, Science 167, 463.

ALEXANDER, P., 1957, Atomic radiations and life. (Pelican Press, London).

ARRHENIUS, S., 1908, Worlds in the making: The evolution of the universe. (Harper and Bros., New York).

BARGHOORN, E. S. and J. W. SCHOPF, 1966, Science 152, 758.

BARTH, C. A., W. G. FASTIE, C. W. HORD, J. B. PEARCE, K. K. KELLY, A. I. STEWART, G. E. THOMAS, G. P. ANDERSON and O. F. RAPER, 1969, In: Mariner-Mars, 1969, A preliminary report, NASA SP-225. (Washington, D.C.) pp. 97.

BELTON, M. J. S., A. L. BROADFOOT and D. M. HUNTEN, 1968, J. Geophys. Res. 73, 4795.

BELTON, M. J. S., D. M. HUNTEN and M. B. McELROY, 1967, The Astrophys. J. 150, 1111.

BROCK, T. D., 1967, Science 158, 1012.

BROWN, H., 1964, Science 145, 1177.

BRUESCHKE, E. E., R. H. SUESS and M. WILLARD, 1961, Planet. Space Sci. 8, 30.

CATTELL, M., 1936, Biol. Ref. Cambridge Phil. Soc. 11, 441.

CHAPMAN, C. R., J. B. POLLOCK and G. SAGAN, 1969, Astronomical J. 74, 1039.

DRAKE, F. D., 1962, Intelligent life in space. (MacMillan Co., New York).

ENGEL, A., 1968, Conference on: The environment of the primitive earth. (Harvard University, Cambridge, Mass.).

GLASSTONE, S., 1968, The book of Mars, NASA SP-179. (Washington, D.C.).

HENDERSON, L. J., 1927, The fitness of the environment: An inquiry into the biological significance of the properties of matter. (MacMillan Co., New York).

HERR, K. C. and G. C. PIMENTEL, 1969, In: Mariner-Mars 1969; A preliminary report, NASA SP-225. (Washington, D.C.) pp. 83.

HUANG, S., 1959. Scientist 47, 397.

JASTROW, R., 1968, Science 160, 1403.

KLIORE, A., D. L. CAIN, G. S. LEVY, V. R. ESHELMAN, G. FJELDBO and F. D. DRAKE, 1965. Science 149, 1243.

KLIORE, A., G. FJELDBO, B. SEIDEL and S. I. RASSOOL, 1969, In: Mariner-Mars 1969; A preliminary report, NASA SP-225. (Washington, D.C.) pp. 111.

LARKIN, J. M. and L. J. STOKES, 1968, Can. J. Microbiol. 14, 97.

LEDERBERG, J. and D. B. COWIE, 1958, Science 127, 1473.

LEIGHTON, R. B., N. H. HOROWITZ, B. C. MURRAY, R. P. SHARP, A. H. HERRIMAN, A. T. YOUNG, B. A. SMITH, M. E. DAVIES and C. B. LEOVY, 1969, Science 166, 49.

MARSHALL, N. B., 1954, Aspects of deep sea biology. (Philosophical Library, New York).

MICHAUX, C. M., 1967, Handbook of the physical properties of the planet Jupiter, NASA SP-3031. (Washington, D.C.).

MOROWITZ, H. and C. SAGAN, 1967, Nature 216, 1259.

NEUGEBAUER, G., G. MUNCH, S. C. CHASE, Jr., H. HATZENBELER, E. MINER and D. SCHOFIELD, 1969, In: Mariner-Mars 1969; A preliminary report, NASA SP-225. (Washington, D.C.) pp. 105.

OLIVER, B. M., 1966, Technological approaches to interstellar communication; Amer. Astronautical Soc., 12th Ann. Meeting, Anaheim, California.

ORO, J., W. S. UPDEGROVE, J. GIBERT, J. MCREYNOLDS, E. GIL-AV, J. IBANEZ, A. ZLATKIS, D. A. FLORY, R. L. LEVY and C. WOLF, 1970, Science 167, 765.

OWEN, T. and H. P. MASON, 1969, Science 165, 893.

OYAMA, V. I., E. L. MEREK and M. P. SILVERMAN, 1970, Science 167, 773.

PIMENTEL, G. C., K. C. ATWOOD, H. GAFFRON, H. K. HARTLINE, T. H. JUKES, E. C. POLLARD and C. SAGAN, 1966, In: C. S. Pittendrigh, W. Vishniac and J. P. T. Pearman, eds., Biology and the exploration of Mars; Natl. Acad. Sci. (NRC) Publ. 1296. (Washington, D.C.) pp. 243.

POLLARD, E. C., 1965, J. Theoret. Biol. 8, 113.

PONNAMPERUMA C. K. KVENVOLDEN, S. CHANG, R. JOHNSON, G. POLLOCK, D. PHILPOTT, I. KAPLAN, J. SMITH, J. W. SCHOPF, C. GEHRKE, G. HODGSON, I. A. BREGER, B. HALPERN, A. DUFFIELD, K. KRAUSKOPF, E. BARGHOORN, H. HOLLAND and K. KEIL, 1970, Science 167, 760.

PULLMAN, B. and A. PULLMAN, 1962, Nature 196, 1137.

PURCELL, E. M., 1960, Radioastronomy and communication through space; Brookhaven Natl. Lab. Lectures, No. 1.

RASMUSSEN, R. and R. B. PAINTER, 1964, Nature 203, 1360.

SAGAN, C., 1961, Organic matter and the moon; Natl. Acad. Sci. (NRC) Publ. 757. (Washington, D.C.).

SAGAN, C., 1964, In: M. Florkin and A. Dollfus, eds., Life sciences and space research II. (North-Holland Publ. Co., Amsterdam) pp. 35.

SAGAN, C., 1966, In: C. S. Pittendrigh, W. Vishniac and J. P. T. Pearman, eds., Biology and the exploration of Mars; Natl. Acad. Sci. (NRC) Publ. 1296. (Washington, D.C.) pp. 73.

SAGAN, C., 1967, Nature 216, 1198.

SCHOPF, J. W. and E. S. BARGHOORN, 1967, Science 156, 508.

SHKLOVSKII, I. S. and C. SAGAN, 1966, Intelligent life in the universe. (Holden-Day, Inc., San Francisco)

SIEGEL, S. M. and C. GIUMARRO, 1965, Icarus 4, 37.

TASS, 4 June 1969, An important step in understanding the universe. The Soviet 'Venera 5' and 'Venera 6' Interplanetary Stations. (Isvestiya, Moscow) pp. 1, 3, 4.

UREY, H. C., 1959, In: S. Flugge, ed., Handbuch der Physik, vol. 52. (Springer-Verlag, Berlin) pp. 363.

VON HOERNER, S., 1962, Science 137, 18.

WOELLER, F. and C. PONNAMPERUMA, 1969, Icarus 10, 386.

ZOBELL, C. E. and R. Y. MORITA, 1957, J. Bacteriol. 73, 563.

C. Ponnamperuma (ed.), Exobiology. © North-Holland Publishing Company

CHAPTER 15

Life beyond the solar system

CARL SAGAN

From our very limited terrestrial perspective the evolution of intelligence appears to be a natural outcome of cosmic evolution. Modern astrophysics and biology have converged on a scheme of almost breathtaking scope, which begins $\sim 10^{10}$ years ago in a universe made entirely of hydrogen and helium. Inhomogeneities in this medium led to gravitational instabilities and the condensation of gas into first generation stars. Thermonuclear reactions, particularly in the later red giant and supernova evolutionary phases, generated carbon, nitrogen, oxygen, silicon, iron and the rest of the elements in the periodic table; and then ejected these newly synthesized elements back into the interstellar medium. Subsequent generations of star formation then occurred out of a medium enriched with heavy elements.

Accompanying second and later generations of star formation was the origin of two categories of planets, massive gas giants preserving to a significant degree the hydrogen and helium abundance of their origins, and smaller planets, formed closer to their primaries, from which the primordial hydrogen and helium has been largely depleted. In our solar system these two categories of bodies are known as the Jovian and terrestial planets respectively.

On at least one of these planets, our own, as detailed elsewhere in this volume, a progression of chemical events occurred beginning with the synthesis of small organic molecules in the primitive atmosphere, and followed by the aggregation of larger polymers; the production of a self-replicating molecular system capable of evolution by natural selection; and the long evolutionary chain from molecule to man. Somewhere in this chain a quality which we characterize by the word intelligence developed – one of many ways to control the environment for the benefit of the organism. In some fundamental sense we can consider ourselves as ambulatory collections of self-replicating molecules, and many of our actions, thoughts and emotions are clearly of high selective advantage for these molecules which we contain.

Assuming that the selective advantage of intelligence continues, and that we do not destroy ourselves – an assumption by no means obvious, as even a casual glance at the daily newspapers indicates – we cannot consider ourselves and the increasingly global civilization which we have erected as the culmination of this evolutionary sequence. If we avoid our self-destruction there must in some sense be successor civilizations to our own, distant thousands, millions and perhaps even larger intervals of time into the remote future.

This grand perspective of cosmic evolution, supported as it is by a wide range of recent scientific findings (see, for example, Shklovskii and Sagan 1966), is not proved. It is only a likely story. What we seek is the generality of these events.

At first sight the continuous trend from disorder to order appears to fly in the face of the spirit if not the letter of the second law of thermodynamics. We may be prepared to believe that the increase in order represented by the origin of life or the development of intelligence on a planetary surface is more than compensated for by the increasing entropy in the local star attendant to these processes. But why should there be any tendency to increase order at all? A very general answer to this question – so general that it applies both to chemical, biological and social evolution – may be contained in a theorem by Prigogine on open system thermodynamics. Prigogine has shown that in an open system, which is in contact with its environment by the flow of energy through it, the order may increase without limit. (A cative discussion of these results can be found in the book by Morowitz (1968).

The search for extraterrestrial intelligence has received considerable attention in recent years. I wish in this brief discussion to touch only on a few of the highlights and some of the more recent developments. The interested reader can find a more detailed discussion in Shklovskii and Sagan (1966), Cameron (1965), Tovmasyan (1965), and in Kaplan (1969). The latter two are books published in the Soviet Union and unfortunately not generally known in the West, despite the fact that one has been translated into English. (Appendix A reproduces the Tables of Contents of these two books.)

The first major problem is one of the generality of the formation of planetary systems. Quite independent of models of the origins of such systems, we have close at hand examples of not one but four such systems: the solar system proper and its three small scale versions: the satellite systems of Jupiter, Saturn and Uranus. Whatever the mechanism of formation of such systems the existence of four of them right at hand suggests that they are not difficult to produce.

More than half the stars in the sky are members of double or multiple star systems. The mean separation of double stars is on the order of 10 astronomical units, or about the distance from the Sun to Saturn, which is somewhere near the center of mass of the planets of our solar system – a result which suggests that, with remarkable generality, massive stars tend to be formed with less massive companions at a common mean distance. If the small companion has a mass of several Jupiter masses or smaller, its interior temperatures remain low and it never can initiate thermonuclear reactions and become a star. If the less massive companion however has a mass some tens of times that of Jupiter, it will reach interior temperatures larger than a few million degrees Kelvin and thermonuclear reactions will occur; it will turn on, and become a small star. Such considerations suggest that the distribution of planets, in absolute distance from the central star, has a tendency, by no means well understood, to be comparable to those in our solar system.

When we look at the nearest stars which are not members of binary or multiple systems and which are close enough for methods of gravitational perturbation to indicate the presence of a dark companion, we find that fully half have companions of planetary mass. The nearest single star, Barnard's star, has two companions of roughly Jovian mass and semi-major axis, although the orbits appear to be much more elliptical than the Jovian planets in our own solar system (Van de Kamp 1969). Planets of terrestrial mass would not be detectable by these techniques even at the distance of the nearest star. Kukarkin (1965) suggests that the linear superposition of a system of planets in circular orbits can produce effects in gravitation perturbation which would be interpreted as a planet in a highly eccentric orbit.

A very provocative theoretical result, although based on a model which is certainly too simple for the actual physics of planetary formation, is contained in a recent paper by Dole (1970). A by now conventional solar nebula of gas and dust surrounding the forming sun is assumed, and aggregation nuclei are injected into the nebula with semi-major axes and orbital eccentricities determined by a random number generator. The nuclei grow by accretion and sweep out lanes which are free of dust; beyond a certain critical mass they can gravitationally aggregate gas as well. Collisions of growing nuclei lead to further aggregation. Dole's computer program continues this process until all of the gas and dust has been swept free. This very simple procedure produces for a wide range of cases planetary systems with masses and semi-major axes quite comparable to those in our own solar system and with a distribution of distances reminiscent of the 'Bode law' relation for our solar system. The distribution of planets in our solar system is shown in fig. 1. The numbers above each planet indicate the planetary mass in units of the Earth's and the trident markings below give the orbital

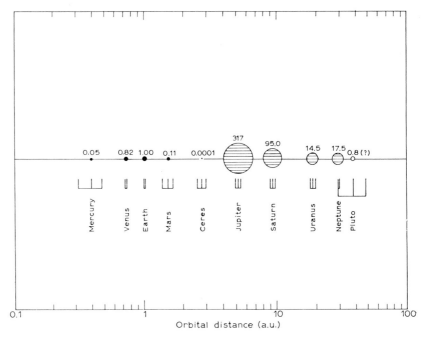

Fig. 1. Representation of the masses and orbital semi-major axes of the planets in the solar system. The tridents indicate orbital eccentricities and horizontal shading distinguishes Jovian from terrestrial planets. (After Dole 1970).

eccentricities. Fig. 2 and 3 give some of Dole's results; they differ only in the random numbers generated which are indicated by the index number x_0 on the left. The large planets with horizontal cross hatchings will be Jovians, and the planets represented by small filled circles will be terrestrial planets. In some of these models, the eccentricity, even of Jovian planets, is high and in most cases there are only two or three large Jovian planets. There seems to be no problem in accounting for the Barnard's star system as determined by Van de Kamp with this category of solutions.

The evidence both observational and theoretical seems strongly to imply that planetary systems more or less like our own are a very frequent accompaniment of stars in our galaxy.

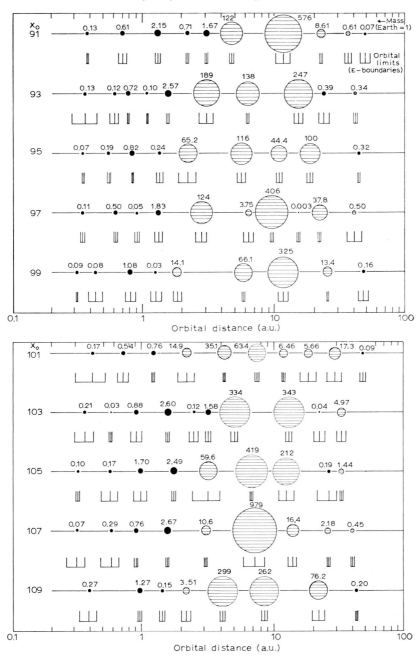

Figs. 2, 3. Model solar systems generated by Dole's (1970) program. The symbols are as in fig. 1 with the mass of the Earth having unit value. The numbers marked x_0 down the left hand margin are random number indices for the injection of accretion nuclei.

The next question is how likely it is that appropriate prebiological organic chemistry occurs on such planets. The laboratory experience which we have to date shows that under quite general primitive planetary conditions – essentially cosmic composition at low temperatures and cosmically high densities – essentially all the organic compounds of interest for the origin of terrestrial life occur. The results are remarkably independent of fair variation in pressure and temperature, composition (as long as the net conditions are reducing) and energy source. Such experiments on the origin of life are of course biassed to detect the molecules of contemporary biochemical interest, and it is by no means out of the question that other molecules not used in terrestrial biology are made in high yield in such experiments and might be of importance for extraterrestrial biochemistry. Nevertheless it is striking how high the yield is in the production of molecules of undisputed provincial terrestrial interest. For example in recent work on high temperature pulse production of organic compounds in a shock tube, Bar-Nun et al. (1970) found that 30% of the ammonia was converted into amino acids in a single shock. Such results suggest that at least some of the same biochemicals will be prominent in extraterrestrial as in terrestrial biochemistry and that the laboratory experience garnered in studying the origin of life on earth will be of relevance in evaluating the question of the origin of life elsewhere.

In addition there is information from a variety of other sources. Largely because of the carbonaceous chondrites, something like 1% by mass of the meteorites which fall on the Earth, and are of presumptive asteroidal origin, are composed of organic compounds. The spectra of comet tails show the presence of CH, CN, OH, C_2, C_3 and so on – almost surely fragments of organic compounds contained in the comet heads. Infrared spectra of cool stars and optical and microwave line studies, both in absorption and in emission, of the interstellar medium show similar fragments. In addition, the cool stars display signs of H_2O, HCN, and C_2H_2 (see, for example, Spinrad and Wing 1969); and the interstellar microwave spectrum shows unambiguously the presence of H_2O, OH, NH_3, CO, CN, HCN, HCHO, and most recently cyanoacetylene, CNC_2H. These molecules, found exclusively in dense interstellar clouds, clearly indicate, in my view, the presence of a field of interstellar organic chemistry (Sagan 1970, 1971). Claims have also been made from studies of the optical interstellar spectrum of diffuse lines by Johnson (1970) for the presence of a magnesium-chelated pyridinated benzoporphyrin. While I have indicated elsewhere my reasons for skepticism in accepting this identification (Sagan 1970, 1971) there is every reason to think that quite complex organic molecules may exist in interstellar space. It is also possible that the interstellar grains have a significant organic component. Some evidence for the presence of graphite has been adduced, but from the point of view of the absorption spectrum the difference between graphite and large polycyclic aromatic hydrocarbons – probably even more likely in the hydrogen-rich interstellar environment – is not very great. It is, incidentally, by no means obvious that a planet condensing out of a solar nebula rich in organic matter is any further along towards the origin of life than a planet condensing out of a solar nebula in which no carbon molecule more complex than methane is present. The gravitational energy of accretion, the impact of final planetesimals in the formation process, and the presence of extinct and uncommon radionuclides are together likely to volatilize all organic compounds near the surface of a new planet, even if the accretion rate is slow.

There is a debate on the question of whether organic compounds are present in the Jovian planets. I have argued that the Jovian coloration is most likely explained by organic chromophores in the Jovian clouds (Sagen et al. 1967); and have speculated that an enigmatic and apparently transient absorption feature in the rocket ultra-violet was due either to benzene and its derivatives or to purines and pyrimidines and their derivatives (Sagan 1968). Woeller and Ponnamperuma (1969) and Sagan and Khare (1971) have shown directly that chromophores can be produced in large quantities under simulated Jovian conditions. Shocks, for example from thunder, are to be expected in the Jovian clouds in very high yield. Other organic compounds can be expected from such processes (Bar-Nun et al. 1970), and calculations modeling the shock front (Bar-Nun and Sagan 1971) show that the general run of organic compounds produced should be insensitive to the molecular hydrogen abundance over a range of many orders of magnitude. Lewis and Prinn (1970) have proposed that the Jovian coloration in the visible is due instead to compounds of H_2S and ammonia.

But the presence of organic compounds in meteorites, probably in Jovian planets, in comets, in the interstellar medium and in cool stars implies that the production of organic compounds essential for the origin of life should be pervasive throughout the universe.

The production of organic molecules necessary for the origin of life is not at all the same as the origin of life (see, for example, Sagan 1970b; Horowitz et al. 1970). Nevertheless this range of laboratory and astronomical information on prebiological organic chemistry makes it seem very likely that the origin of life is probable under quite general primitive planetary conditions. It therefore appears that planets suitably situated from the local star, prebiological organic chemistry and the origin of life are all very common.

Data on the likelihood of the evolution of intelligence, and of a technical civilization are of course restricted to our own planet, at least until now; and any attempt to extrapolate from terrestrial to extraterrestrial cases in these areas must be fraught with uncertainty. I have discussed these issues elsewhere (Shklovskii and Sagan 1966). An estimate which seems reasonable to me and which is derived in the cited reference is that the number of extant technical civilizations in the galaxy today is approximately one tenth the mean lifetime of a technical civilization in years. By a technical civilization, I mean one capable of interstellar radio communication. Other workers have come up with a similar number but the skeptical reader is invited to perform an alternative calculation.

If, then, a small fraction, say 1%, of technical civilizations learn to deal with weapons of mass destruction and so have lifetimes comparable to geological time or stellar evolutionary time scales it follows that there are approximately 10^6 extant technical civilizations in the Galaxy. A simple geometric argument, assuming such civilizations to be randomly spaced within the Galaxy, then shows that the distance to the nearest such civilization is several hundred light years. Also, from the relation just quoted, we see that technical civilizations in the sense defined are emerging in the Galaxy about once a decade. Since we are ourselves only a decade or two after having achieved the bare capability for interstellar radio communication, it follows that there is no technical civilization within the Galaxy, with whom we might be able to communicate by radio, which is as backward as we. This means we might have a great deal to receive but very little to send.

Now there are great advantages to interstellar radio communication. The cost per bit is low. There are a few natural frequencies known both to senders and receivers. Even radio facilities existing at the present time on the Earth are capable of transmitting and receiving over a distance of a thousand light years or more. And there seems to be no serious problem in devising a message which can be decoded by a distant civilization, even though the two civilizations share no common language except the common knowledge of physics which enabled them to achieve radio contact. These matters are discussed in the cited references.

Nevertheless, radio is a recent technique for the planet Earth. It is less than a century old and it seems clear that over a very long time scale – say periods of millions of years – not only will better techniques for interstellar communication be invented based upon laws of physics now known, but also entirely new laws of physics will be discovered, laws at which we can now only vaguely guess. It is therefore possible that a typical extraterrestrial technical civilization employs means of interstellar communication which are entirely beyond our present abilities. But a truly advanced civilization will of course know our limitations, and understand that we are restricted to such primitive techniques as radio communication. If they wish to contact us it will surely be possible for them to do so. At the present time, however, there is no systematic effort on our planet aimed at detecting interstellar radio signals from advanced technical civilizations, although it is entirely possible that an international cooperative search for such signals might be initiated in the coming decade.

Appendix A

EXTRATERRESTRIAL CIVILIZATIONS

Proceedings of the First All-Union Conference on Extra-terrestrial Civilizations and Interstellar Communication

Byurakan, 20–23 May 1964

V. A. Ambartsumyan, Ya. B. Zel'dovich, V. A. Kotel'nikov, A. S. Pistol'kors, V. I. Siforov, N. L. Kaidanovskii, B. N. Kukarkin, D. Ya. Martynov, V. S. Troitskii, S. E. Khaikin, I. S. Shklovskii, and other participants.

Edited by G. M. Tovmasyan

Izdatel'stvo
Akademii Nauk Armyanskoi SSR
Erevan 1965

Table of Contents

EXTRATERRESTRIAL CIVILIZATION:
PROBLEMS OF INTERSTELLAR CONTACT

L. M. Gindilis, S. A. Kaplan, N. S. Kardashev, B. N. Panovkin, B. V. Sukotin, G. M.
Hovanov

Under the General Editorship of
Professor S. A. Kaplan

Science Publishing House
Moscow 1969

Table of Contents

Introduction. Exosociology – the quest for signals from extraterrestrial civilization (S. A. Kaplan, page 7). Theory of the development of civilization (11). Problem of the quest for signals from extraterrestrial civilizations (14). The decoding aspect of the quest for signals from extraterrestrial civilizations. References p. 24.

Chapter I. Astrophysical aspect of the quest for signals from extraterrestrial civilizations (N. S. Kardashev, page 25). § 1. Introduction. § 2. The basic dilemma (page 29). § 3. The plenitude and credibility of contemporary astrophysical data (page 30). § 4. The idea of civilizations, the basic laws and character of their growth (page 43). § 5. The quest for the practical manifestation of stellar civilizations. § 6. The quest for the communication of information. § 7. About the program for the quest for stellar civilizations (page 97). References p. 100.

Acknowledgement

This research was supported in part by NASA Grant NGR 33–010–101.

References

BAR-NUN, A., N. BAR-NUN, S. BAUER and C. SAGAN, 1970, Science 168. 470.

BAR-NUN. A. and C. SAGAN, 1971, to be published.

CAMERON, A. G. W., ed., 1965, Interstellar Communication. (Benjamin, New York).

DOLE, S., 1970, Icarus 13, 494.

HOROWITZ, N. H., F. D. DRAKE, S. L. MILLER, L. E. ORGEL and C. SAGAN, 1970, In: P. Handler, ed., Biology and the future of man. (Oxford University Press).

JOHNSON, F., 1970, Bulletin of the American Astronomical Society, 2, 323.

KAPLAN, S. A., ed., 1969, Extraterrestrial civilizations: problems of interstellar communication. (Moscow, in Russian).

KUKARKIN, B. V., 1965, In: Tovmasyan, 1965.

LEWIS, J. and R. G. PRINN, 1970, Science 169, 472.

MOROWITZ, H. J., 1968, Energy flow in biology. (Academic Press).

SAGAN, C., 1968, Science 159, 448.

SAGAN, C., 1970, Trans. Intern. Astron. Union XIV B, in press.

SAGAN, C., 1970b, Life. (Encyclopaedia Britannica).

SAGAN, C., 1971, in preparation.

SAGAN, C., M. O. DAYHOFF, E. R. LIPPINCOTT and R. ECK, 1967, Nature 213, 273; also Priroda (Moscow), 9, 73.

SAGAN, C. and B. KHARE, 1971, Astrophys. J., in press.

SHKLOVSKII, I. S. and C. SAGAN, 1966, Intelligent Life in the Universe. (Holden Day, San Francisco).

SPINRAD, H. and R. WING, 1969, Ann. Rev. Astron. Astrophys. 7, 249.

TOVMASYAN, G. M., ed., 1965, Extraterrestrial Civilizations. (Armenian Academy of Sciences, Erevan). English translation by Israel Program for Scientific Translations, Jerusalem. Z. Lerman, translator, IPST 1823, 1967.

VAN DE KAMP, P., 1969, Private communication; also 1963, Astron. J. 68, 515.

WOELLER, F. and C. PONNAMPERUMA, 1969, Icarus 10, 386.

Subject index